Climate Change and Agriculture

Edited by Saddam Hussain

Published in London, United Kingdom

IntechOpen

Supporting open minds since 2005

Climate Change and Agriculture
http://dx.doi.org/10.5772/intechopen.78427
Edited by Saddam Hussain

Contributors
Mark Matsa, Saddam Hussain, Sadam Hussain, Abdul Khaliq, Umer Mehmood, Tauqeer Qadir, Muhammad Saqib, Muhammad Amjed Iqbal, Shilpi Kumari, Moonmoon Hiloidhari, S N Naik, R P Dahiya, Muhammad Mumtaz, Jose A. Puppim De Oliveira, Saleem H. Ali, Asmat Ullah, Ishfaq Ahmad, Muhammad Habib Ur Rahaman, Umer Saeed, Ashfaq Ahmad, Abid Mehmood, Gerrit Hoogenboom, Reazul Ahsan, Elliot Zwane, Saifullah Khan, Surendra N. Kulshreshtha, Sajid Hussain, Muhammad Shaukat, Muhammad Ashraf, Qianyu Jin, Junhua Zhang, Chunquan Zhu, Ghulam Abbas, Zartash Fatima, Syed Aftab Wajid, Shakeel Ahmad, Burhan Ahmad, Alexander Ruzmaikin, Joan Feynman, Sumiani Yusoff, Zameri Mohamed

First published in London, United Kingdom, 2019 by IntechOpen
IntechOpen is the global imprint of INTECHOPEN LIMITED, registered in England and Wales, registration number: 11086078, 7th floor, 10 Lower Thames Street, London, EC3R 6AF, United Kingdom
Printed in Croatia

British Library Cataloguing-in-Publication Data
A catalogue record for this book is available from the British Library

Additional hard and PDF copies can be obtained from orders@intechopen.com

Climate Change and Agriculture
Edited by Saddam Hussain
p. cm.
Print ISBN 978-1-78985-667-5
Online ISBN 978-1-78985-668-2
eBook (PDF) ISBN 978-1-83880-039-0

For EU product safety concerns:
IN TECH d.o.o., Prolaz Marije Krucifikse Kozulić 3, 51000 Rijeka, Croatia, info@intechopen.com or visit our website at intechopen.com.

Meet the editor

Dr. Saddam Hussain is an Assistant Professor in Plant Stress Physiology in the Department of Agronomy, University of Agriculture, Faisalabad. He completed his Ph.D degree at the Huazhong Agricultural University, China. He has been recognized with several international distinctions/awards in recent years. He has published over 100 refereed journal papers (IF:>370, h-index: 29, citations:>3000), many of which have sought to understand the basis of crop responses (rice in particular) to individual and concurrent abiotic stresses. He has devised promising strategies for improving the crop performance under sub-optimum conditions in the context of changing climatic optima, and he has suggested various novel indicators for augmenting stress tolerance in plants. He is also on the editorial board of several international journals.

Contents

Preface

Agriculture is essential to the livelihood of all people and nations, especially in the developing world. Climate change is likely to have an extensive impact on agriculture around the world through changes in temperature, precipitation, concentrations of carbon dioxide, and available water flows. This book entitled "Climate Change and Agriculture" provides the most recent research on the interaction between climate change and the agriculture sector. With contributions from internationally recognized scientists, this volume contains 13 chapters covering the key topics related to climate change hazards, risk assessment, mitigation strategies, and climate-smart agriculture innovations. It offers a solid foundation for the discussion of climate resilience in agricultural systems and the requirements to keep improving agricultural production in the face of mounting climate challenge.

I wish to express my gratitude to the contributing scientists in this edition for their overwhelming response and for readily accepting my invitation. All the contributors not only shared their knowledge, but admirably integrated the scattered information from diverse fields in composing the chapters and efficiently incorporated the editorial suggestions finally to produce this venture. I greatly appreciate their commitment. I am also thankful to the IntechOpen team (especially Ms. Dajana Pemac) for their generous cooperation at every stage of the book production. I hope this volume will be useful to all agriculturists, environmentalists, climate change specialists, policy makers, research scholars, and others concerned with climate change and agriculture.

Dr. Saddam Hussain
Assistant Professor,
Department of Agronomy,
University of Agriculture,
Faisalabad, Pakistan

Chapter 1

Climate Stability and the Origin of Agriculture

Joan Feynman and Alexander Ruzmaikin

Abstract

Although modern man had developed long before the migration from Africa began ~ 55,000 years ago, no agricultural societies developed until about ~ 10,000 years ago. But in the next 5000 years, agricultures developed in several unrelated regions of the world. It was not a chance occurrence that new agricultures independently appeared in the same 5000 years. The question is what inhibited agriculture worldwide for 44,000 years and what changed ~ 10,000 years ago? We suggest that a major factor influencing the development of agricultural societies was *climate stability*. From the experience of several independent cultures, we estimate that the development of agriculture needed about 2000 years of climate free from significant climate variations on time scales of a few centuries.

Keywords: origin of agriculture, climate stability, paleoclimate data, younger Dryas

1. Introduction

One of the most important events in human history was the establishment of agriculturally based societies, that is, societies with fully developed agriculture. Modern human beings (*Homo sapiens sapiens*, Hss) had developed in East Africa by about 195,000 ybp (years before present) [1]. [Note that the dates throughout the chapter are the calibrated by ^{14}C years before 1950.] However, no agriculture appeared during the first 100,000 years after the development of modern man [2]. Even after migrations out of Africa began about 55,000 years ago, no agricultural societies developed during the next 44,000 years. Although before 20,000 ybp, the cave walls in the south of France were being painted so beautifully that we can understand the art [3], there was no agriculture. But then around 10,000 ybp agricultural societies were independently established in many regions during the same few thousand years.

The relationship between climate and the development of agriculture has been widely discussed for many years by both the anthropology and climate scientific communities. Many competing views have been developed primarily based on studies of the archeology of the Near East (for a review see [4]). For example, it had been suggested [5] that agriculture appeared as a result of technological advances that gradually increased man's ability to exploit the environment after man had occupied vast areas of the Earth. Certain conditions were found necessary for the development of agriculture [4] such as the technology for collection, processing, and storage of agricultural products and the presence of potential domesticates in the local environment. Examples included development of improved hunting technology by one group and perhaps experimentation with agriculture by another. The

IntechOpen

increased efficiency of hunting failed as a survival technique, but the experimenting with agriculture may have had more success, and when the stresses of the YD were removed, agricultural development was accelerated. Bar-Yosef [6] emphasized that the initiating event in this view was a response to the environmental stress during the YD. This same idea has been applied to the development of agriculture in China [7].

Here, we discuss a different explanation for the origin and development of agriculture. We examine the proposition that the agricultural development depended on the stability of the climate. It was the decrease of climate variability at the Pleistocene/Holocene boundary (i.e., at the termination of the YD) that allowed the establishment of societies fully based on agriculture [8, 9]. Paleoclimate data from Greenland ice cores and ocean climate proxies show that the last glacial climates were extremely unfavorable for development of agriculture-dry, low in atmospheric CO2 [8], and extremely unfavorable for development of agriculture on short time scales. We hypothesize that agriculture was impossible under the last glacial conditions. The quite favorable for agriculture climate conditions appeared in the Holocene. Note that Rehfeld et al. [10] argued that although glacial-interglacial changes in variability have been quantified for Greenland, a global view may remain elusive. However, the Greenland ice core records faithfully reflect the timing and relative magnitudes of climate variability before and after the start of Holocene and, we believe, still can be used for the study of culture development such as the origin of agriculture.

The current consensus is that agriculture arose independently in several regions of the world located in Asia, South America, Europe, and the Fertile Crescent (an area near the Tigris and Euphrates rivers that spans modern-day Iraq and Syria) after the termination of the YD [11–13]. A probability that agriculture would appear by chance in these independent regions during the same 5000-year period after man left Africa is very small [9]. The first factor is the time when human evolution had progressed to the point where mankind was essentially the same as we are now, that is, the point at which Hss almost certainly had the mental and physical capabilities required for agriculture. A conservative time estimate may be made by considering the Aurignacian people who drew pictures of horses on the walls of the Chauvet cave in Southern France 30,000 ybp (see **Figure 1**) [2]. These ancestors of modern Europeans [14] not only produced art but also were apparently highly organized socially. Since then,there have been roughly six periods of 5000 years each, it is certain that the development of so many independent agricultures in the same 5000-year period did not occur by chance. There must have been something special about that period of time. It seems implausible that it was the release of the stress of the YD and the sudden increase in global mean temperature because during the last 40,000 years, there have been nine sudden increases in temperature, in so-called Dansgaard-Oeschger (DO) events, in addition to the YD termination [15–17], but agriculture developed only after the most recent one, that is, after the transition to the Holocene.

The question is what prevented agriculture for more than 40,000 years after the exodus of man from Africa and what changed after YD about 11,000 years ago? We propose that until the end of the last Ice Age, frequent climate change inhibited the transition from the hunter-gatherer way of life to an agricultural way of life, which became possible due to more stable climate conditions after the end of the YD [8, 9]. We will first give four examples of plant domestication that took place early in that transition (see the next section). Then, we estimate the time required for the transition to be completed based on the information on the most extensively investigated case, the Levant. The paleoclimate data lead us to suggest that transition from a hunter-gatherer-based society to the agriculture-based society required an extended period of the climate stability on characteristic times of centuries. The Greenland ice core sodium ion and oxygen isotope climate proxy records for the wind and temperature from 50,000 ybp until the present show that the climate

Figure 1.
Chauvet cave (35,000–22,000 ybp). Panel of the four horses [3].

variability changed when the Younger Dryas cold period ceased at 11,570 ± 200 ybp [18] and the magnitude of the climate variations on time scales relevant to the development of agriculture decreased markedly. The same conclusion is supported by an analysis of a lower latitude climate proxy data set record taken from the Cariaco sea sediment.

Richerson et al. [8] pointed out to another factor the plant productivity limited by lower atmospheric CO2 during the last glacial that may prevented the development of agriculture, because the CO2 content of the atmosphere was only about 190 ppm during the last glacial, compared to about 250 ppm at the beginning of the Holocene (c.f. http://cdiac.essdive.lbl.gov/trends/co2/ice_core_co2.html).

2. First steps toward agricultural societies

Let us briefly describe the current state of knowledge about the earliest domestication of plants focusing on the four best-studied examples of agricultural societies that had developed independently: the Levant (Middle East), China, MesoAmerica, and the Andes-Amazon area of America.

3. The Levant

The first definite evidence of cultivated cereals in the Levant has been dated to about 10,600–10,000 ybp [19, 20]. Wild cereals were extensively gathered even before domestication took place; their remains are found at various settlement sites. In order to properly date the time of domestication, a distinct marker is needed to distinguish domesticated plants from their wild progenitors. In the case of cereals, such as the wheat, domesticated in the Levant, there are such distinguishing characteristics [19]. For example, in the wild cereals, the seeds ripen over a period of time and leave stems when ripe. In domesticated cereals, all seeds ripen at the same time and are retained on the stems until harvested, which is necessary if a farmer wants to have control of the crop. For dating the onset of agriculture, grain has the advantage that the plant material is not easily destroyed; both wild and domesticated remains of wheat have been recovered from ancient sites in the Levant. The wild grain inadvertently goes via genetic changes required for the domestication. The estimated time required for these changes is of the order of a few centuries [19]. Hence, the climate must remain stable on time scales of a few centuries for even the first step in the development of an agricultural society to take place.

4. China

The main plant in this case is rice. Various types of wild and domesticated rice have a wide distribution in Asia and beyond. Domesticated rice is descended from the wild plant *Oryza rufipogon*, and at least two major types of cultivated rice, *Oryza sativa Japonica* and *Oryza sativa indica*, are currently major crops in Asia, although it is still unknown whether or not these crops are due to one or separate domestications (c.f. [21]). Many authors suggested the sites of domestication in Asia (see [22] and references there in). The best-studied and oldest rice known is from the middle Yangtze Basin from the Diaotonghuan Cave in that region [20]. The time to be assigned to the domestication of the rice is also uncertain and depends on the definition of "domestication." The presence of the double-peaked glume is a good characteristic distinguishing between domesticated and wild rice [20]. Wild rice grew in the Yangtze Basin and was harvested by local people by ~12,000 ybp, that is, before the climate emergence from the Younger Dryas. Domesticated rice is dated to 9000–10,000 ybp. If this latter date is correct, then the remains found in the Diaotonghuan Cave site were the earliest domesticated rice remains found to date. This time agrees with the termination of the YD (see [22] and references therein). As in the case of the Levant, although the wild cereal grain was clearly being utilized before the sudden end of the YD, the evidence for the domesticated counterpart seems to appear shortly after the YD ended.

5. MesoAmerica

Three major crops were domesticated in the early history of agriculture in MesoAmerica: the maize, the common bean, and the squash. The cultivation of maize was widespread, and it exhibits a very large morphological and genetic diversity. Maize samples were studied from its entire pre-Columbian range, which is extending from eastern Canada to northern Chile [23]. From the genetic makeup of these samples, Matsuoka et al. [23] constructed a map showing the relation of the maize types in North and South America and concluded that all of the many types of maize were derived from a single domestication of the wild grass *Teosinte* in the highlands of Mexico about 9000 years ago. Their molecular data are consistent with the date of 6250 ybp for the oldest known fossil maize [13] and with archeological estimates that crop domestication in Mexico did not precede 10,000 ybp. There is some evidence that squash (*Cucurbita pepo*) was also domesticated in the Mexican highlands between 10,000 and 8000 ybp [24].

6. Andean-Amazon region of America

A recent study of domestication of plants in the Andean-Amazon region indicates that squash and gourds were domesticated there very early [25] and

Region	Domesticated plant	Dates of the oldest remains
Levant	Wheat (emmer and einkorn)	10,600–10,000 ybp
China	Rice (*Oryza sativa japonica*)	10,000–9000 ybp
MesoAmerica	Maize	~ 9000 ybp
Andean-Amazonian	*Cucurbita pepo* *Cucurbita*	9000–7000 ybp 10,000–9000 ybp

Table 1.
Early domesticated plants.

independently from their domestication in Mexico. *Phytoliths* recovered from two sites in southwest Ecuador have been dated to 10,100–9300 ybp [25]. As in the cases of wheat, rice, and maize, there is evidence that the wild precursors were exploited earlier. The highly cultured societies, which were present when the Europeans arrived, were strongly dependent on the cultivation of the potato in the Andean highlands. The research, which indicates that Peru was the only site of potato domestication [26], gives no reliable time estimate for the development of that crop.

Table 1 gives the cited regions, the domesticated plants, and the time known for domestication.

Table 1 indicates that the initial domestications in the four regions took place in the same period. The development of independent agricultures in these regions must have been due to something that occurred in each of the regions at the same time, which we identify with the onset of relative stability in the climate. Climate instability can be expected to strongly inhibit agriculture since agricultural societies are dependent on a relatively few species compared to simple foraging societies. For example, six species of large prey animals have been reported in the Levant before the termination of the YD ([27] but only sheep and goats are domesticates). As far as plants are concerned, each plant, whether domestic or wild, thrives best in a specific growing environment. If the environment changes so that it becomes too far from the plant's required conditions, the crop fails. For an agricultural society, this can be catastrophic because of the limited number of plants utilized and the permanence of the settlement site. It is interesting to note that the plants domesticated differed from region to region and the climates of the four regions differed widely. This supports a hypothesis that it was not the specific values of the local climate parameters (annual rainfall or mean temperature) that were of foremost importance in the inhibition of agriculture but the stability of these parameters.

7. Time required for establishing an agricultural society: the Levant

The regions listed in **Table 1** went on to complete the establishment of the four independent agricultural societies [28] are compared in **Table 2**. Some of the main changes to be accomplished to become an established agricultural society are listed in **Table 3**.

The independence of the four agricultures is borne out by the diversity of species involved in each case as shown in column 2. Each area developed its own distinct constellation of domesticated species. The last column gives the date at which the complex agriculture-based society appears to have been well established and shows that they were all established within the same 5-millennium time period.

Each of the changes indicated in **Tables 2** and **3** has its own problems, and the order in which they are accomplished may differ from place to place, but all of them must be carried out to complete the development of an agricultural way of life. In

Location	Species	Development accomplished
Levant	Wheat, barley, chickpeas, flax, sheep, goats	~ 9000 ybp
China	Rice, millet, pigs, silkworms	by 9000 ybp
Mesoamerica	Corn, beans, squash, turkey	by 5500 ybp
Andean-Amazonian	Potato, manioc, guinea pig, llama	by 5500 ybp

Table 2.
Four agricultural societies developed independently.

Item	Hunter-gatherers	Agricultural society
Food plants	Wild grains, fruit, tubers	Domesticated counterparts
Animals	Many species of wild prey	A few domesticated species
Clothing	Wild animals, vegetable fibers	Domesticated counterparts
Tools	Projectiles and traps	Farming tools, cereal preparation, food storage techniques
Housing	Temporary, easy to erect	Permanent structures
Settlement patterns	Small bands	Villages, towns

Table 3.
Main steps in transition from hunter-gatherer to agricultural societies.

addition to the genetic changes required to produce a domesticated plant or animal, fundamental changes in technology and social structure are needed. Although the time to domesticate one particular plant may be of the order of a few centuries [19], the total time it takes to change from a hunter-gatherer society to an agricultural one is much longer. Here, we estimate this time scale for the best-studied case, the Middle East, using archeological investigations and studies of wild progenitors of cultivated plants [19]. There is also information on animal domestication that comes from bones found in archeological sites [4, 29]; housing, tool development, and settlement patterns are also available from archeological studies [4, 20]. However, much of the material from earlier archeological studies, while provocative, is of limited use because reliable dating methods are still not perfect.

The earliest agricultural society was developed in the southern Levant, an area that includes southern Syria and Lebanon, Israel, Palestine, Jordan, and the Sinai Peninsula during the cultural-historical time known as the Pre-Pottery Neolithic (PPN) that spanned the years between about 11,700 and 8250 ybp [6]. In the well-studied Israeli section of the Levant, much of the area was an open grassland with wild cereals and pistachio trees [6]. Local Levant climate history has been derived from speleothems in Israeli caves (at 32°N) [29] that have recorded a proxy signal of the Eastern Mediterranean region. An analysis of the data since 18,000 ybp shows several markedly different climates. The most interesting time for our purposes is the ~3000-year period that has a rainfall of 675–950 mm, almost twice the present-day values, and the Dead Sea reached its maximum level. The onset of this wet episode determined within the accuracy of the time determinations corresponds to the end of the YD. After 8000 ybp, the temperature and rainfall in the Levant became more similar to the current values. The plant remains were combinations of cereals, pulses, and flax and were similar to the plants that appeared later all over the Near East. Three cereals were cultivated: emmer wheat, einkorn wheat, and barley. Cereals alone, however, do not provide all of the nutrition required by man. In the case of the Levant, the cereals were added by legumes (lentils, peas, bitter vetch, and chickpeas) [20]. Sheep and goats were also domesticated and replace gazelles, and other wild games have previously been hunted [19]. During the PPN, the size and complexity of settlements increased by a factor of 14 [20], but the earliest of the PPN sites show some cultural continuity with the preceding final phase of the preagricultural Natufian period (12,500–12,000 ybp). The agriculture was not brought to the region by an invading force due to total absence of evidence for interpersonal or intercommunity aggression or violence during the PPN.

The earliest evidence of cultivated cereals in the Levant has been dated to ~ 10,600–10,000 ybp when the remains of emmer and einkorn wheats show the telltale signs of domestication. The fact that such remains were found in several sites during this period implied that the actual beginning of wheat cultivation in this area was earlier, perhaps as early as the first part of the Pre-Pottery Neolithic (PPNA) (11,700–10,500 ybp) [6]. Kislev et al. [30] presented evidence that during this same period figs appear to have been gathered from trees grown intentionally from planted branches. The development of agriculture so soon after the termination of the YD-led Bar-Yosef et al. [6] to suggest that the first experiments in systematic cultivation may have occurred during the YD which ended abruptly ~11,750 ybp. However, they report that no remains of domesticated plants have yet been recovered from the YD itself [6]. Although there is some uncertainty in the date of the beginning of wheat farming, ~11,000$ ybp is a reasonably conservative estimate of that date.

Thus, the development of the earliest agricultural societies would be encouraged by the absence of large century-scale climate variability during a period of at least 2000 years.

8. When climate was stable on these time scales

Here, we discuss the variations in the Earth's climate derived from two of climate proxy records (CPR): the polar ice cores in Greenland and the sea sediments from the Cariaco Basin of the Northern coast of South America.

The best and the most discussed of these data banks is from the polar ice cores in Greenland, which contains climate memory of diverse climate variables [31]. The oxygen isotopes in snow characterize temperatures [16, 32], while the dust blown from the deserts and the sea salt blown from the ocean characterize the atmospheric wind.

The records from the Greenland Ice Sheet Project 2 (GISP2) cover the time period from 110,000 ybp to the present, although with differing sampling rates and accuracy. The variability found in these records indicates a nonlinear nature of the climate variations. For example, using a composite of the time series of the Ca, Na, Cl, SO_4, K, and Mg ions and a narrowband filtering technique, Mayewski et al. [31] found that between 110,000 and 11,000 years ago there was a variation with a persistent period of 1450 years but with time-varying amplitude.

To take into account the nonstationary and nonlinear character of the climate paleo records, we apply the empirical mode decomposition (EMD) techniques [33], which are especially designed for analyses of nonlinear and nonstationary time series. The EMD represents the data as the sum of a small number of empirical orthogonal modes that have time-variable amplitudes and instantaneous frequencies capturing the nonstationary spectral content of the data. This method employs empirical basis that is changing in time to adapt to the actual variability of the data so that the selection of the modes is equivalent to locally adaptive filtering of the signal. The lack of leakage from one power to another in the EMD method presents an advantage over narrowband filtering and many other techniques. A set of the EMD modes that have mean periods in the range shorter than about 300 years provides a detailed characterization of the climate variability in the frequency range relevant to the development of agriculture.

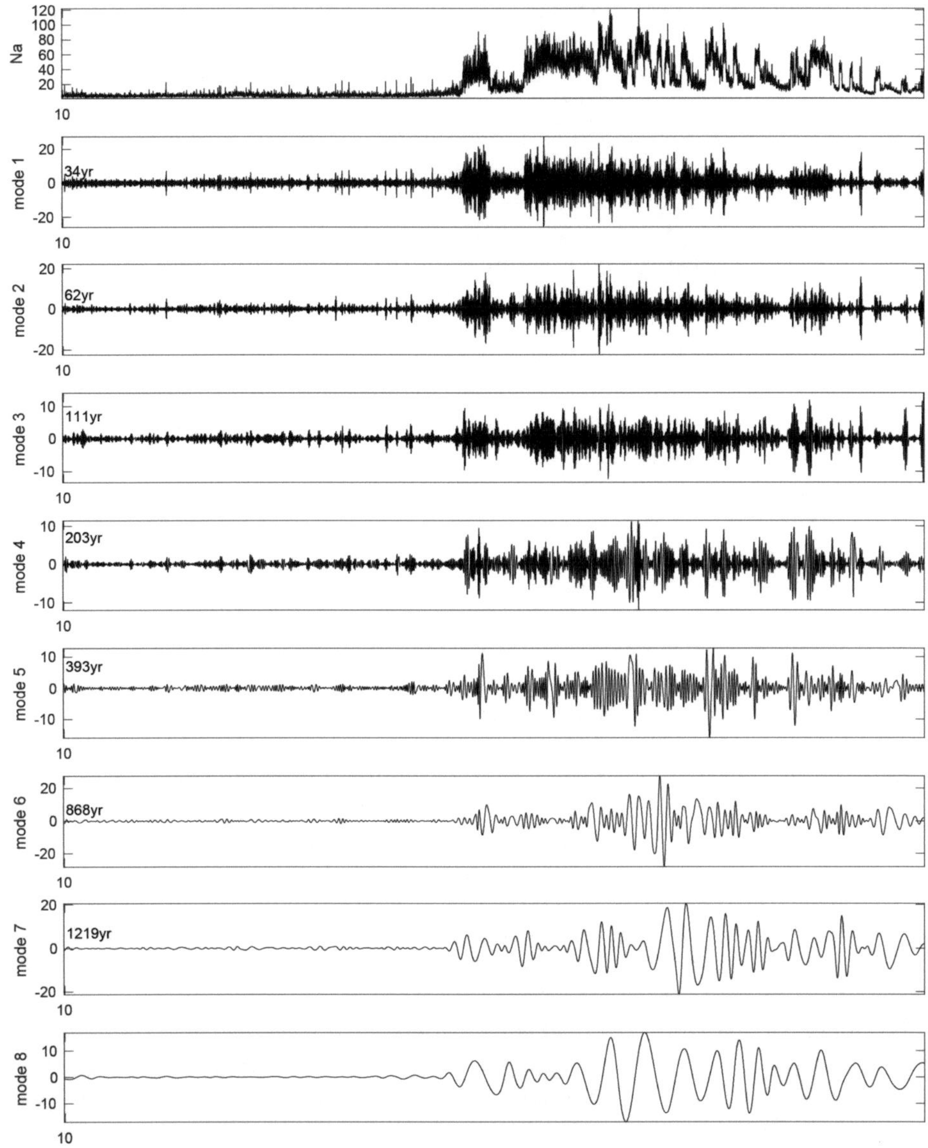

Figure 2.
The GISP2 ice core data for the Na ion (upper panel). To obtain a uniformly sampled Na data set, we interpolated between the gaps in the original GISP2 unevenly spaced data and resampled with a time cadence of 10 years using the piecewise cubic Hermite interpolating polynomial, which preserves the shape of the data and respects monotonicity. This procedure works well for the most recent 50,000 years of the data set because there are no large unevenly spaced data gaps. The eight lower panels show the decomposition of the data variations into EMD modes. Each EMD mode varies in amplitude (seen in this figure) and frequency (can be estimated by the inverse quasi-period between zero crossings). The inverse frequencies of the EMD modes are shown in the left upper corner for each mode. At any given time, the sum of these modes equalizes the data. The mode amplitudes are scaled to the maximum of the data. The decrease in the amplitudes of variations at about 11,000 ybp is evident in all panels.

First, we apply the EMD to analyze the behavior of the concentration of the Na ion, which characterizes the meridional winds transporting ions toward the North pole [31, 34]. The resulting data time series is shown in the top panel of **Figure 2**.

Figure 2 clearly shows strong variations in all of the modes between 50,000 ybp and 11,000 ybp (with the possible exception of a short period at about 44,000 ybp). There is a sharp decrease in the amplitude of all modes at the termination of the Younger Dryas.

To characterize the Greenland temperature (**Figure 3**), we use the GISP2 bi-decadal ^{18}O record based on measurements done at the Quaternary Isotope Laboratory, University of Washington, and the calibration given in Ref. [34]:

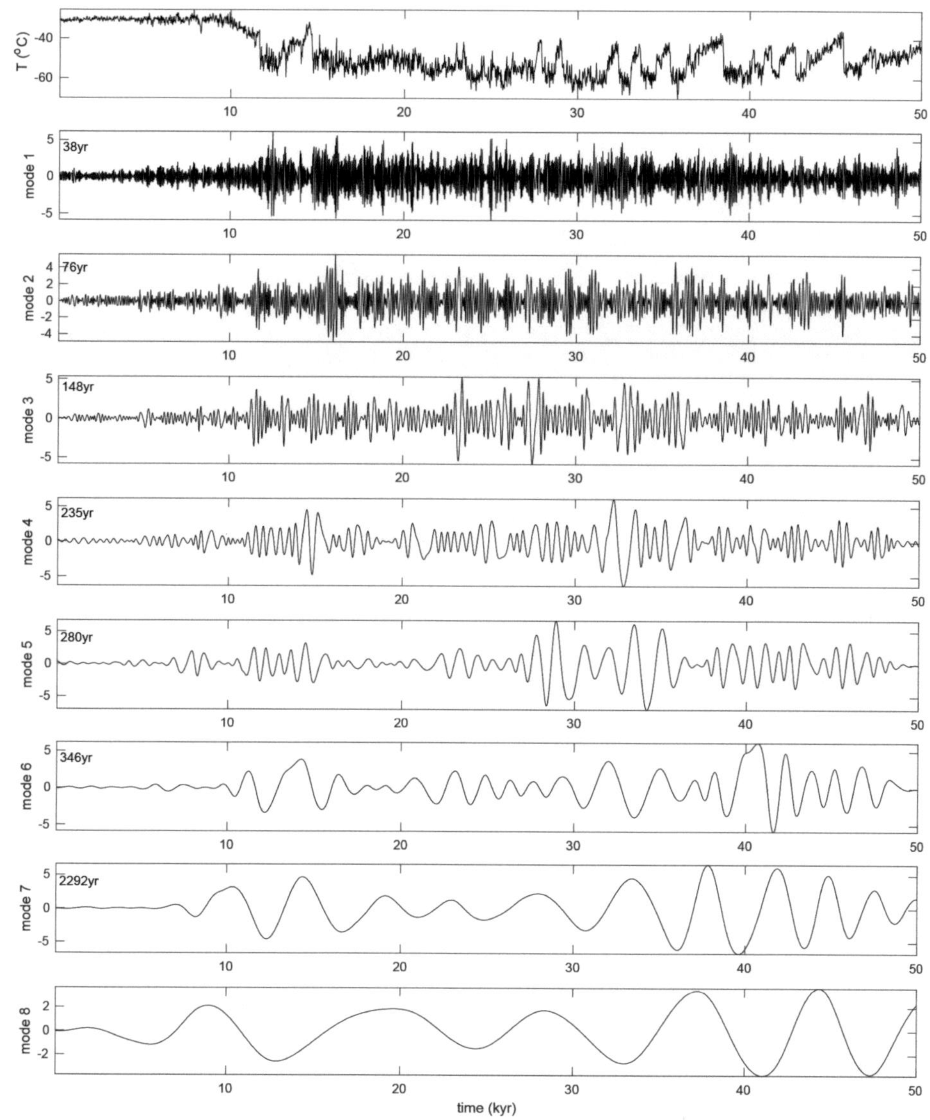

Figure 3.
*(Upper panel) The air temperature estimated from the GISP2 $\Delta^{18}O$ data. The evolution of the temperature, which is given at a 20-year cadence, its EMD modes in the relevant frequency range, and the characteristic quasi-periods are shown (see the upper left side of the figure). In accordance with the results for the Na ion shown in **Figure** 2, we see that the amplitudes of the variations of temperature were much greater before the Holocene. The warming at the termination of the YD was very abrupt and may have been accomplished in only a single decade at 11,570 ± 200 ybp [18].*

$$T = a\Delta^{18}O + b \tag{1}$$

where (a,b) = (2.15, 43.4) for t < 0.5 kyr, (a,b) = (3.99, 108.0) for 0.5 < t < 3 kyr, (a,b) = (3.98, 207.7) for 3 < t < 8 kyr, and (a,b) = (3.05, 75.4) for t > 8 kyr.

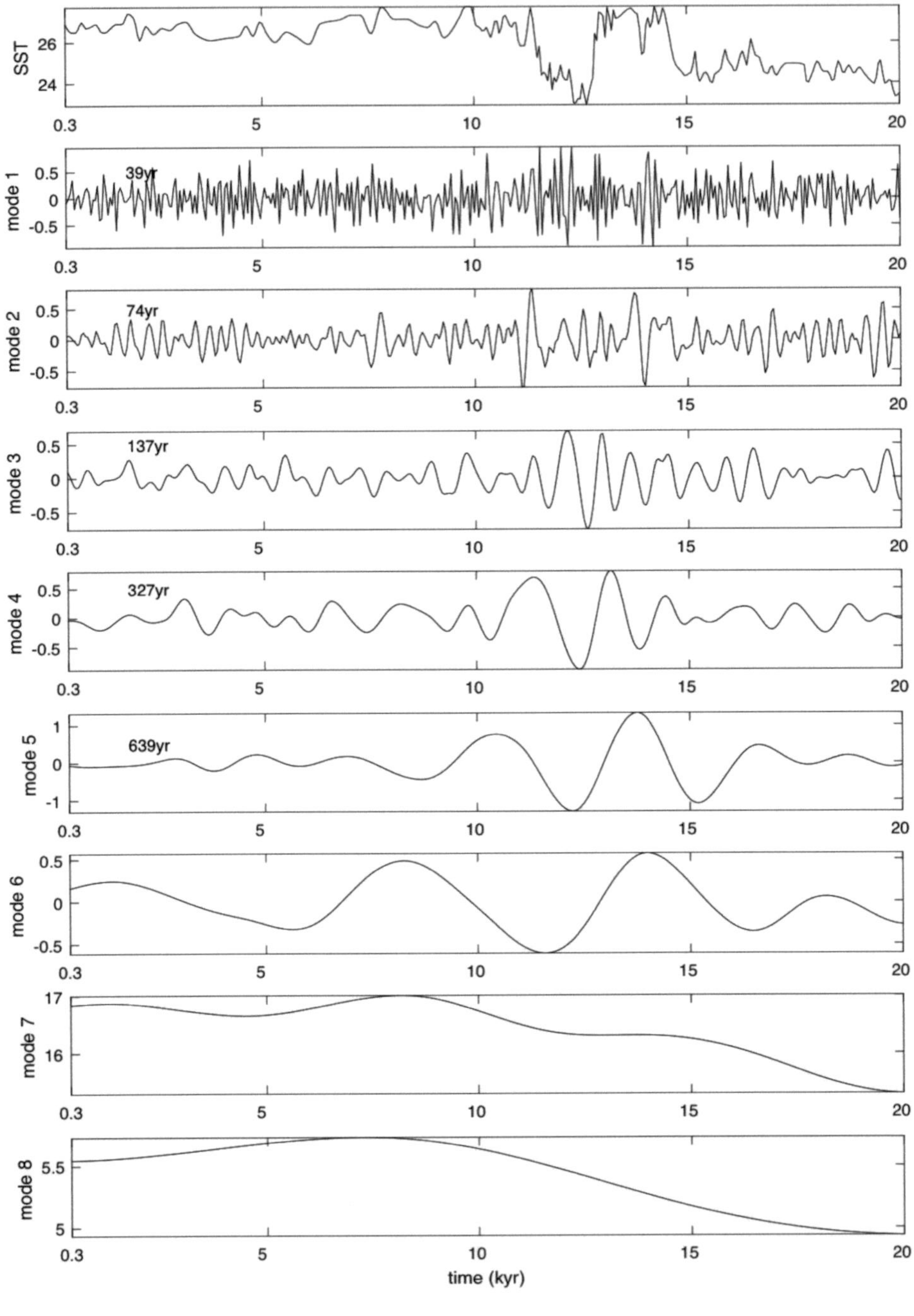

Figure 4.
(Left rows) Top most panel shows the SST data in the Cariaco Basin, and the lower panels show the amplitudes of the EMD modes with the mean quasi-periods marked on the left side. Compared with GISP2 (see ***Figures*** *2 and 3), these data have lower resolution, and the time interval without major data gaps is more limited. We interpolated the data with a 50-year time cadence in the time interval from 300 ybp to 20,000 ybp. Note the presence of ~1500-year period in these data as well as in the GISP2 data. The amplitudes of all modes are large before and during the YD and decrease when the YD ends.*

Thus, in from 50,000 ybp until the recovery from the YD ~11,500 ybp, there were always (except perhaps about 44,000 ybp) large amplitude variations in both the winds and the temperature. The high amplitude of the variations continued throughout the YD, but the amplitude dropped precipitously at the end of that period. The figures also show that data meet the stability requirements we indicated above in our discussion of the development of agriculture (i.e., 2000 years of stability against variations with periods ≤300 years) after the YD termination and not before.

As an example of climate variability found at low latitudes, we show the EMD (**Figure 4**) of the sea surface temperature (SST) from sediments of the Cariaco Basin (~10°N) off the coast of Venezuela [35]. This record shows essentially the same variations as the $\Delta^{18}O$ from the GISP2 record [35, 18] and directly characterizes the MesoAmerican climate during the summer and early fall [56]. Compared with GISP2, these data have lower resolution than the GISP2 data, and the time interval without major data gaps is more limited. The data from 300 to 20,000 ybp are utilized here. We interpolated the data with a 50-year time cadence. Note that the ~1500-year period is present in these data as well as in the GISP2 data. The amplitudes of all modes are large before and during the YD and decrease when the YD ends. We also found that the spectral content of the EMD modes obtained from Cariaco data resembles the content of the modes obtained from Greenland ice core data smoothed with similar 50-year cadence in that both data sets show large high-amplitude, high-frequency changes in climate when there was no agriculture and that these variations became much smaller before agriculture appeared.

9. On relationship between climate variability found in widely separated regions

Current studies show that climate variations recorded in geographically widely separated regions are strongly interrelated. In addition to the climate records analyzed above, there are many other climate proxy records (CPR) obtained from widespread regions of the Earth including coral cores from several oceans [36] and stalagmites from Soreq caves in the area of the ancient Levant [29]. Many CPR have been intercalibrated to obtain well-dated records of tracers of worldwide patterns in climate change [36, 32]. Comparisons of these records demonstrate the Northern Hemisphere wide extent of both millennium scale [38] and more rapid climate changes [39]. Thus, the abrupt termination of the YD cold-dry period has been detected throughout the Northern Hemisphere in Greenland, Western Europe, North America, and Central America, off the Venezuelan coast, in the Middle East, with some evidence from Central Africa and the Indian Ocean [35, 40–42].

The agreement of the climate variability found from the GISP2 data and the Cariaco data is an example of the climate tele-connections between distant regions of the Earth [43, 31] that have been demonstrated by ocean and atmospheric observational and modeling studies for both the Pleistocene [44] and Holocene [45]. Intercomparison of data shows that climate variations in the Arctic are linked to variations in Antarctic and to lower latitudes [37, 39, 46]. Here are some examples: Grootes and Stuiver [16] compared the ice core records with the deep ocean $\Delta^{18}O$ records from Atlantic and Pacific Oceans and with tree-pollen land records in North America and Europe; Bond et al. [47] established correlations between climate records from North Atlantic Ocean and Greenland ice; Barlow et al. [34] established a link between stable isotope ratios (for deuterium) found from GISP2 ice cores data and the North Atlantic Oscillation (NAO) for the time period 1840–1970; and Wang et al. [48] linked the Asian monsoon in Southern China with the climate in the North

Atlantic. Century-scale temperature variations in the Greenland Ice record [17] have also been related to the NAO [49]. The NAO is associated with a sea surface temperature anomaly having a meridional average tripole pattern: cool north of 55°N, warm in 20°–55°N latitudinal band, and cool south of 20°N in the positive phase of the NAO [50]. The Red Sea coral cores in an area close to the initial domestication of wheat also currently reflect variations in the North Atlantic Oscillation [36]. The Cariaco Basin data directly reflect the MesoAmerican rainfall during the summer and early fall in the Yucatan peninsula [50] near where corn was first domesticated. The NAO is one of several large-scale climate circulation patterns (such as Pacific Decadal Oscillation, Northern Annular Mode, Aleutian-Iceland Seesaw, and Cold Ocean-Warm Land patterns) that can be expressed as made up of different combinations of the current first and second Empirical Orthogonal Functions of the Northern Hemisphere winter sea-level pressure [45]. In these large-scale patterns, a climate variable, such as temperature, is above its mean value in one area of the coherent climate pattern and below its mean value in another geographically remote part of the pattern (and vice versa). The global mean is only weakly, if at all, affected, and the pattern temperature variability is not necessarily related to a change in global temperature. The pattern variability sometimes manifests itself as a prolonged statistical preference of one of the two basic states. For example, during the Little Ice Age, the temperature pattern tended to be in a state with a cold northern Eurasia more frequently than during more recent times [51, 52].

Studies such as these indicate that the climate variations observed in the GISP2 and Cariaco data reflect worldwide patterns that mankind would have experienced during the Pleistocene as well as the Holocene. It is not important for the development of agriculture, if these Pleistocene patterns are exactly the same as those we are familiar with from the Holocene. In fact, modeling suggests that they were not identical, but that the large-scale worldwide coherence was maintained [29, 42]. The large-scale coherence is an important feature because it strongly implies that the century-scale variations seen in Greenland and Cariaco Basin are parts of interrelated worldwide climate variations.

10. Conclusion

When an agricultural society is developing, it may not be important if the local climate tends to be colder or warmer and dryer or wetter. What is important is that the local climate remains stable enough so that the crops and the livestock being domesticated continue to thrive.

It appears that from the time man left Africa about 50,000 ybp until 11,750 ybp there was essentially continuous climate variability in the period range of a century to a few centuries and no agriculture-based societies developed. These climate variations quieted at the beginning of the Holocene after the Younger Dryas terminated and were quickly followed by the development of several agricultural societies. It was the intense Pleistocene *climate variability* that prevented agriculture from developing until the onset of the relatively stable Holocene [8, 9]. This suggestion is supported by studies of the responses of already well-established agricultural societies to the relatively mild periods of climate variability that have taken place during the Holocene. For example, there was a weakly variable climate event at about 4000 years ago. This event strongly disrupted the Neolithic culture of Central China [48] as well as destroying the Egyptian Old Kingdom circa 4250–3950 ybp [53] and Akkadian in Mesopotamia 4170 ± 150 ybp [54]. A later period of climate variability observed in the Cariaco Basin sediments was accompanied by the fall of the classical Maya civilization in the Yucatan during the ninth century [55].

We argued that the conditions required for the development of agricultural societies include about a millennium or more periods during which there are no large-century-scale climate variations. We have presented evidence that this has been the case in the Northern Hemisphere since the end of the Younger Dryas but not during the last several tens of millennium of the preceding Pleistocene and suggested that this resulted in the failure to develop agricultural-based societies until after the termination of the Younger Dryas. We conclude that there is considerable evidence that climate variability inhibited the development of agriculture until ~11,000 ybp when relative climate stability was established and many independent agricultural systems were developed.

List of abbreviations

~	approximate
ybp	year before present (1950)
Hss	*Homo sapiens sapiens*
YD	younger Dryas
DO events	Dansgaard-Oeschger climate events
PPN	pre-pottery neolithic
CPR	climate proxy records
GISP2	Greenland Ice Sheet Project 2
EMD	empirical mode decomposition
SST	sea surface temperature
NAO	North Atlantic Oscillation

Author details

Joan Feynman and Alexander Ruzmaikin*
Jet Propulsion Laboratory, California Institute of Technology, Pasadena, CA, USA

*Address all correspondence to: alexander.ruzmaikin@jpl.nasa.gov

References

[1] The Oldest *Homo sapiens*: Fossils Push Human Emergence Back To 195,000 Years Ago [ScienceDaily]. Available from: www.sciencedaily.com/releases/2005/02/050223122209.htm [Accessed: 28 February 2005]

[2] Wells S. The Journey of Man: A Genetic Odyssey. Princeton University Press; 2003. ISBN: 0-8129-7146-9

[3] Clottes J. La Grotte Chauvel, L'art Des Origine. Paris: Editions du Seuil; 2001

[4] Henry DO. From Foraging to Agriculture: The Levant at the End of the Ice Age. Philadelphia: University Pennsylvania Press; 1989

[5] Braidwood R. The agricultural revolution. Scientific American. 1960;**203**:130-148

[6] Bar-Yosef O. The Natufian culture in the Levant, threshold to the origins of agriculture. Evolutionary Anthropology. 1998;**6**:159-177

[7] Cohen DJ. The origin of domesticated cereals and the Pleistocene-Holocene transition in eastern Asia. The Review of Archaeology. 1998;**19**:22-29

[8] Richerson PJ, Boyd R, Bettinger RL. Was agriculture impossible during the Pleistocene but mandatory during the Holocene? A climate change hypothesis. American Antiquity. 2001;**66**(3):387-341

[9] Feynman J, Ruzmaikin A. Climate stability and the development of agricultural societies. Climatic Change. 2007;**84**:295-311. DOI: 10.1007/s10584-007-9248-1

[10] Rehfeld K, Münch T, Ho SL, Laepple T. Global patterns of declining temperature variability from the last glacial maximum to the Holocene. Nature. 2018;**354**. DOI: 10.1038/nature25454

[11] Piperno DR, Pearsall DM. The Origins of Agriculture in the Lowland Neotropics. San Diego, CA: Academic Press; 1998

[12] Piperno DR. On maize and the sunflower. Science. 2001;**292**:2260-2261

[13] Piperno DR, Flannery KV. The earliest archaeological maize (*Zea mays* L.) from highland Mexico: New accelerator mass spectrometry dates and their implications. Proceedings of the National Academy of Sciences of the United States of America. 2001;**98**:2101-2003

[14] Semino O et al. The genetic legacy of Paleolithic Homo sapiens sapiens in extant Europeans: A Y chromosome perspective. Science. 2000;**290**:1155-1159

[15] Dansgaard W, Jonsen SJ, Clausen HB, Dahl-Jensenm NS, Gundstrup NS, Hammer CU, et al. Evidence for general instability of past climate from a 250-kyr ice-core record. Nature. 1993;**364**:218-220

[16] Grootes PM, Stuiver M. Oxygen 18/16 variability in Greenland snow and ice with 10^{-3} to 10^{5} -year time resolution. Journal of Geophysical Research. 1997;**102**:26455-26470

[17] Ditlevsen PD, Kristensen MS, Andersen KK. The recurrence time of Dansgaard-Oeschger events and limits on the possible periodic component. Journal of Climate. 2005;**18**:2594-2603

[18] Hughen KA, Southon JR, Lehman SJ, Overpeck JT. Synchronous radiocarbon and climate shifts during the last deglaciation. Science. 2000;**290**:1951-1954

[19] Zohary D, Hopf M. Domestication of Plants in the Old World. 3rd ed. Oxford University Press; 2000

[20] Kuijt I, Goring-Morris N. Foraging, farming, and social complexity in the pre-pottery neolithic of the southern Levant: A review and synthesis. Journal of World Prehistory. 2002;**16**:361-439

[21] Morishima H. Evolution and domestication of rice, rice genetics IV. In: Khush GS et al. Proc. of 4th Intern Rice Genet Symp; 2000; India & IRRI Philippines: Science Publishers; 2001

[22] Zhao Z. The middle Yangtze region in China is one place where rice was domesticated: Phytolith evidence from the Diaotonghuan cave, northern Jiangxi. Antiquity. 1998;**72**:885-897

[23] Matsuoka Y, Vigouroux Y, Goodman MM, Sanchez J, Buckler ES, Doebley JF. A single domestication for maize shown by multilocus microsatellite genotyping. Proceedings of the National Academy of Sciences of the United States of America. 2002;**99**:6080-6084

[24] Smith BD. The initial domestication of Cucurbita pepo in the Americas 10,000 years ago. Science. 1997;**276**:932-934

[25] Piperno DR, Stothert KE. Phytolith evidence for early Holocene Cucurbita domestication in Southwest Ecuador. Science. 2003;**299**:1054-1057

[26] Spooner DM, McLean K, Ramsey G, Waugh R, Bryan GJ. A single domestication for potato based on multilocus fragment length polymorphism genotyping. Proceedings of the National Academy of Sciences of the United States of America. 2005;**12**(41):14694-14699

[27] Turnbull PF, Reed CA. The fauna from the terminal Pleistocene of Palegawra cave. Fieldiana Anthropology. 1974;**63**:81-146

[28] Diamond J. Evolution, consequences and future plant and animal domestication. Nature. 2002;**418**:700

[29] Bar-Matthews M, Ayalon A, Kaufman A. Late quaternary paleoclimate in the eastern Mediterranean region from stable isotope analysis of speleothems at Soreq cave. Quaternary Research. 1997;**47**:155-168

[30] Kislev ME, Hartmann A, Bar-Yosef O. Early domesticated fig in the Jordan valley. Science. 2006;**312**:1372-1375

[31] Mayewski PA, Meeker LD, Twickler MS, Whitlow SI, Yang Q, Lyons WB, et al. Major features and forcing of high latitude northern hemisphere atmospheric circulation using a 110,000 yearlong glaciochemical series. Journal of Geophysical Research. 1997;**102**:26345-26366

[32] Cuffey KM, Clow GD. Temperature, accumulation, and ice sheet elevation in Central Greenland through the last deglacial transition. Journal of Geophysical Research. 1997;**102**:26383-26396

[33] Huang NE, Shen Z, Long SR, Wu MC, Shih HH, Zheng Q, et al. The empirical mode decomposition and the Hilbert spectrum for nonlinear and non-stationary time series analysis. Proceedings of the Royal Society of London. 1998;**A45**:903-995

[34] Barlow LK, White JWC, Barry RC, Rogers JC, Grootes PM. The North Atlantic oscillation signature in deuterium and deuterium excess signals in the Greenland ICE CORE sheet project 2 core, 1840—1970. Geophysical Research Letters. 1993;**20**:2901-2904

[35] Lea DW, Pak DK, Peterson LC, Hughen K. Synchroneity of tropical and high-latitude Atlantic temperatures over the last glacial termination. Science. 2003;**301**:1361-1364

[36] Rimbu N, Lohmann G, Felis T, Pätzold J. Arctic oscillation signature in a Red Sea coral. Geophysical Research Letters. 2001;**28**:2959-2962

[37] De Angelis MJ, Steffensen P, Legrand M, Clausen H, Hammer C. Primary aerosol (sea salt and soil dust) deposited in Greenland ice during the last climatic cycle. Comparison with East Antarctic records. Journal of Geophysical Research. 1997;**102**:26598-26681

[38] Clark PU, Webb RS, Keigwin LD, editors. Mechanisms of Global Climate Change at Millennial Time Scales. Vol. 112. Washington, D. C.: Geophysical Monograph Series, American Geophysical Union; 1999

[39] Peterson LC, Haug GH, Hughen KA, Rohl U. Rapid changes in the hydrologic cycle of the tropical Atlantic during the last glacial. Science. 2000;**290**:1947-1951

[40] Schulz H, Von Rad U, Erlenkeuser H. Correlation between Arabian Sea and Greenland climate oscillations for the past 110,000 years. Nature. 1998;**393**:54-57

[41] National Research Council. Abrupt Climate Change: Inevitable Surprises, National Academy of Sciences Report NAS2002. Washington, D.C.: National Academy Press; 2002

[42] Genty D, Blamart D, Ouahdl R, Gilmour M, Baker A, Jouzel J, et al. Precise dating of Dansgaard-Oeschger climate oscillations in Western Europe from stalagmite data. Nature. 2003;**421**:833-837

[43] Rind D et al. Components of the ice age circulation. Journal of Geophysical Research. 1987;**92**:4241-4281

[44] Toracinta ER, Oglesby RJ, Bromwich DH. Atmospheric response to modified CLIMAP ocean boundary conditions during the last glacial maximum. Journal of Climate. 2004;**17**:504-522

[45] Quadrelli R, Wallace JM. A simplified linear framework for interpreting patterns of northern hemisphere wintertime climate variability. Journal of Climate. 2004;**17**:3728-3744

[46] Yiou F, Fuhrer PK, Meeker LD, Jouzel J, Johnsen S, Mayewski PA. Paleoclimate variability inferred from the spectral analysis of Greenland and Antarctic ice-core data. Journal of Geophysical Research. 1997;**102**:26441-26454

[47] Bond G, Broecker W, Johnsen S, McManus J, Labeyrie L, Jouzel J, et al. Correlations between climate records from North Atlantic sediments and Greenland ice. Nature. 1993;**365**:143-147

[48] Wang Y et al. The Holocene Asian monsoon: Links to solar changes and North Atlantic climate. Science. 2005;**308**:854-857

[49] White JWC, Barlow LK, Fisher D, Grpptes P, Jouzel J, Johnsen SJ, et al. The climate signal in the stable isotopes of snow from summit, Greenland: Results of comparisons with modern climate observations. Journal of Geophysical Research. 1997;**102**(C12):26425-26439

[50] Seager R, Kushnir Y, Visbeck M, Naik N, Miller J, Krahmann G, et al. Causes of Atlantic Ocean climate variability between 1958 and 1998. Journal of Climate. 2000;**13**:2845-2862

[51] Shindell DT, Schmidt GA, Mann ME, Rind D, Waple A. Solar forcing of regional climate change during the maunder minimum. Science. 2001;**294**:2149-2152

[52] Ruzmaikin AJ, Feynman XJ, Noone DC, Waple AM, Yung YL. The pattern of northern hemisphere surface air temperature during prolonged periods of low solar output. Geophysical Research Letters. 2004;**31**:L12201. DOI: 10.1029/2004GL019955

[53] Butzer KW. Early Hydraulic Civilization in Egypt. Chicago, London: University of Chicago Press; 1976. p. 54

[54] Cullen HM, deMenocal PB, Hemming S, Hemmiing G, Brown F-H, Guilderson T, et al. Climate change and the collapse of the Akkadian empire: Evidence from the deep sea. Geology. 2000;**28**:379-382

[55] Hodell DA, Brenner M, Curtis JH, Guilderson T. Solar forcing of drought frequency in the Maya lowlands. Science. 2001;**292**:1367-1370

[56] Haug G, Günther D, Peterson LC, Sigman DM, Hughen KA, Aeschlimann B. Climate and the collapse of maya civilization. Science. 2003;**299**:1731-1735

Chapter 2

Climate Smart Interventions of Small-Holder Farming Systems

Asmat Ullah, Ishfaq Ahmad, Habib-ur-Rehman, Umer Saeed, Ashfaq Ahmad, Abid Mahmood and Gerrit Hoogenboom

Abstract

Agriculture is very vulnerable to temperature and drought in semi-arid and arid regions. Farming communities are especially vulnerable to the potential impact of climate change on crop and livestock. For Pakistan, a potential increase of 2.8°C for the maximum day temperature and 2.2°C decrease in night temperature by the mid-century has been reported. The goal of this chapter is to introduce climate-smart interventions as mitigation and adaptation strategies coupled with crop diversification through the introduction of climate resilient crops in existing cropping systems. Firstly, it describes the impacts of climate change in context to current food security situation in Pakistan and, secondly, potential climate smart interventions to combat changes in the country. Crop models, their application for developing adaptations, modeling technique and its integration with breeding, remote sensing and its application, policy interventions and resource smart interventions in context to changing climate are imperative means to favor the farming community in future farming. Introducing climate resilient crops can be rescued and recognized in dry and hot areas of Pakistan using climate smart interventions and resource use efficiency may be determined with the aid of computer and decision support IT tools in resource inefficient areas.

Keywords: climate change adaptations, strategies, farming systems, small farmers, Pakistan

1. Introduction

Agriculture is significantly affected by temperature variability and other climatic variables (events) all over the world including Pakistan. Farmers are feeling the worst impressions of climate change and variability on produces and livings. We must ensure nutritious food for all through increasing production even up to 60% in 2050 [1], while fighting changes in climate in the world.

In such conditions, it is important for a country to make its agriculture sector efficient to ensure and enhance food security. In recent decades, high temperature has been recorded in most parts of Asia and Pacific regions. According to the global climate risk index (CRI) 2015, Pakistan is the 10th most vulnerable country to climate change with a CRI score of 31.50, 3989 million US$ loss and 141 events during 1994–2013 [2]. The other countries are Honduras, Myanmar, Haiti, Nicaragua, Philippines, Bangladesh, Vietnam, Dominican Republic and Guatemala.

Less developed countries (LDCs) are generally more vulnerable to climate change than industrialized countries. In context to future climate change, CRI may assist as an alarm for previously current vulnerability that may further extended in regions where risky events will become more recurrent or more severe due to climate change. The Fifth Assessment Report of IPCC stressed that the risks associated with extreme weather events will further increase regularly. These risks are likely to get worse and uneven distributed in tendency.

There are two main growing seasons of crops in Pakistan, namely winter (Rabi) and summer (Kharif) season. The performance of winter season crops (wheat, chickpea, barley, sugar beet, etc.) and summer season crops (cotton, pearl millet, etc.) depends upon specific weather conditions. Each crop has its own tendency to face climate change impacts which depends on its genetic makeup. While management options (irrigation, fertilization, planting geometry and density, etc.) applied to get good yields by a farmer. The response of each crop to change in climate is different. Climate change commonly affects crops through changes in temperature and precipitation. In Pakistan, it is assessed that temperature will increase by 3°C by 2040 and 4–6°C by the end of this century. Due to this scenario, Asia can lose 50% of its wheat production [3].

Wheat, rice and maize are important cereals bearing huge population pressure in context of food demand all over the world including Pakistan. But these cereals are at more risks to climate change. The wheat and rice production may decrease up to 15 and 17%, respectively by midcentury (2040–2060) due to changes in temperature in Pakistan. It has been projected that day temperature of 2.8°C will significantly increase with 2.2°C decrease during nights by the midcentury (2040–2060) [3]. In such harsh conditions, growth and production of crops will be affected severely. The scientists must anticipate the possible solutions of climate calamities through introducing climate resilient crops and recognizing climate smarts production options for safe and secure food.

Climate resilience improvement is in common use now a day to inform crop management options. In view of current and future climate change and variability, interest among researchers to apply such technique is increasing to strengthen the climate resilience in crops of hot and dry areas [4, 5]. Climate resilience is quite resemblance to vulnerability and commonly defined as "the ability to bounce back after an external shock or stress". Resilience of a system can also be illustrated through its components including system disturbance, maintaining system & control and returning to stable state [6, 7]. Agriculture system is affected by extreme weather events associate with climate change, therefore adaptative measures are needed to mitigate the negative impacts of climate change. Keeping in views, the current study was planned to adapt climate change by developing climate smart practices for sustaining the agriculture productivity.

1.1 Land use and patterns

Pakistan is rich in natural resources, with half of the total area (36 million ha) lies under agricultural. About 84% of its land is arable, while remaining 14% is permanent pasture. The country is very poor in forests reserves which are 2% of total land as compare to world average ~30%. Uncontrol deforestation is further reducing its area at the rate of 0.2–0.5% per year. The area under agricultural production systems has remained stable over the last few decades. The high cropping intensity and use of fertilizer are main components to get high production in the country. Approximately 160 kg/ha fertilizers are used in Pakistan across all cropping systems. The land holding classification showed that 43% farmers are small scale (<1 ha) as compare to 36% large scale (>2 ha) [8].

Agroecological based agricultural production systems are different in different zones in Pakistan. However, the redefining of agroecological zones in Pakistan is importantly needed in the context of changing climate. Sugarcane, rice, cotton and wheat are more dominated along the Indus River. This zone is highly productive and lies in Punjab province. Chickpea crop is usually cultivated in Thal desert region of southern Punjab. It contributes >80% of chickpea production in country. As this crop is rainfed, hence the most affected by the climate variability in the region.

The rainfall patterns determine the cropping seasons in the country. These are "Kharif" (April to June), suitable for rice, sugarcane, cotton, maize, millet, mung bean, and the "Rabi" (October to December) season is suitable for wheat, barley, lentil, rapeseed, mustard and sugar beet. Tunnel farming in vegetables has got importance in the country due to high economic return. Climate variability, climate change, value chain, market monopoly and pest pressure are the main challenges to crop productivity [9].

1.2 Food supply and security

Natural disasters and economic instability and malnutrition are main challenges to country food security over the past decades. Pakistan ranks 78th out of 113 countries in the global food security index. It is alarming that 60% of population is experiencing food insecurity. The average food supply of 2440 kcal/person/day in the country is yet insufficient to meet the demand [5]. In Pakistan 62% of total energy is meet through cereals (wheat > rice > maize) after milk and vegetables in terms of calories consumed. In Pakistan, 44% of household's income is spent on their food, which is higher than any other commodity. The ratio is quite higher in villages (48%) [10].

1.3 Challenges fostering climate change

Although agriculture is the main stay to the country's economy, but still this sector is facing many challenges including population growth, rapid urbanization and reduction in water resource availability. The population of Pakistan has more than doubled in the past two decades. It is growing very fast, approximately at the rate of 2% yearly and expected to be 244 million in 2030 and 300 million in 2050. This ever-increasing population will ultimately put horrible strains on already vulnerable agricultural production system of Pakistan [10].

Industrial extension and low priority to village life are the main reasons for rapid urbanization in Pakistan. These activities seriously affect the agricultural areas by deteriorating the quality and safety of food products through harmful chemicals. The crops near these areas are often irrigated with waste water, which heavily absorb heavy metal and become part of food chain.

Water resources are limited all over the world; availability of water and its management are main problems in Pakistan. The per capita availability of water was 5300 m^3 in 1950, which is reduced to <1000 m^3 now a day in Pakistan [4]. This availability is projected to go down (800 m^3 per capita) in 2025. While a country is declared water scarce if its per capita water availability is <1000 m^3 and absolute scarce in <500 m^3 water availability [11]. These finite conditions are big threats to Pakistan. Changing climate conditions such as erratic rainfall patterns, variable temperature and humidity multiply these threats.

1.4 Climate change profile of Pakistan

It is obvious that agriculture is significantly affected by the climate variability and climate change. The country is ranked 7th among the top ten climate vulnerable countries in the world in global climate risk index and have diversified geography. The stress (heat & drought) followed by devastating floods are common in the country and have contributed to low crop yields, and disturbing food chain. The main reason for climate change is reduction in rainfall in the semiarid and arid region of Pakistan. The mean temperature across the country has increased by 0.5°C during the historical period (last 30 years) [4].

The projections show an increase of 1.4–3.7°C by mid-century in Pakistan. The temperature is expected more increase in winter than summer in Pakistan. Projections for rainfall have some unclear results due to uncertainties in performance of climate models; however, the trends are decreasing rainfall in future. The increase in temperature altered the phenology of crop i.e. leaf development, anthesis, harvest, fruit production and in asynchrony between anthesis and pollinators [7]. The rise in temperature also resulted in high respiration rates, reduction in pollen germination, shorter grain filling period, lesser biomass production and low yields. Pre and post flowering heat waves at 35°C led to yield loss in wheat, barley and triticale. High temperature above 35°C in combination with high humidity and low wind speed caused a 4°C increase in temperature, resulting in floret sterility in cereals [12].

In view of above, we must adapt such interventions which may create resilience in agricultural production systems to combat the climate variability and long-term climate change in the country.

2. Climate smart interventions

Global food production could be doubled by 2050 to contest the population and income growth in Asia and Africa. Pearl millet, sorghum are the important component of global food security in both continents. Owing to challenging anxieties for land, water, labor and capital, there is need to improve crop production per unit of land and water, as these resources are dwindling. Keeping in view the condition, sustainable increase in agricultural productivity is vital to the future of food security of Pakistan. Practices to adapt agriculture to climatic risks, take time to root and become effective for each. In such cases, approaches that enhance climate smart agriculture are the most apt starting point for sustainable agriculture. Climate smart intervention approaches are derived from the CSA-climate smart agriculture (**Figure 1**). CSA is usually defined as the "the agriculture that agriculture that sustainably increases productivity, enhances resilience (adaptation), reduces or removes greenhouse gases where possible and enhances achievement of national food security and development goals [1]. CSA aims to strengthen livelihoods and food security, especially of smallholders, by improving the management and use of natural resources and adopting appropriate methods and technologies for the production, processing and marketing of agricultural goods. The agricultural productivity, adaptations and mitigations are the main pillars of CSA.

2.1 Adaptations through crop modeling interventions

Crops models have capability to frame adaptation packages with in no time as a strategy for further implementation in the fields. However, experimentation through field setting is time consuming and utilize much resources unless the results are applicable or not. Climatic adaptations deal with effective investments/changes in technologies/policies in response to future climate change. While non-climatic adaptations focus on agronomic management options. Crop model frame comprehensive, cost effective and reliable adaptation packages by changing

DOI: http://dx.doi.org/10.5772/intechopen.82872

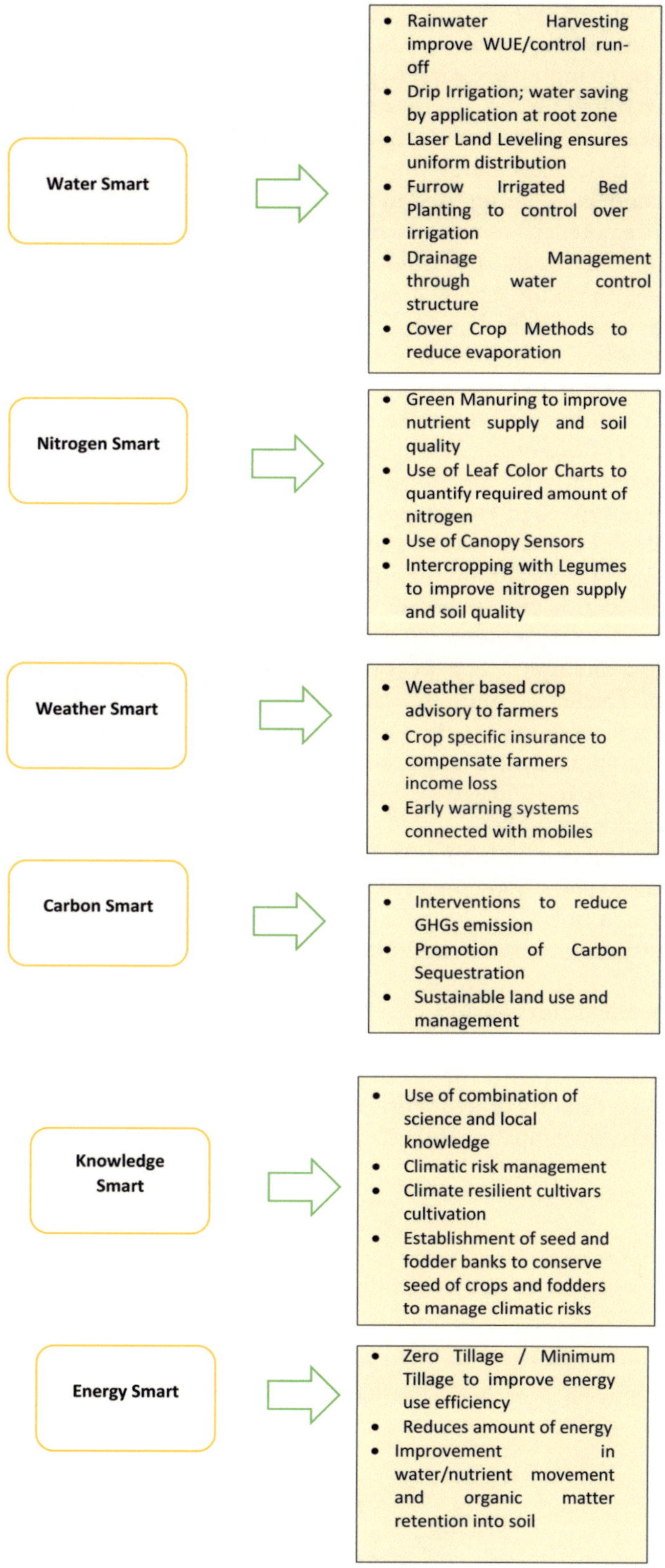

Figure 1.
Climate smart interventions in agriculture sectors.

the planting time, fertilizer dose, planting geometry, cultivar/hybrid and residue management. Increasing number of grains and crop growth rate, re-fitting crop season length by changing growing degree days (GDD) to anthesis and maturity, adjusting grain filling period, decreasing root length of crop are important elements of adaptations in crop model configuration. These elements of adaptations give sound genetic concepts as an important intervention in designing cultivar [4]. Various crop models are being used in improving natural resources to evaluate the impact of future potential climate on crop production [5]. Crop simulation models are appropriate tools for the assessment of crop production options for an environment, including inorganic fertilization levels, plant spacing, planting times and others management options [13, 14].

Development in crop improvement through plant breeding on molecular basis is inadequate by our skill to predict phenotype of plant which based on its genotype, specifically for multifaceted traits [15, 16]. In addition to this, there has been an extended history of designing and application of crop growth and development models for prediction in crop management [17]. The use of such modeling interventions for genotype to phenotype prediction are at beginning [18, 19]. Current studies encouraged that use of crop models have considerable potential to face the genotype-phenotype prediction for application in plant breeding. However, the competence of existing crop models for these type of applications is uncertain [20, 21] and need improvements. The intervention of integration of simulation with plant breeding is an important aspect to design a "virtual cultivar" which can guide to breeders and further recommendation for general cultivation in area. This type of simulation can assemble a virtual variety with acclimatize characteristics for site specific cultivation. Hence, this approach can also elevate the farming to the extent of revolution and have ability to feed the world in safe and healthy style (**Figure 2**). DSSAT (crop modeling) [22], APSIM (crop modeling) are used to assist all type of stakeholders in decision making [23]. PLABSIM (marker-assisted backcrossing) [24], PLABSOFT (plant breeding) [25], QU-GENE (genotype-by-environment interaction) [26] and E-CELL (whole cell simulation) [27] are useful tools being used in computer simulation in breeding programs.

2.2 Weather smart intervention

The risks associated with change in rainfall and temperature at different crop growth stages are directly linked with increase or decrease in pearl millet production. The stakeholders must be linked to automated weather stations and site

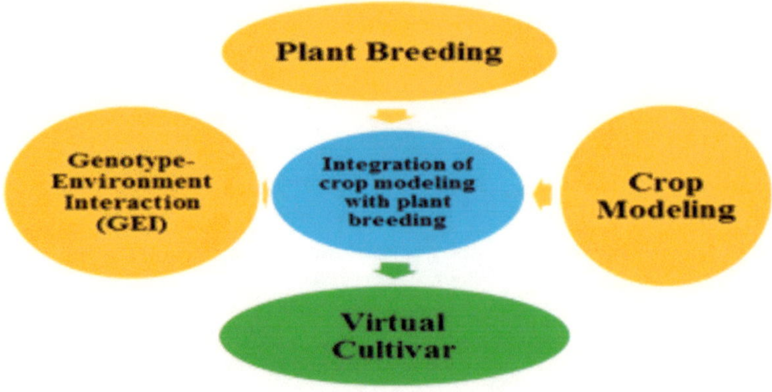

Figure 2.
Integrated approach of crop modeling and breeding.

specific agro-advisories through radio shows, television, newspapers and mobile phone voice messages. Agricultural Weather Network (AgWeatherNet) is a good example of this intervention. The current and historical weather data along with a range of models and decision aids can be accessed through this system. The weather data, advisories, weather data products and decision support systems provided by AgWeatherNet is very helpful to improve production and product quality, efficient resource use and reduce environmental impact on crops. Early warning system is also the important component of AgWeatherNet [28].

2.3 Remote sensing and its application

The role of geo-spatial technology is vital in precision agriculture, crop monitoring and yield forecasting. Remote sensing is a potential tool to monitoring crop health and condition. Abiotic stresses; high temperature, insects attack, diseases, moisture deficiencies, fertilizer stress and area affected by these stresses can be detected earlier for quick mitigation. Time series data and maps using remote sensing help to understand the spatial and temporal changes and their drivers. Climate change, biotic and abiotic stresses, and their impacts on crop can be monitored using remote sensing. It depicts and reveals area under crop and its trend over the years. There are many people who are involved in selling, purchasing and pricing of their farm produce. Crop yield forecasting using satellite imagery well before the time of harvest helps to devise policy for crop import and export. Remotely sensed data can be assimilated in crop models to forecast yield on regional scale and for site specific crop production technology [29]. The current innovation in remote sensing sciences is use of higher spatial resolution satellites imagery for precision agriculture and to monitor within field variability. Crop production and management practices can be improved and modified by obtaining information using higher resolution multispectral and hyperspectral data. Use of same input for larger area is wastage of resources and causes crop yield reduction because significance within field variability exists. Pearl millet is grown with diverse distribution in country. Estimation of area, production and average yield through conventional method is a tough job. Remote sensing figures out all these aspects quickly before maturity of crops. Using this type of intervention, we can point out historically, the potential areas, low and high yielding areas of pearl millet with in Pakistan for its retrieving using better management options. Future trends of area and production of field crops can also be predicted using various techniques of remote sensing such as normalize difference vegetation index (NDVI) and random forest as statistical tool which not only assist in decision making for policy makers but also help to evade food insecurity in country [29].

2.4 Policy interventions

Changes in climatic conditions are affecting agricultural produce and would become more calamitous in future. We need to be proactive to figure out solutions and make Pakistan a food secure country through better policy interventions. Mitigation and adaptation strategies to reduce climate risks are the best policy interventions if the policy makers strictly regularize those polices. Reducing greenhouse gases (GHGs) emissions to mitigate climate change should be encouraged as a measure through different schemes that convince farming communities without loss to their crops and production. Afforestation and reforestation are healthy activity to improve carbon sinks and should be an important component of policy intervention to mitigate the impact of climate change. Use of renewable energy resources should be accentuated to mitigate the impact of changing climate. Climate compatible agriculture should be endorsed as an adaptation strategy in country. Genetic divergence and biotechnology can also play an imperative role to develop crop varieties which can tolerate drought, heat stress and submerge conditions. Use of information technology (IT) in agriculture has become crucial to quantify the impact of climate change. Monetary protection of farmers by insurance companies and government should be promoted in the country against natural catastrophes. An index-based insurance schemes should be introduced to cover risks associated with changes in rainfall and temperature at the different stages of crop growth. Early warning systems and automatic weather stations to alert farmers with weather data are good tools those must be installed at district level [10].

Globally, all research institutes are non-profitable and funded with sufficient finance to carry smooth research activities. Contrarily to the facts, most of the research institutes in Pakistan are considered profit-oriented by fixing financial targets each year. Hence, as a policy intervention, research institutes in Pakistan should be declared as non-profitable. This approach enables scientists to work without restrictions rather than making profit. In addition to this, investment in research and development must be increased in country to develop site specific adaptation strategies for changing climate. These adaptations must be fit in new designed agro-ecological zones.

3. Resource smart interventions

3.1 Water smart

Water is limiting resource throughout the world and must be used professionally to ensure a safe and plentiful food supply. Explanations to water shortage glitches will necessarily include adoption of innovative water conservation measures, flexible water delivery systems and precision irrigation. Better control and management of water applications are important components for any crop in this action [20]. Irrigation efficiency of system is 40% which is quite poor in Pakistan. Smart water management practices aim to enhance the efficiency and productivity of water which are very important interventions. These might be comprised of water course improvement, precision land leveling, bed planting, furrow planting of row crops and high efficiency irrigation systems, aquifer recharge, community management of water and water conservation [30] which have potential to save water in the range of 20–70%. These interventions can be resulted into increase in crop yields by 20–30% and increase in net income to the farmers by 20% under changing climate scenarios of Pakistan. In climate smart water, irrigation plays a major role in stabilizing agricultural production. Climate smart irrigation is good irrigation approach for the given agro-climatic and societal context that may result directly or indirectly from the different aspects of climate change, it aims to increase per unit production and income from irrigated cropping systems. These interventions through climate smart agriculture, aim to reduce the exposure of farmers and their irrigation systems to short term risks. Adaptations are generally developed to strengthen resilience by developing their capacity in the face of shock and other stresses.

Irrigation is applied usually to overcome the stress due to spatial and temporal variability in rainfall on crop growth and its quality. Characteristically, climate smart irrigation practices are founded in the mix of technical and non-technical measures. These may be included;

- The source of water must be reliable; however, the reliability of water sources largely depends on the policy interventions for regulating, protecting and sustainability of natural resources.

- Building of sound infrastructure to extract and convey water from the source to where it is needed. The policy interventions again play an important role in managing and maintaining all types of irrigation conveyance systems.

- The role of irrigation management systems and water user association is important to implement the innovations like; weather or soil sensors, control devices etc. according to certain schedule.

- Drainage infrastructure is often needed to reduce the risks of waterlogging and soil salinization.

- Farmers must be given land and water tenure that gives right to use both resources and, the incentives to invest in activities aimed at improving the productivity and resilience of their irrigation systems.

- The agriculture production system is declared to be climate smart, when the cropping systems are well adapted to relevant biophysical factors (agro-climatic conditions, soils, extreme events) and societal factors (markets, labor availability). Policy interventions to support farmers and farming systems, credit services, suppliers of seeds and irrigation related infrastructure, suppliers of irrigation and agricultural equipment, etc.

- Value chains are the most important aspects that connect farmers to markets. They ensure returns on investment and minimize risks of post-harvest losses. In these context, the interventions are needed to support farmers and traders in trading related legislation.

3.2 Carbon smart

Total organic carbon is the amount of carbon related to living organisms or derived from them. Increasing carbon content in the soil may reduce atmospheric carbon dioxide and improve soil quality. Recent interests in carbon sequestration have raised many questions about how much carbon can be stored in the soil. The organic carbon content in soil can be improved through agricultural practices such as agroforestry, livestock, manure management, conservation tillage, diversified land use systems and residue management. As millet crop is grown on mostly marginal land with low organic matter (OM). Improving OM using different interventions can boost the pearl millet yields in Pakistan where soils of most areas are low in organic matter [31].

3.3 Nitrogen smart

Conventionally, climate resilient crops (pearl millet, sorghum, quinoa etc.) are grown on marginal soils (low fertility) without application of nitrogen. Nitrogen application should be considered as an important component of dryland, rain-fed and irrigated systems. Soil fertility along with climatic risks are obstructing crop productivity in arid and semi-arid environments of Pakistan. Nitrogen smart interventions are those strategies in precision agriculture where most of the people do not focus before growing crops, especially the ignoring

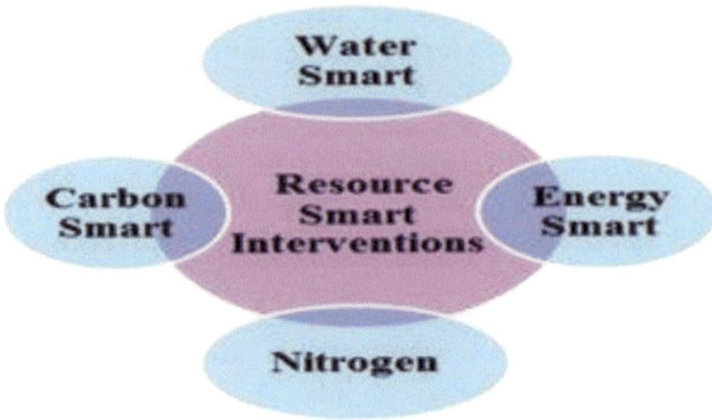

Figure 3.
Resource (water, carbon, nitrogen and energy) smart interventions.

crops (pearl millet, sorghum, quinoa etc.) and started to prepare their lands for cultivation. Applying the right source of N, at the right rate, at the right time, and in the right place is another important aspect in nitrogen management. In climate smart interventions, farmers must use leaf color charts, handheld crop sensors, and nutrient decision-maker tools to decide the most appropriate dosage of nitrogen fertilizers for crops. The use of crop canopy sensors is a good example of the interventions and an important precision agriculture tool into the decision-making process. These interventions not only save costs, but also cause decrease in greenhouse gas (GHG) emissions especially from puddled rice fields [32].

3.4 Energy smart

Energy is also a limited resource which must be used sensibly to secure the safe, healthy and nutritious food for the growing population in country. Promotion of biofuels, high efficient agro machineries, residue management and conservation practices (reduce tillage) are the important energy smart interventions to conserve energy and reduce GHGs [33]. Use of biogas systems through manure slurry from intensive dairy initiatives is also appreciated as energy smart approach in most part of the world (**Figure 3**).

4. Knowledge gaps and the way forward

- The effect of land use and land cover patterns on food production and food security may be focused in studies in context of current changing climate.
- Agro ecological based temporal and spatial models to be developed coupling with climate smart interventions.
- Integrated impact assessment on different agro ecosystems still needed to improve agricultural production system's efficiency.
- Impact of shocks/extreme weather events on crop plants require more attention.
- Food safety and food security in the countries like Pakistan may be linked with incentives. The mechanism of incentives must be properly examined.

- The researchers and extension workers may be fully equipped with nutrients and SOC kits.
- The mobile applications related to weather predictions and early warning systems may be developed locally.
- Subsidies may be provided to those technologies that support more diverse climate adapted agricultural production systems.
- The impact of climate change related projects may be evaluated or reviewed from the agro climatologists.
- The new adapted species may be integrated with conservation practices and supported with favorable policy interventions.
- The regions for research on climate change may be selected based on the most vulnerability and risk within country.

5. Conclusions

Agriculture is significantly hit by temperature variability and other variable events all over the world including Pakistan. Farming communities are feeling the worst impacts of climate change and variability on crops and livelihoods. In such conditions, climate resilient crops like pearl millet prove to be a robust, climate smart grain crop. It endures to out yield more nutritious grain sustainably, thereby cheering the fight against poverty and food insecurity due to its resilience. Because it is selected as water saving, drought tolerant and climate change complaint crop. Crop modeling and its application, computer simulation in plant breeding and resource smart approaches supported with policy interventions can present better picture under changing climate scenarios of Pakistan.

Acknowledgements

The authors acknowledge the critical review of this chapter from K. Grace Crummer, Coordinator, Institute for Sustainable Food Systems, University of Florida, USA.

Conflict of interest

There is no 'conflict of interest' among the authors.

Author details

Asmat Ullah[1,2,3*], Ishfaq Ahmad[4], Habib-ur-Rehman[5], Umer Saeed[6], Ashfaq Ahmad[2], Abid Mahmood[1] and Gerrit Hoogenboom[2]

1 Ayub Agricultural Research Institute (AARI), Faisalabad, Punjab, Pakistan

2 Institute for Sustainable Food Systems, University of Florida, USA

3 Agro Climatology Laboratory, University of Agriculture, Faisalabad, Punjab, Pakistan

4 Centre for Climate Research and Development, COMSATS University Islamabad, Pakistan

5 MNS-University of Agriculture, Multan, Punjab, Pakistan

6 Independent Researcher, Punjab, Pakistan

*Address all correspondence to: drasmatu@gmail.com

References

[1] Alexandratos N, Bruinsma J. World agriculture towards 2030/2050: The 2012 revision. In: ESA Working Paper. FAO, Rome; 2012. Available from: http://environmentportal.in/files/file/World%20agriculture%20towards%202030.pdf

[2] Kreft SD, Eckstein L, Junghans L, Kerestan C, Hagen U. Global Climate Risk Index 2015. Bonn Office: Germanwatch e.V.; 2014. pp. 2-6. Available at: http://germanwatch.org/de/download/10333.pdf

[3] Ahmad A, Ashfaq M, Rasul G, Wajid SA, Khaliq T, Rasul F, et al. Impact of climate change on the rice-wheat cropping system of Pakistan. In: Rosenzweig C, Hillel D, editors. The Agricultural Model Inter-comparison and Improvement Project (AgMIP): Integrated Crop and Economic Assessments, Part 2. ICP Series on Climate Change Impact, Adaptation, and Mitigation, Chapter 7. Vol. 3. American Society of Agronomy, Crop Science Society of America, Soil Science Society of America, Imperial College Press; 2015. pp. 219-258

[4] Ullah A, Ahmad A, Khaliq T, Akhtar J. Recognizing production options for pearl millet in Pakistan under changing climate scenarios. Journal of Integrative Agriculture. 2017;**16**(4):762-773. DOI: 10.1016/S2095-3119(16)61450-8

[5] Ahmed I, ur Rahman MH, Ahmed S, Hussain J, Ullah A, Judge J. Assessing the impact of climate variability on maize using simulation modeling under semi-arid environment of Punjab, Pakistan. Environmental Science and Pollution Research. 2018;**25**(28):28413-28430

[6] Ahmad I, Wajid SA, Ahmad A, Cheema MJM, Judge J. Assessing the impact of thermo-temporal changes on the productivity of spring maize under semi-arid environment. International Journal of Agriculture and Biology. 2018;**20**(10):2203-2210

[7] Rahman MH et al. Multi-model projections of future climate and climate change impacts uncertainty assessment for cotton production in Pakistan. Agricultural and Forest Meteorology. 2018;**253**:94-113

[8] Ali G, Nitivattananon V. Exercising multidisciplinary approach to assess interrelationship between energy use, carbon emission and land use change in a metropolitan city of Pakistan. Renewable and Sustainable Energy Reviews. 2012;**16**(1):775-786

[9] Iqbal MF, Khan IA. Spatiotemporal land use land cover change analysis and erosion risk mapping of Azad Jammu and Kashmir, Pakistan. The Egyptian Journal of Remote Sensing and Space Sciences. 2014;**17**(2):209-229

[10] Alderman H, Garcia M. Poverty, household food security, and nutrition in rural Pakistan. International Food Policy Research Institute. 1993;**96**:1-108

[11] Ahmad I, Wajid SA, Ahmad A, Cheema MJM, Judge J. Optimizing irrigation and nitrogen requirements for maize through empirical modeling in semi-arid environment. Environmental Science and Pollution Research. 2018;**26**:1227-1237

[12] Barlow KM et al. Simulating the impact of extreme heat and frost events on wheat crop production: A review. Field Crops Research. 2015;**171**:109-119

[13] Saseendran SA, Nielsen DC, Vigil MF, Ahuja LR. Simulating planting date effects on corn production using RZWQM and CERES-Maize models. Agronomy Journal. 2005;**97**:58-71

[14] Tsuji GY, Uehara G, Balas S. DSSAT Version 3. Vol. 4. Honolulu, HI, USA: University of Hawaii; 1993

[15] Cooper M, Chapman SC, Podlich DW, Hammer GL. The GP problem: Quantifying gene-to-phenotype relationships. In Silico Biology. 2002;**2**:151-164. Available from: http://www.bioinfo.de/journals.html

[16] Cooper M, Podlich DW, Smith OS. Gene-to-phenotype models and complex trait genetics. Australian Journal of Agricultural Research. 2005;**56**:895-918

[17] Sinclair TR, Seligman NG. Crop modeling: From infancy to maturity. Agronomy Journal. 1996;**88**:698-703

[18] Hammer GL, Jordan DR. An integrated systems approach to crop improvement. In: Spiertz JHJ, Struik PC, van Laar HH, editors. Scale and complexity in plant systems research: Gene-plant-crop relations. Wageningen UR, Frontis Series No. 21. Dordrecht, The Netherlands: Springer; 2007. pp. 45-61

[19] Hammer GL, Kropff MJ, Sinclair TR, Porter JR. Future contributions of crop modelling: From heuristics and supporting decision-making to understanding genetic regulation and aiding crop improvement. European Journal of Agronomy. 2002;**18**:15-31

[20] Hammer GL, Cooper M, Tardieu F, Welch S, Walsh B, van Eeuwijk F, et al. Models for navigating biological complexity in breeding improved crop plants. Trends in Plant Science. 2006;**11**:587-593

[21] Messina CD, Jones JW, Boote KT, Vallejos CE. A gene based model to simulate soybean development and yield responses to environment. Crop Science. 2006;**46**:456-466

[22] Hoogenboom G, Jones JW, Wilkens PW, Porter CH, Boote KJ, Hunt LA, et al. Decision Support System for Agrotechnology Transfer (DSSAT) Version 4.6. Prosser, Washington: DSSAT Foundation; 2015. Available from: www.DSSAT.net

[23] Keating BA, Carberry PS, Hammer GL. An overview of APSIM, a model designed for farming systems simulation. European Journal of Agronomy. 2003;**18**:267-288

[24] Frisch C, Bohn M, Melchinger A. Computer note. PLABSIM: Software for simulation of marker-assisted backcrossing. The Journal of Heredity. 2000;**91**(1):86

[25] Maurer HP, Melchinger AE, Frisch M. Population genetic simulation and data analysis with Plabsoft. Euphytica. 2008;**161**(1-2):133-139

[26] Podlich DW, Cooper M. QU-GENE: A simulation platform for quantitative analysis of genetic models. Bioinformatics. 1998;**14**(7):632-653

[27] Tomita M, Hashimoto K, Takahashi K, Shimizu TS, Matsuzaki Y, Miyoshi F. E-CELL: Software environment for whole-cell simulation. Bioinformatics. 1999;**15**(1):72-84

[28] Agricultural Weather Network [Internet]. 2016. Available from: http://weather.wsu.edu

[29] Ahmad I, Saeed U, Fahad M, Ullah A, ur Rahman M, Ahmad A, et al. Yield forecasting of spring maize using remote sensing and crop modeling in Faisalabad-Punjab Pakistan. Journal of the Indian Society of Remote Sensing. 2018a;**46**:1701-1711

[30] Evans RG, Iversen WM, Stevens WB, Jabro JD. Development of combined site-specific mesa and lepa methods on a linear move sprinkler irrigation system. Applied Engineering in Agriculture. 2010;**26**:883-895. DOI: 10.13031/2013.34951

[31] Santacana E et al. Getting smart. IEEE Power and Energy Magazine. 2010;**8**(2):41-48

[32] Scherr SJ, Shames S, Friedman R. From climate-smart agriculture to climate-smart landscapes. Agriculture & Food Security. 2012;**1**(1):12

[33] Scholz G, Quinton JN, Strauss P. Soil erosion from sugar beet in Central Europe in response to climate change induced seasonal precipitation variations. Catena. 2008;**72**:91-105. DOI: 10.1016/j.catena

Chapter 3

Beitbridge Minority Farmer Communities and Climate Change: Prospects for Sustainability

Mark Matsa and Beauty Dzawanda

Abstract

Indigenous minority farmer communities in Beitbridge district of Zimbabwe are on the cutting edge of climate change and climate vulnerability. This chapter assesses through questionnaires, interviews and focus group discussions how these communities are triangulating their indigenous knowledge systems, government and NGOs initiatives to achieve sustainability. Results reveal that although the farmers are appreciative of external assistance through government and NGOs assistance, such assistance can only be sustainable provided it is built around their indigenous knowledge systems which they hold sacrosanct. The study therefore recommends more use of the abundant natural resources in Beitbridge and The district's competitive advantage is a rich livestock district. The community identifies itself with these resources, so all developmental endeavours should coalesce around these resources for sustained social, economic and environmental growth as a cushion against the climate change phenomenon and associated threats. All such efforts should be community driven rather than being imported from central government or NGO headquarters or country offices. The resilient and hardworking qualities of these communities need not be destroyed by food aid and free farming input hand-outs. Instead, these qualities should be utilised to drive community development initiatives for household livelihood sustainability.

Keywords: climate change, Beitbridge district, indigenous minority farmer communities, Maramani, community sustainability, CAMPFIRE

1. Background

Climate change through natural and anthropogenic forces has drastically changed the earth's climate over the past century worsening key challenges for global food production [1]. Climate change impacts, which are expected to be mainly negative, are likely to be felt mostly by the already vulnerable communities in economically less-developed countries. Most developing countries in Africa, Asia, the Americas outside the United States of America and Canada are generally poorly equipped both financially and infrastructurally which makes them more vulnerable to climate change impacts [2]. The effects of climate change are alarming enough in themselves, for instance, droughts, burning temperatures, more frequent hurricanes, worse floods and new plagues of diseases.

Climate change defines alterations in the long-term average conditions of the climate, persisting for unusually longer periods, which can be decades or generations [3]. These changes may be due to natural or persistent anthropogenic alterations in the composition of the atmosphere or in land use. Climate variability, on the other hand, refers to unusual changes in the spatial and temporal state of climatic variables. Food is an incessant priority for subsistence for many indigenous minorities who are already vulnerable to changing environmental conditions. Due to a continuous dependence on agriculture for most livelihoods, the effects of climate change on productive croplands are likely to drastically threaten the well-being of the population [4]. The close relationship of minorities with their natural environments makes them mainly sensitive to the effects of climate change. One of the most outrageous examples of minorities' greater exposure to climate change is in India where indigenous minority groups known as Dalits, Adivass and Muslims were economically, physically and socially excluded from the rest of society [4]. As a result, they were worst hit by the abnormally severe monsoon floods in 2007.

Minority people tend to live close to nature, in relatively natural environments, rather than in cities, growing and making much of the food and other products that they need to survive [5]. This gives them an exceptionally intimate knowledge of local weather, plant and animal life. Customary wisdom on issues such as where to hunt for food or when to plant crops has been accrued over many generations, but now that the climate is shifting, some of those understandings are proving to be no longer applicable [6]. Masud et al. [7] also argue that some rainfall patterns have changed in line with what climate change scientists are predicting and, as a result, people's customary knowledge about when to plant crops is no longer consistent. Hence, the capability to accurately recognise the rainy season has suffered leading them to plant crops impulsively.

Coping strategies means actions that reduce the actual and expected effects of climate change making people adapt to prevailing conditions. These coping strategies can actually take place at a local level where people make changes they can, independently of government. Coping strategies can also be introduced by governments and NGOs to indigenous minority people. For example, in the Arctic Sam reindeer, herders transport food to the reindeer in winters when the animals cannot reach the lichen [8]. They also reverse their traditional pattern and take their animals inland during the summer and to the coast in winter, where there is no snow and so grazing is less. However, the author further asserts that their ability to adapt is limited by lack of financial resources and technical expertise. There is so much that they cannot perform without government support, and as a result, this affects the sustainability of their own introduced coping strategies. In Kenya, some pastoralists have adapted to climate change by growing livestock fodder crops in wetter areas near rivers, selling some of their livestock rather than allowing them to die during droughts [8]. But their insufficient representation in national politics has smashed their capacity to cope with the increasingly harsh climate thereby affecting the sustainability of their own coping strategies.

Climate variability has always been experienced in Southern Africa. During the 1991–1992 drought in Zimbabwe, average annual precipitation fluctuated from 335 to 1004 mm and averaged 640 mm [9]. Recurrent droughts and sporadic seasonal floods that have been experienced in the region have resulted in the loss of human life, livestock and property. They have also caused severe localised shortages of the main cereal crops like maize and other food items. Implications of climate change for Zimbabwe are serious. The number of years with below average rainfall is increasing. A survey on farmers' perceptions of climate change in Zimbabwe by Masendeke [10] indicates that farmers have noticed changes in the quantity, quality and efficacy of rainfall. There is a general decline in the amount of rainfall, which is more pronounced in the semi-arid tropics, the largest part of which is constituted by south-west Zimbabwe, the study area. Zimbabwe has an agricultural economy which is generally rain-fed. It has a rural population of more than 70% which depends on subsistence agriculture for livelihood [11]. This makes most sectors of the national economy sensitive to extreme changes or shifts to weather and climatic patterns.

The lack of research into the ways in which minorities are being affected by climate, how they are coping with effects of climate change and the sustainability of the coping strategies only aggravate their disadvantage and susceptibility. For minorities to get the assistance they need, their circumstances must first be documented and acknowledged by academics, development and environmental NGOs, governments and intergovernmental organisations. Climate change is a serious issue affecting the world, but seldom does its impact on minorities get a mention, even though they are among the worst affected. Despite the high susceptibility of Zimbabwe to climatic fluctuations, very little research has been carried out on climate change, particularly on coping strategies of minority communities, most of whom occupy marginal, remote, hot, dry regions of the country. There is a serious lack of community-specific and household-specific data demonstrating their vulnerability. Communities, however, have been witnessing the gradual changes in climate over the years, and have been attempting to cope, albeit with mixed success. Few studies have tried to highlight and interrogate these coping strategies by communities for possible development and improvement, despite the fact that this is what has made them resilient to this day. This study therefore aims to cover this gap in knowledge.

2. The study area

Beitbridge lies in the Limpopo Valley, a paragneiss zone that stretches from the east of Chiredzi to the Border with Botswana. Mean annual rainfall for the district is between 300 and 600 mm [12]. Most of the rainfall is experienced in summer from October to March. Mean annual temperature is between 25 and 27.5°C. Soils in the district are varied, depending on the parent materials and age. On sedimentary formations, soils that occur in younger deposits are deep and often stratified. On levee deposits, soils are relatively light-textured with a high proportion of coarse sand of granitic origin. Basin areas have heavy-textured soils derived from fine materials deposited during floods. Vegetation varies from the savanna on deep fertile soils to shrub savanna on shallower ones. It is of lower stature of 2–6 m high with a sparse grass cover of mainly *Sporobolus* spp (love grass) and *Cynodon dactylon*. Common trees in this region include mopani, which is prevalent on salt-rich soils, baobab, *marula* and various species of combretum and acacia [13]. Apart from the urban setting of Beitbridge as a border town with South Africa, the west of Beitbridge district is semi-arid, remote and marginal. Farmers are generally sedentary pastoralists who practice dry land farming which concentrates on drought-tolerant small grains like sorghum, millet and rapoko. Communities in Beitbridge traditionally keep large heads of the indigenous thuli cattle and other breeds. They also keep large flocks of sheep and goats. The district has some natural salt pans where salt extraction takes place. Beitbridge has a heterogeneous population of marginalised minority farmer communities which include the Venda, Shangani and Suthu. There is no evidence of a concerted study which disaggregates the age-sex classes of each of these minority ethnic communities and this emphasises their social and economic vulnerability. However, Beitbridge district has a population of 80,083 (14) comprised mainly of these three ethnic groups whose combined population, together with other Bantu-speaking minority ethnic communities like the Tonga (Binga district) and Kalanga (Bulilima and Mangwe districts) is slightly over 1% of Zimbabwe's total population

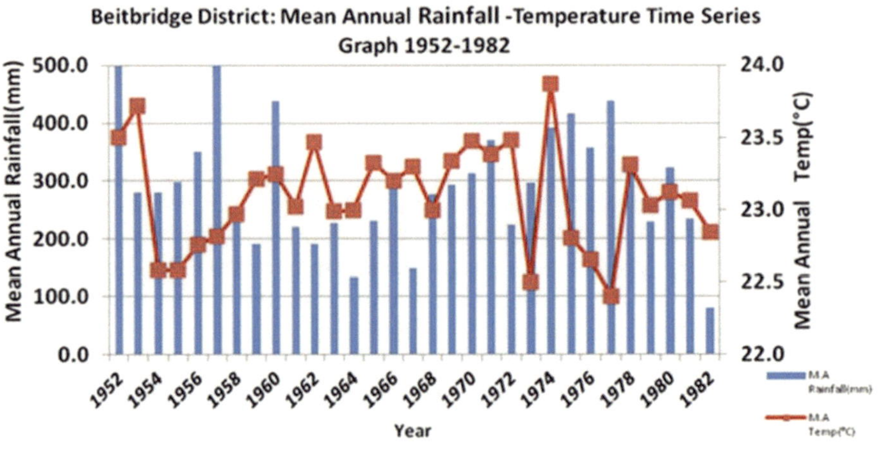

Figure 1.
Mean annual rainfall-temperature graph for Beitbridge district (1952–1982). Data source: Meteorological Department, Harare.

of about 17 million. The population density of Beitbridge district is between 4.42 and 10.61 [14]. The district has 15 wards for administrative and developmental purposes (**Figure 1**).

3. Methodology

The study used qualitative ethnographic research design to investigate coping strategies to climate change and climate variability by specific minority farmer communities. Communities borrow indigenous knowledge from their culture to adapt to their environment. Fraenkel and Wallen [15] submit that ethnography describes social groups or situations; delineating behaviour and shared beliefs of a particular group of people and in the process gaining an understanding of how and why the participants function and behave as they do in the context of their culture. Qualitative ethnographic design enabled the researcher to obtain in-depth data concerning minority farmer communities through interviews, focus group discussions and direct observations. It embraced their experiences, feelings and behaviour as they respond to climate change and climate variability issues.

An analysis of relevant documents in text format was done to gain preliminary knowledge of the minority communities in Beitbridge district. Documents analysed included scientific work published on the district and documents outlining the district's developmental plans. Secondary rainfall and temperature data from the Meteorological Department was also collected for analysis. Document analysis enabled the researcher to examine records and documents and to get an idea of past and present life of these communities in an unobtrusive manner [16]. In this study, interest was on determining coping strategies of minority communities to climate change and variability.

A total of 20 purposively selected farmers were subjected to in-depth interviews focusing on answering the research questions like environmental evidence which point to a changing climate; and community knowledge-based initiatives in place to cope with climate change impacts. The farmers are geographically intermingled, incorporating Venda, Suthu and Shangani communities and thus geographical differentiation was not an important variable in their selection. The snowball technique was used to ethnically but purposively select the 20 farmers (8 Venda, 6 Suthu and 6 Shangani) to give each ethnic group a voice. Chief

Executive Officer for Beitbridge Rural District Council was interviewed to get insight into government's involvement in climate change and climate variability issues in the district. A randomly selected traditional leader (local Chief) from the district was interviewed to get the traditional perception of climate change issues in relation to their respective areas of jurisdiction The International Crops Research Institute for the Semi-Arid Tropics (ICRISAT) country representative was also interviewed to find out the activities it is involved in which are aimed at climate change mitigation/adaptation as well as communities' response to their initiatives. ICRISAT is based in Zimbabwe's south-west district of Matobo (Matopos), some 40 km to the south-east of Zimbabwe's second largest city, Bulawayo.

Focus group discussions were used to give independent voice mainly to women, who because of patriarchal domination, do not usually narrate their experiences as they experience them, but instead answer questions as expected of them by society. Discussions on climate change issues and how communities are coping with them were initiated at social gatherings like food for-work programmes, boreholes, school development gatherings and similar gatherings to get communities' views. Focus group discussions, therefore, provided some quality controls on data collection in that participants were able to provide checks and balances on each other. This helped in weeding false or extreme views, and it was fairly easy for the researcher to assess the extent to which there was a relatively consistent or shared view among the participants. Focus groups allowed insight into the values, beliefs, fears and aspirations of the different communities regarding climate change and variability.

Structured observation was used to verify communities coping strategies in situ. It was used to verify findings from interviews, questionnaires, focus group discussions and ethnography. Climate change and coping strategies-related evidence like drying of wetlands, state of pastures, traditionally preserved foods, etc., were observed. The researcher observed events or phenomena in the natural and social spheres without any manipulation, interference or intervention. All relevant observations were noted on a coding sheet or checklist.

The questionnaire method was applied and analysed as a quantitative research method. A questionnaire with closed items which focussed on facts on evidence and impacts of climate change was developed for this study. The questionnaire survey was meant to collect as much data as possible from respondents on their coping strategies to climate change and climate variability. To ensure a 100% response rate and also to save on time, the questionnaire was administered in the form of an interview. Questionnaires were distributed to 10 randomly selected households per each of Beitbridge district's 15 wards to confirm or refute information obtained from in-depth interviews and other methods. This means 150 questionnaires were randomly administered across the district.

To verify whether rainfall and temperature were changing over time in Beitbridge district, secondary data from the Meteorological Department were analysed in Microsoft Excel. Sum monthly rainfall totals from September to August were calculated and averaged. The sums of monthly maximum and minimum temperature were also calculated and averages computed in Microsoft Excel. The district's rainfall and temperature data were divided into three generations for the period 1922–2012.The data were obtained from the Meteorological Services Department in Harare. However, the first generation, 1922–1952, had inadequate data to make meaningful comparisons, so only data from generations 2 and 3 were used. Meteorological data were thus divided into 1952–1982 and 1982–2012 generations and rainfall-temperature compound time-series graphs were computed in order to compare rainfall and temperature variations which impact minority farmer communities' livelihoods in Beitbridge.

4. Results

4.1 Socio-demographic characteristics of respondents

In Beitbridge district, a total of 100 respondents successfully answered questionnaires, 50 were males and another 50 females. Respondents' age-groups ranged between 26 and 105 years who had lived within the district for 30 years or more and were believed to have witnessed considerable climatic and environmental changes. Like in other border districts in the south-western part of Zimbabwe, it was, however, easier to find female respondents than male ones. Most males and the youth go to neighbouring South Africa and Botswana to look for better employment and livelihood opportunities [17]. About 68% of the respondents were married, 18% were single and 10% were widowed, while 4% were divorcees. The level of education in the district is generally low, with 30% of the respondents having acquired only primary education, another 30% lower secondary level and 28% basic secondary education. Only 2% had advanced secondary level of education and 10% had tertiary training, 8% of whom being teachers and 2% agricultural extension officers. Average family size in Beitbridge was 6. About 52% of the respondents were unemployed and were typical agro-pastoralists who relied mainly on livestock and subsistence crop farming for their day-to-day household livelihood. About 26% were in part-time employment which mainly involved cross-border trading in household goods between Zimbabwe and South Africa. This group would shuttle weekly or monthly, by hook or by crook, legally or illegally between the two countries for their household livelihood. Only 22% were in formal employment, most of them being teachers or employed in other government departments like agriculture or Beitbridge Rural District Council employees.

4.1.1 Evidence attributable to climate change in south-west Zimbabwe

In Beitbridge district, there is a high realisation that the rainy season is now starting late (100%) and yet ending early (94%) (**Table 1**). This could be linked to the change in wind patterns which 80% of the respondents identified as having changed. Summer season is becoming hotter (80%), while winter is becoming warmer (64%). The apparent rise in temperature could partly be the reason for the drying of perennial

Ward number	Ward name	Ward number	Ward name
1	Chipise	9	Machuchuta
2	Dite 1	10	Dendele
3	Dite 2	11	Siyoka 2
4	Mtetengwe 1	12	Siyoka 1
5	Mtetengwe 2	13	Leasnth
6	Mtetengwe 3	14a	Bishopstone
7	Masera	14b	Bishopstone
8	Maramani	15	Old Tuli

Source: generated by author.
N = 100.

Table 1.
Environmental evidence indicative of climate change in Beitbridge district.

rivers and springs (78%) through increased evapotranspiration. Increased temperature can also help explain diminishing pastures (88%) through depletion in soil moisture.

In an interview, Beitbridge Rural District Council Officer in charge of the central wards of Mtetengwe I, II and III, Mr. Sibanda (65) observed that early summer rains which used to start in September/October are now starting in December. June and July which are supposedly the coldest months of the year are no longer very cold. Instead, February, which is characteristically a very rainy month, is turning out to be very cold. Drought is now more frequent and temperatures are no longer predictable. Mr. Mudawu (70), a Shashe village elder in the western ward of Maramani, added that when the rains come, they are uncharacteristically erratic and cause lots of damages to bridges, schools, homes and even kill people and livestock through very violent and strong winds, flooding, lightning and thunder. Mrs. Simuta (64) of Chaswingo village in Dite I ward concurs that seasons have changed but added that this confuses farmers as to when to plant. Pastures in the eastern wards of Beitbridge comprising Dite I, Dite II and the Shangani-dominated Chipise are now few even if it rains. This has resulted in many livestock deaths in recent years.

In the Suthu-dominated north-western wards of Siyoka I, Siyoka II and Dendele, environmental evidence attributable to climate change include heavy siltation of Umzingwane River which used to be perennial with multiple permanent pools. The river no longer has any pools. According to Mr. Siziba (70) of Vuturura village (Dendele ward), harvests are now very poor yet in yesteryears grain used *ukubola eziphaleni ngobunengi babo* (grain used to rot in granaries because of huge harvests).

Places like Lutumba village in central Beitbridge which used to successfully produce more dry land maize than pearl millet and sorghum no longer produce maize because of unreliable rainfall. Amacimbi (mopane worms) used to be plentiful in the whole district but because of low and variable amounts of rainfall, amacimbi have become fewer, smaller and as a result fetch much less income for the farmers than before. In very dry years, amacimbi die before maturity. Amarula juice, the most common fruit product in Lutumba, has become less common.

Harvests used to take place up to August but now they end in May. Wildlife, which used to be a common sight in the wards, has also disappeared partly because of negative variations in climate change-related environmental modification.

4.1.2 Meteorological evidence of climate change in Beitbridge district

Figures 2 and **3** show some descriptive evidence that Beitbridge is generally a dry district. Only 2 years (1952 and 1957) received annual rainfall above 500 mm in the generation 1952–1982. Comparatively, the third generation (1982–2012) had only 1 year (2000) receiving annual rainfall above 500 mm. The second generation had 19 years which received mean annual rainfall figures of 300 mm and below compared to the third generation's 21 years. The two climate graphs (**Figures 2** and **3**) also show that temperature range for Beitbridge increased from 1.4°C in the second generation (1952–1982) to 3.3°C in the third generation (1982–2012). Beitbridge district has, therefore, become drier and hotter in recent years.

Rainfall is the most important climate variable in south-western Zimbabwe because of minority farmers' reliance on semi-nomadic pastoralism and rain-fed crop farming. Mean annual rainfall coefficient of variation was calculated in SPSS version 20 to determine the percentage variation from one generation to the next.

Beitbridge district's mean annual rainfall coefficient of variation for the second generation (**Figure 3**) varied from as low as 35% in the second generation to 53% in the third generation (**Figure 4**).

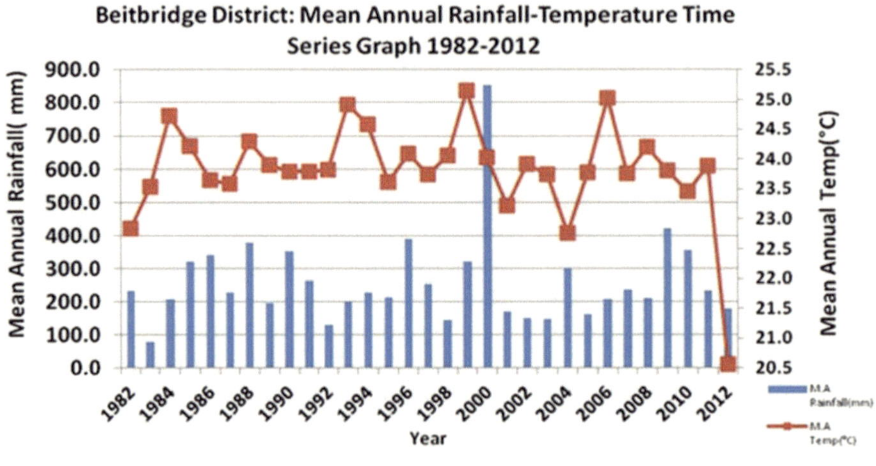

Figure 2.
Mean annual rainfall-temperature graph for Beitbridge district (1982–2012). Data source: Meteorological Department, Harare.

4.2 Indigenous household strategies for coping with climate change

In Beitbridge district, common indigenous coping strategies for coping with food insecurity associated with climate change include planting drought-resistant crops (94%) like sorghum and millet which are quite popular (**Table 2**). Rapoko is grown by few farmers. The semi-arid conditions of the district and the unpredictable and variable nature of the rainfall pattern forces farmers to practise probability planting which farmers gave a 72% frequency rating. Traditional adaptation mechanisms like collecting and drying wild fruits for future use, drying some crops for future use and eating wild fruits as household meals all have suppressed frequency ratings of 56 and 52%, respectively. This could be because crops like sweet reeds, groundnuts and watermelon which used to be dried for future use are no longer productively grown in the district because of inadequately distributed

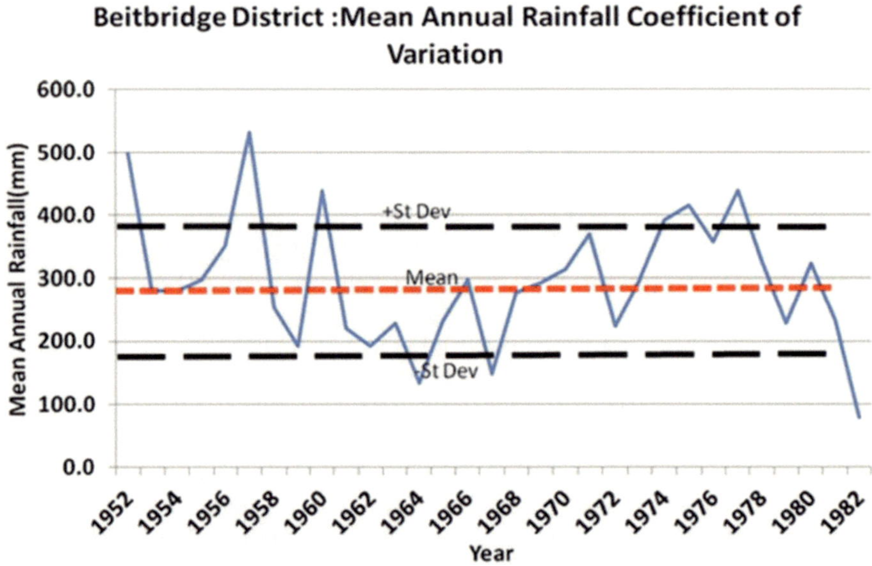

Figure 3.
Mean annual coefficient of variation for Beitbridge district (1952–1982).

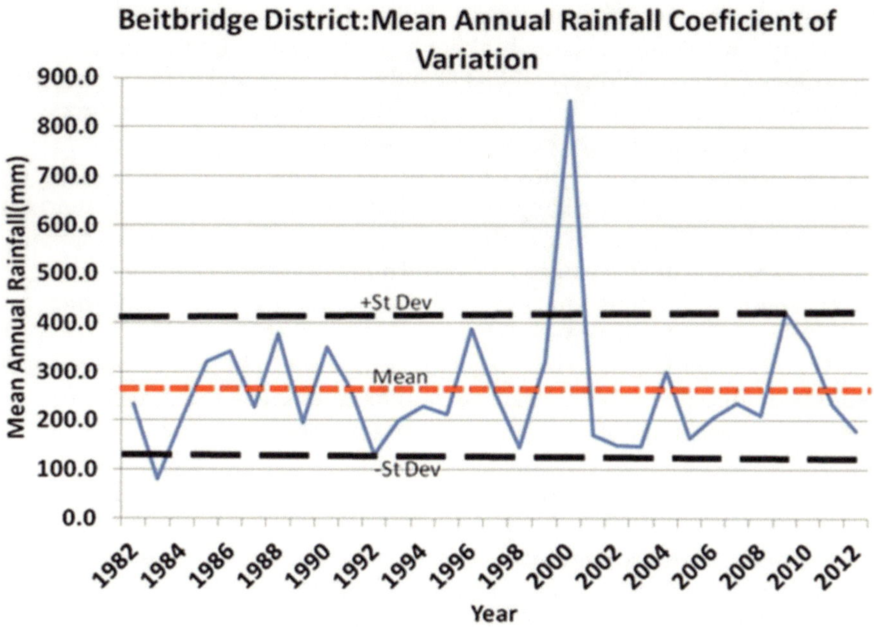

Figure 4.
Mean annual coefficient of variation for Beitbridge district (1982–2012).

soil moisture levels. Traditional fruit trees like *umkhomo, uxakuxaku, umgano* are now very few in Beitbridge because the trees no longer observe their normal reproductive cycle because of rainfall, temperature and soil moisture variability related to climate change. About 68% of the farmers, however, believe planting early maturing varieties of the national staple maize is still helpful in the district. This could be because in climatically favourable years, some farmers get decent maize yields.

The Venda, Shangani and Suthu also cope with climate change through diversification of economic activities. This mixed economy strategy involves pastoralism, cultivation, hunting, fishing, barter trading, cross-border trading, formal and informal employment and remittances from siblings in the regional Diaspora, mainly South Africa and Botswana. These diverse sources of livelihood help communities remain hopeful even when the climatic regime becomes bleak.

Environmental evidence	Frequency			
	Evident	%	Not evident	%
Drying of perennial rivers and springs	78	78	22	22
Late onset of the rains	100	100	0	0
Early cessation of the rains	94	94	6	6
Change in wind patterns	80	80	20	20
Diminishing pastures	88	88	12	12
Cold season warmer	64	64	36	36
Hot season hotter	80	80	20	20

N = 100.

Table 2.
Indigenous household strategies for coping with climate change.

When in season, minority women, girls and small boys gather fruits like umviyo, umqokolo, mtshwankela, amaganu, umkhomo, umkhemeswane and umhali as household food supplements. Wild vegetables like imbuya, idelele and ulude are important relish during the wet season. These, together with bean leaves, pumpkin leaves and garden vegetables are dried and preserved for use during drought or famine periods.

Locusts, inyeza (cicadas), inhlwa and mopane worms (amacimbi) are important among the Venda, Suthu and Shangani communities. Although mopane worms can be eaten while fresh, for future use, they are usually dried and roasted. Mopane worms are a regional delicacy in south-western Zimbabwe; hence, they are an important source of income. The income is used to purchase other important food items like mealie-meal and grain.

The Venda also gather roots like mutobhi, mukwikwi and mudzamoyo and add to their diet in drought periods. These onion-like roots are found in wet areas which because of climate change are now under severe threat of extinction.

4.3 Coping strategies for livestock sustenance under climate change

Indigenous minority communities in Beitbridge are sedentary pastoralists who depend heavily on livestock for their sustenance (**Table 3**). Their various types of livestock which include goats, sheep, donkeys and cattle are vulnerable to both climate change and environmental change.

In Beitbridge, with livestock being their major fallback cushion against poverty, communities have adopted means of sustaining their animals albeit with limited success. In the Shangani-dominated eastern wards of Dite I, Dite II and Chipise, farmers collect both wet and dry mopane, *mutsingidzi* and *murabva* leaves which they sprinkle with salt solution and give to cattle. Others mix dry leaves with molasses which they buy from Lutumba and feed their cattle.

In the Venda-dominated central wards of Mtetengwe I, Mtetengwe II and Mtetengwe III where pastures are now a problem, natural cattle husbandry is no longer practised because of heavily depleted pastures partly due to climate change. Farmers characteristically stock pile crop stover to feed cattle during drought periods. Such practice is new in a district known to be a natural pastoral region. Some farmers now sell part of their herd to raise money to buy stock feed for their cattle so that they survive the more frequent drought periods. Beitbridge experienced drought periods in the seasons 1982/1983, 1987/1988, 1991/1992, 1992/1993, 1996/1997, 1997/1998, 2001/2002, 2002/2003, 2004/2006, 2005/2006, 2011/2012 [17].

Strategy	Frequency			
	Helpful	%	Not helpful	%
Planting early maturing varieties of staple maize	68	68	32	32
Planting drought-resistant crops	94	94	6	6
Collecting and drying wild fruits for future use	56	56	44	44
Drying crops for future use	62	62	38	38
Probability planting	72	72	28	28
Eating wild fruits as household meals	52	52	48	48

N = 100.

Table 3.
Indigenous strategies for livestock sustenance under climate change.

Mr. Sibanda highlighted that in western wards (Machuchuta, Masera and Maramani) and north-western wards (Dendele, Siyoka I and Siyoka II), cattle deaths during drought periods are relatively lower. During the fast-track land reform programme communities generally agreed to repossess land from former white commercial farmers. However, they chose not to go and settle in most of those farms but reserved them as grazing areas (miraga) for their cattle during drought periods. Communities of Machuchuta, Masera and Maramani, for example, send their cattle to Shobi Block farm during dire periods. This explains why central and eastern wards which chose to go and settle in newly acquired farms record high figures of cattle losses to drought. These areas no longer have typical *miraga* (reserved natural pastures).

4.4 Constraints encountered by indigenous minority farmers in coping with climate change and variability

Attempts by farmers to cope with climate change in Beitbridge district are being hindered by a number of constraints. Both planting time (94%) and harvesting time (92%) are no longer predictable (**Table 4**). This is mainly because of the unpredictability and unreliability of seasonal rainfall patterns. A large section of the district, including places like Lutumba and Chaswingo, are generally dry with few rivers. This is the reason why 70% of the respondents describe water sources as being few and far from their homesteads. The national radio and television broadcaster, The Zimbabwe Broadcasting Corporation, covers only a small section of the district, in and around Beitbridge town. The larger area of the district, for example, Shashe, Chaswingo, Masera and Maramani either do not have this service or the signal is erratic and unreliable. This explains why 74% of the respondents do not have access to weather forecasts. Of the 26% who receive the signal, 24% of them describe weather forecasts provided as unreliable. This could be because the meteorological department itself is also experiencing challenges with changes in climatic phenomena.

Mr. Siziba of Vuturura village argued that although monthly council-administered cattle sales are a good idea which is supposed to help farmers creep out of poverty, the fact that council insist on farmers having temporary trading licences to conduct business at such sales renders the endeavour futile. Farmers do not raise much from the sales because their livestock will generally be in bad shape and yet goods sold by other traders are expensive. In the eastern wards (Dite I, Dite II and Chipise), for example, a mature cow or bull (or ox) could fetch as little as US$90 according to Mrs. Simuta of Chaswingo village in Dite I ward. In group discussions, farmers from Chaswingo village thanked one white who they said 'helped' them by buying their very thin cattle which he would feed first before transporting them to his Matengeni farm in Mwenezi district.

Strategy	Frequency			
	Helpful	%	Not helpful	%
Sending cattle to far-off places for better pastures	96	96	4	4
Preserving crop residue after harvest	58	58	42	42
Sourcing tree leaves/twigs/fruits for livestock	80	80	20	20
Reducing the size of livestock herds during drought periods	80	80	20	20

N = 100.

Table 4.
Constraints in coping with climate change.

In the western wards, farmers at Shashe and Toporo complained that although they sell cattle and goats every month, private buyers cheat them by buying their livestock at less than a dollar per kilogramme which is unsustainable. They appealed to government to re-introduce the Cold Storage Company which pays more sustainable prices. Even weavers at Maramani Craft Centre complained of unfair markets for their products. Despite all the time and effort invested, one broom would cost 2 South African rands and yet the buyer would sell the same broom at 20 rands. Buyers take advantage of the remote location of these wards in relation to Beitbridge town centre. Roads in rural Beitbridge are barely navigable. Feeding cattle with tree leaves requires considerable amount of labour which is in short supply in the district given that most young people migrate to South Africa. It also requires scotch-carts to transport leaves home. Families who do not have scotch-carts usually suffer heavier losses of livestock during drought periods. It is important to note that feeding livestock with leaves is only a stop gap measure to reduce livestock mortality. It does not prepare them for the market, neither is it sustainable if drought periods persist beyond a year.

All areas of the district bemoaned the prevalence of cattle rustlers who steal cattle, goats and sheep at night. They target mainly female-headed households where they know women would not go out for fear of possible harm or even death.

Although *miraga* (extensive natural rangelands) reduce cattle mortalities, farmers complained that they are expensive. Where these *miraga* are owned by private farmers, community farmers have to pay a single herd of cattle per every 10 herd of cattle. This is despite the fact that during severe drought years, cattle die even there and yet miraga owners will insist on their charge. Mrs. Muleya of Chaswingo, for example, lost 50 herds of cattle at the *muraga yakholomo* (cattle-dominated resettlement areas) during the severe 2012 drought. She only received cattle bells as evidence of cattle deaths. Cattle usually go to *miraga* in June/July when pastures have completely depleted and watering points have dried up.

In Beitbridge, minority farmer communities still have considerable faith in their traditional staples like sorghum and millet (**Figure 5**). A frequency of 50 for sorghum and 36 for millet show that the household granary provides most of the seeds grown, which are traditionally treated and carefully stored for replanting. Maize, the national staple crop, generally does not do well in Beitbridge district because of erratic and unreliable rainfall. Maize requires mean annual rainfall of between 600 and 800 mm which the district rarely gets. Farmers, however, insist on planting maize despite low yields because of its palatability which is more than that

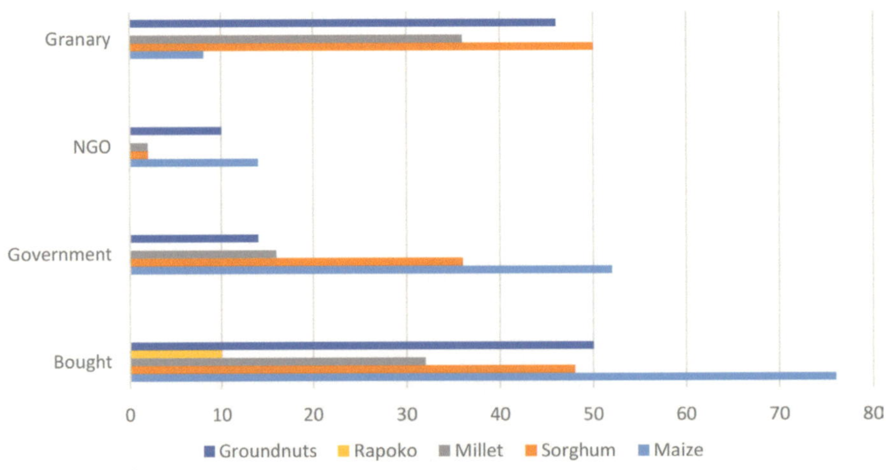

Figure 5.
Frequency of household sources of seed in Beitbridge district (%).

of sorghum and millet. Most of the maize seed (frequency of 76%) is either bought or supplied by government through its grain loan scheme where farmers are given seeds and are expected to repay with maize grain after harvesting. The loans are rarely repaid because of persistent poor harvests.

Maize is not a traditional Venda, Suthu or Shangani crop and hence these communities do not have traditionally time-tested ways of maize seed preservation as with sorghum and millet. This explains the granary's contribution to maize seed in the district. Though groundnuts require significant rainfall amounts for favourable harvests, households in Beitbridge grow them because of their multiple uses as relish, as relish enhancers or as body lotion. Farmers prefer traditionally prepared and preserved seed from the granary but if this is not available or is not enough, they buy from seed houses or from other households.

4.5 Government initiatives to help farmers cope with the effects of climate change

In Beitbridge district, questionnaire results show that government has not done much to help Venda, Suthu and Shangani farmers cope with climate change. Government initiatives cited include borehole drilling in wards 2, 3, 5 and 9. This stretched from the 1980 decade to the 1990s. In 2012, the severe drought which hit the district forced government to initiate a supplementary feeding scheme for cattle. This, however, came after many households had lost large herds of cattle.

Through the Livestock Development Programme (LDP), government advises farmers to sell some of their cattle to serve others. Government provides inputs like maize, sorghum and pearl millet seeds and fertilisers which farmers claim are common during election campaign periods at political rallies. The authorities also give drought relief at subsidised prices to farmers. Through AGRITEX, government educates farmers on helpful techniques to cope with climate change. For example, they are encouraged to practise dry planting to make maximum use of the early rains. During the 2012 drought, government availed supplementary feeding at subsidised prices where a 50 kg bag of 'beef survival' feed was sold at US$6.50 instead of US$14 from National Foods Limited. This assistance, however, was not very helpful since most farmers could still not afford.

Farmers around Zhove Dam thanked government for the dam from which they get an alternative source of livelihood through fish projects. They sell fish to clients from as far afield as Gwanda, Bulawayo and to middlemen who sell along highways and in villages.

Beitbridge has abundant wildlife which most rural farmer communities manage under Communal Areas Management Programme for Indigenous Resources (CAMPFIRE) projects which can be very useful in helping communities cope with climate change and climate variability through giving hunting quotas to professional hunters. CAMPFIRE is a government-initiated strategy designed to help rural communities manage their resources for the development of their localities. It was initiated in 1986. CAMPFIRE programmes are run by local authorities (rural district councils) and are therefore an arm of central government.

Assistant to Chief Executive Officer (Administration) Ms. Ponela revealed that all CAMPFIRE projects in the district face the danger of being rendered unviable by climate change. The signs are already showing in that the district is always hit by all forms of drought, that is, meteorological, agricultural, hydrological and socio-economic droughts. When there is no surface water in rivers and pools, wildlife migrate to South Africa or Mozambique. This results in reduced safari hunting and consequently reduced dividends to communities.

4.6 Initiatives by NGOs to help farmers cope with the impacts of climate change

In Beitbridge district, initiatives by NGOs to help communities cope with climate change and variability were non-existent between 1980 and 1999. The 2000–2009 decade, however, saw NGOs like Lutheran Development Society (LDS) and CARE International building earth dams to harness rain water for communities. They also introduced community gardens to help reduce malnutrition and food insecurity in general. LDS, CEZVI, Red Cross and World Vision sunk boreholes in various wards across the district and this improved water access for both domestic use and livestock watering. In an attempt to reduce deaths of livestock due to drought and improve the value of cattle for subsistence farmers, LDS introduced cattle fattening. The programme only became successful during the inception period but farmers could not sustain it once the NGO had left due to high running costs.

Community gardens and earth dam construction by CARE and LDS continued into the 2010–2013 period. During the same period, CARE, Southern Alliance for Indigenous Resources (SAFIRE), EU and LED collaboration with government initiated community orchards at Shashe Irrigation Scheme. These are expected to provide economic relief to the Shashe community once the trees start to bear fruits.

Lutheran Development Services (LDS) has rehabilitated boreholes in north-western wards of Dendele, Siyoka I and Siyoka II. World Vision and LDS drilled boreholes at primary schools which include Dendele, Madali and Vuturura. LDS also helped villagers initiate Phondongoma community garden in Vuturura village and Malusungane garden in Dendele ward. Villagers grow different types of vegetables and fruits like oranges, mangoes, peaches and guavas in an attempt to circumvent the effects of climate change and variability. LDS pays school fees for orphans from primary school up to Advanced Level as a way to increase their chances of creeping out of poverty.

In the central wards of Mtetengwe I, II and III, CARE International is the most active. It has helped villagers start community gardens and small irrigation schemes by providing fence, poles and seed packs. The gardening is mainly done by women who are organised in groups of 10. The NGO advise farmers on what to grow taking into account ambient climatic variability. Farmers in these wards, however, complained that although the NGO helps them cope with climate change, it is not a permanent solution because it only stays in an area for two or three seasons and leaves at critical times when they are mastering the new technologies introduced. During the time of the study, CARE had shifted to Masvingo Province and its projects in Beitbridge were already showing signs of collapse.

In the western wards of Machuchuta, Masera and Maramani, CARE International, ICRISAT and ORAP established sale pens for livestock. This has helped farmers realise fairly reasonable prices from their livestock before they become very thin from lack of pasture, browse and water. During severe years, CARE International and World Food Programme provide food packs like cooking oil, soya beans, maize and samp to alleviate starvation. The most notable contribution to these western wards is Shashe Irrigation Scheme where government partnered SAFIRE and CEZVI to initiate the growing of citrus which when complete will cover 140 hectares. About 80 hectares are currently under irrigation. Farmers also grow wheat to cover for their family needs as well as for sale. World Vision built three small earth dams in ward 9 which helped farmers initiate community gardens. The dams also provide water for livestock.

In the eastern wards, CARE International supplied cooking oil and a bag of maize per household per month for 7 months from September 2012 to March 2013, while World Vision provided supplementary feeding for cattle, donkeys and goats. LDS, World Vision, Red Cross and CEZVI drilled boreholes.

4.7 Perspectives by communities in Beitbridge on sustainability of interventions

Minority farmer communities in Beitbridge district comprising the Venda, Suthu and Shangani generally believe their indigenous knowledge is least sustainable at 28% (**Figure 6**). Only 52% of the respondents believe indigenous knowledge is between sustainable to very sustainable. This could be because of environmental change resulting from climate change and variability. NGOs are the best rated in providing interventions against climate change and variability. Respondents gave NGOs a combined rating of 78% between sustainable and very sustainable against a government rating of 56%. This could mainly be because NGOs come to communities' rescue during the hyperinflationary period of the 2000–2009 decade which, combined with famine from recurrent droughts and floods during that period, threatened to wipe out these communities.

4.8 Discussion

In Beitbridge district, climate change is causing failed agricultural produce and reduced yields. These result from climate extremes like droughts, strong winds and floods which also destroy infrastructure like homes, bridges and schools. They also kill people and livestock. Many irrigation schemes in the district are not functional because of the 2000 and 2003 destruction by cyclones Eline and Japhet, respectively. This has negatively impacted households' livelihoods. Climate-dependent environmental endowments like the mopane worm (amacimbi), a southern district delicacy harvested from mopane trees, has become vulnerable because of climate variability. Mopane worms are an important source of livelihood in Beitbridge district. They are eaten in the household as relish or as a stand-alone meal in difficult times. They are also brisk business when sold in villages, along roadsides, at rural service centres or in towns as communities' source income to cushion themselves against poverty. Another traditionally important household cushion against hunger and starvation, amarula juice, is now only episodically produced as the fruit is also succumbing to climate change. Youths in Beitbridge no longer place their hope on their environment but on neighbouring South Africa. With their drift to South Africa, the youths miss out on cultural initiation which would otherwise equip them with their indigenous knowledge. This points to an imminent erosion of rich cultural values of the Venda, Suthu and Shangani communities.

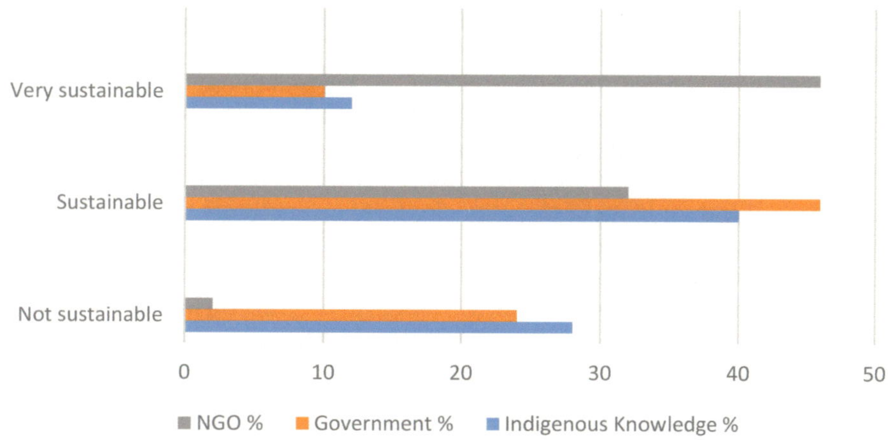

Figure 6.
Interventions sustainability rating.

Climate change is forcing communities to gradually lose confidence and trust in their generations-old indigenous knowledge systems of farming because of erratic rains which lead to reduced agro-ecological yields. Community trust and confidence on indigenous knowledge systems to support food security has now been eroded. Traditional food preparation and preservation are no longer systematic. This leads to loss of these skills by the new generation who traditionally would carry such indigenous knowledge to posterity. Some farmers have since abandoned dry land farming because of persistent droughts. Others have resorted to NGO-sponsored community gardening projects, for example, in Lutumba and Maramani wards. Given the uncertainty of such gardening projects, which usually collapse at the departure of benefactor NGOs, reliance solely on gardens drive households into deeper poverty.

Beitbridge is a livestock district and for most households, livestock is a store of wealth. Climate change induced extreme weather like droughts and floods which kill many animals [17]. Farmers then sell their livestock for very little to people from districts with better pastures. Most of the money that communities get from their livestock is used to buy items and goods from cross-border traders and traders from other districts. This means that even with the large herds of livestock, which can turn them into viable commercial livestock producers, minority community farmers remain poor subsistence livestock herders.

A viable indigenous coping strategy to climate change is, however, manifest in Maramani ward between Shashe and Limpopo Rivers. Both men and women hand-make hats, mats, brooms at Maramani ilala Project Craft Centre using the locally available *ilala* plant which they harvest from wetlands and river banks. They sell their products locally, in towns and even across the border in South Africa. The remote location of both the ward and craft centre, however, negatively affects this local initiative. Clients from outside the ward are few and those who endure the rough drive to the craft centre negotiate for very uneconomical bargains to the weavers. Maramani Ilala project is, however, an example of how communities can use their indigenous knowledge, raw materials from their immediate environment and their skills to cushion themselves against climate change impacts. Largely because of the Maramani Ilala project, climate change effects in Maramani ward are not very dire.

Western wards of Masera, Machuchuta and Maramani, which are rich in wildlife also, have the potential to counter detrimental climate change impacts through CAMPFIRE projects. Wildfire is a promising resource which, if well exploited, can boost tourism in a revamped CAMPFIRE programme. CAMPFIRE proceeds are meant to benefit these wards in infrastructural developments like roads, schools, clinics and bridges are currently inadequate. Communities no longer see any value in conserving wildlife. Some have resorted to poaching in wildlife conservation areas.

These wards are comparatively wetter and sustain more herds of livestock. The tragedy, however, is that they are isolated and have badly damaged roads. This results in poor markets for livestock. Although Beitbridge Rural District Council conducts cattle sales at selected 'central' points within the district, these points are in reality not central for most of the minority farmers in remote parts of the district. The selling points are located close to highways to attract buyers from towns and cities. Besides council-conducted cattle sales being a noble idea, council insists on farmers to have temporary cattle trading licences to conduct business. Most farmers cannot afford the money. Moreover, most of their cattle will be in bad shape and cannot give high returns. To raise the money for a council licence and for household needs, farmers are forced to sell many herd of cattle and at very low prices. In the eastern wards of Dite 1, Dite 2 and Chipise, for example, a mature cow or bull can sell for only US$90. In drought years, government, through the Livestock Development Programme (LDP), advises farmers to sell their livestock and not risk all of them being wiped out by the drought. Minority communities, therefore, have two hard choices for their livestock in drought years; either they sell them very cheaply to people from other districts for fattening and profit-making, or they watch them die miserably.

Even for these comparatively better environmentally endowed western wards, opportunities offered by government in the face of climate change are no longer enough to retain youths. This is because of the strong South African rand pull factor in the more affluent South African cities and farms. The sustainability of indigenous knowledge and local resources as cushions against climate change therefore remain threatened.

Pastures are generally a big problem in Beitbridge district, more so in the Venda-dominated central wards of Mtetengwe1–3 and Eastern Shangani-dominated wards. Although these wards benefitted from the fast track land reform and resettlement programme, beneficiaries preferred to go and settle in the newly acquired areas. This led to quick environmental depletion due to large herds of livestock, clearing of large areas for settlement both of which were not complemented by climate variables, especially rainfall. Some of the resettled areas had over the years been used as *miraga* (winter grazing areas) by these minority farmers. The permanent occupation of such areas, therefore, deprived farmers of fallback options for their livestock in times of drought. Consequently, eastern and central wards lose high numbers of livestock to climate-change-related factors.

By comparison, Suthu-dominated north-western wards of Dendele, Siyoka 1 and 2 and western wards of Machuchuta, Masera and Maramani, livestock deaths are not as devastating. In these areas, instead of occupying the farms they got from the fast track land reform and resettlement programme, communities were prudent enough to reserve most of the farms as *miraga* for their cattle. For example, Maramani, Masera and Machuchuta communities send their cattle to Shobi Block farm which is their winter grazing area.

While *miragas* are generally a viable solution to reduce cattle deaths, they come at a cost to farmers. For every 10 herd of cattle, farmers are obliged to part with one beast as payment to the custodians of respective grazing areas. Besides, at these grazing areas, cattle are not always safe. Cattle rustlers are a problem in the district, with some families losing as many as 50 head of cattle to them at *miraga yakholomo* (winter cattle grazing areas) in one season alone.

Minority Venda, Suthu and Shangani communities in Beitbridge rely on their traditional sorghum and millet varieties as their staple crops. They prepare and treat these seeds and store them in their traditional granaries for replanting. Instead of helping communities improve these local grain varieties through research, government seems determined to change these communities' traditional tastes by introducing maize as an alternative to these small grains. Government provides maize seed loans which farmers rarely pay back because of poor harvests caused mainly by reduced soil moisture content resulting from the changing climate. Government is, therefore, defeating the intentions of the developmental state theory which seeks to promote indigenous initiatives to development.

NGOs are popular in Beitbridge. This is because they attend to the immediate and most pressing needs of the household through, for example, providing food, initiating and funding community gardens, paying fees for school children, fattening cattle during drought periods. NGOs, however, usually do not stay in any ward or district for long. They usually leave after the supposed summer harvest (even if some households do not harvest anything). NGOs usually leave before communities have mastered the art of self-sufficiency in the efficient management of projects. As a result of this pseudo-empowerment, communities fail to sustain themselves.

4.9 Conclusion

All three minority communities (Venda, Suthu and Shangani) that populate Beitbridge are at the mercy of climate change and climate variability. Although they receive some assistance from government and non-governmental organisations, they all have considerable faith in their own indigenous coping strategies to fight the climate change scourge. There is need to recognise, respect and improve their indigenous knowledge systems. A deliberate effort must also be made to integrate this local traditional knowledge with modern technologies to build even stronger resilience and sustainability options.

The case of Beitbridge's minority farmer communities suffering from climate change and climate variability is one of lack of financial and technical resources and to some extent human resources because of an acute outmigration by the youth. Beitbridge is not a poor district from a natural resources' point of view. The district is a livestock region and all minority farmer communities generally have livestock like cattle, goats and sheep. The management of these in the face of climate change is, however, a serious issue as most of them, especially cattle, succumb in large numbers to climate-change-induced droughts and associated hazards like diseases and surface water shortages.

The study acknowledges efforts being initiated by both government and NGOs to reduce effects of climate change in Beitbridge. However, given the rich soils in parts of the district, the palatable pastures and browse, the diverse livestock varieties and abundant surface water potential for both damming and extraction through piping, the minority farmer communities of the district can significantly be better empowered for sustainability. Farmers should not be basket cases but should be helped to develop sustainable strategies and techniques to help them build resilience regardless of climatic hazards.

The interplay of government and NGO assistance on the communities in Beitbridge District risk disorienting the minds of these communities from their traditional cultural beliefs as their political orientation may be influenced by those who feed them. Apart from simply being natural hazardous phenomena, climate change and climate variability may end up being tools of political and economic manipulation of such minority farmer communities by both the rich and humanitarian service providers like NGOs and government through food politics where assistance may only be given to those who comply with the status quo.

4.10 Recommendations

- Relevant government ministries and development partners should conduct workshops to conscientise the community in all 15 wards about climate change and its devastating impact on the environment and their livelihoods in general. This will enable council and the community to work together in coming up with mutually acceptable solutions to problems.

- Beitbridge's comparative advantage in livestock production should be exploited by government through availing tax incentives for livestock-related industries in order to ameliorate climate change impact through employment creation for the locals. This would likely reduce the number of youths who migrate to South Africa. The young minds would thus be taped for the development of the district.

- Beitbridge has/is in the vicinity of perennial rivers like Limpopo, Shashe, Bubi and Runde Rivers which can be dammed to produce large water bodies which can turn the whole district into a green belt. Government and other development partners should work towards the possible realisation of this possible dream.

- Only 20% of the irrigation schemes in Beitbridge district are functional despite the fact that it falls in agro-ecological region 5 with an average rainfall of 300 mm [17]. Lack of fully functional and efficient irrigation schemes has negative implications on food security within the district especially among minority farmer communities who are on the fringes of development. Beitbridge Rural District Council should partner other arms of government like the irrigation department and AGRITEX to resuscitate and improve capacity utilisation of irrigation schemes in the district.

- Emphasis of the district's irrigation schemes is on subsistence farming rather than on commercialisation. Average irrigation plot sizes range between 0.1 and 0.3 hectares which is relatively small [17]. Though this approach used to be sustainable several decades ago, and had stuck in the mindset of communal minority farmers, the advent of climate change and its destructive effects has rendered subsistence farming unsustainable. Government and NGO initiatives have not helped much with their food handouts which are only temporary and do not guarantee food security. Better coordination of the farmers by AGRITEX, NGOs and even among themselves is required to make irrigation schemes commercially successful.

Acknowledgements

The authors would like to thank all research participants for their generous contributions and also for allowing us to use their names in this chapter.

Author details

Mark Matsa* and Beauty Dzawanda
Midlands State University, Gweru, Zimbabwe

*Address all correspondence to: matsam@staff.msu.ac.zw

References

[1] Corbera E, Conway D, Goulden M, Vincent K. Climate Change in Africa: Linking Science and Policy for Adaptation. Norwich and London: The Tyndall Centre and IIED; 2006, Royal Society Workshop Report

[2] Intergovernmental Panel on Climate Change (IPCC). Managing the Risks of Extreme Events and Disasters to Advance Climate Change Adaptation. New York, NY, USA: Cambridge University Press; 2012

[3] USAID. Adaptation to Climate Change. A Guidance Manual for Development Planning. New York: USAID; 2007

[4] Salick J, Byg A. Indigenous Peoples and Climate Change, Report of a Symposium Held on 12-13 April 2007. Oxford: Tyndall Centre for Climate Change Research; 2007

[5] Boko M, Niang I, Nyong A, Vogel C, Githeko A, Medany M, et al. Climate change 2007: Impacts, adaptation and vulnerability. In: Parry ML, Canziani OF, Palutikof JP, van der Linden PJ, Hanson CE, editors. Contribution of Working Group II to the Fourth Assessment Report of the Intergovernmental Panel on Climate Change. Cambridge, UK: Cambridge Univ Press; 2007. pp. 433-467

[6] Abid M, Shilling J, Scheffran J, Zulfiqar F. Climate change vulnerability, adaptation and risk perceptions at farm level in Punjab, Pakistan. Science of the Total Environment. 2016;**547**:447-460

[7] Masud MM, Al-Amin AQ, Junsheng H, Ahmed F, Yahaya SR, Akhtar R, et al. Climate change issue and theory of planned behavior: Relationship by empirical evidence. Journal of Cleaner Production. 2016;**113**:613-623

[8] Aklilu Y. In: JG MP, Little PD, editors. Pastoral Livestock Marketing in Eastern Africa: Research and Policy Challenges. Rugby, UK: Intermediate Technology Publications; 2006. pp. 187-202

[9] Hirji R, Johnson P, Chauta TM, editors. Defining and Mainstreaming Environmental Sustainability in Water Management in Southern Africa. Harare: SADC/IUCN/SARDC; 2002

[10] Masendeke D. Farmers perceptions of climate change in Zimbabwe. In: Building Capacity to Cope with Increasing Vulnerability Due to Climate Change. Bulawayo: ICRISAT; 2008

[11] Chagutah TC. Climate Change Vulnerability and Adaptation Preparedness in Southern Africa. Zimbabwe Country Report. Cape Town: Heinrich Boll Stiftung Southern Africa; 2010

[12] Feresu SB, editor. Zimbabwe Environmental Change: Our Environment, Everybody's Responsibility. Harare; 2010: Government of Zimbabwe's Third State of Environment Report

[13] Chenge M, Sola L, Paleczny D. The State of Zimbabwe's Environment. Harare: Ministry of Mines, Environment and Tourism. Government of Zimbabwe; 1998

[14] ZimStat. Census 2012 Provincial Report. Matabeleland South. Harare: Population Census Office; 2012

[15] Fraenkel JR, Wallen NE. How to Design and Evaluate Research in Education. 3rd ed. New York: McGraw-Hill; 1996

[16] Matsa M, Matsa W. Bulilima's 'look south' policy: Gender and socio-economic implications. Eastern Africa Social Science Research Review. 2011;**xxxvii**(1):85-106

[17] BBRDCSP. Beitbridge Rural District Council Strategic Plan (2011-2015). Beitbridge: Beitbridge Rural District Council; 2010

Chapter 4

Climate Change Impacts and Adaptation Strategies for Agronomic Crops

Ishfaq Ahmed, Asmat Ullah, M. Habib ur Rahman, Burhan Ahmad, Syed Aftab Wajid, Ashfaq Ahmad and Shakeel Ahmed

Abstract

Climate change is a serious threat to agriculture and food security. Extreme weather conditions and changing patterns of precipitation lead to a decrease in the crop productivity. High temperatures and uncertain rainfall decrease the grain yield of crops by reducing the length of growing period. Future projections show that temperature would be increased by 2.5°C up to 2050. The projected rise in temperature would cause the high frequent and prolong heat waves that can decline the crop production. The rise in temperature results in huge reduction in yield of agronomic crops. Sustaining the crop production under changing climate is a key challenge. Therefore, adaptation measures are required to reduce the climate vulnerabilities. The adverse effect of climate change can be mitigated by developing heat tolerant cultivars and some modification in current production technologies. The development of adaptation strategies in context of changing climate provides the useful information for the stakeholders such as researchers, academia, and farmers in mitigating the negative effects of climate change.

Keywords: climate change impacts, climate change projections, adaptation strategies

1. Introduction

Climate change and variability are the real threats to agriculture and food security [1, 2]. Extreme weather events and uncertainty in rainfall patterns are negatively affecting the agricultural crops [3, 4]. The evidences of global trend in rainfall are unclear due to large regional gaps in spatial coverage and temporal shortfalls in the data. Owing to these changes, the drought is more prevailing in many regions of the world including Pakistan [5].

Finding evidences reported that high temperature and uneven distribution of rainfall have negative effects on crop productivity all over the world [6]. These changes in weather and climate are likely to affect the food security of developing world where a large fraction of ever-increasing population is already fronting hunger, insecure and unhealthy food [7]. Warming of weather and climate systems can results in highly corresponding changes in the occurrence of extreme events including rise in temperature, uneven rainfall patterns [4]. These extreme events occur due to shift in their distribution, or to change in the shape of distribution. Various studies suggest that a shift in mean accounts for much the change in observed temperature extremes [8]. Comparisons of various studies showed that both daily maximum (Tmax) and minimum (Tmin) temperatures have shifted toward higher values in all regions. These shift in temperatures and rainfall significantly effected the cropping patterns, crop yields and phenology [9].

The Intergovernmental Panel for Climate Change (IPCC) has found evidences of accelerated global warming, climate variability and change since the early 1990s. The IPCC reported that average global temperature in the last 100–150 years has increased by 0.76°C [10]. The variability in temperature altered the phenology of crop, i.e., leaf development, anthesis, harvest, fruit production and in asynchrony between anthesis and pollinators. The variable temperature range resulted in high respiration rates, reduction in pollen germination, shorter grain filling period, lesser biomass production and low yields [4]. High temperature above 35°C in combination with high humidity and low wind speed caused a 4°C increase in temperature, resulting in floret sterility in cereals and fruits [6]. Climate change impact assessment provides the scientific foundation for the development of adaptations to offset the negative impacts of climate change. Keeping in views, the current study was planned to assess the impacts of climate change and adaptations strategies for agronomics crops.

2. Projections of climate change across the globe especially in ASIA

World faces dreadful challenges due to changing climate as it is indicated by climatic models that global surface temperature is likely to exceed 1.5°C relative to 1850–1900 for all representative concentration pathways (RCP) scenarios for the end of the twenty-first century [11]. It is likely to exceed 2°C for RCP 6.0 and RCP 8.5 and warming will continue beyond 2100 under all RCP scenarios. Increase of global mean surface temperatures for 2081–2100 relative to 1986–2005 is projected to likely be increased 0.3–1.0°C (RCP 2.6), 1.1–2.6°C (RCP 4.5), 1.4–3.1°C (RCP 6.0), 2.6–4.8°C (RCP 8.5) by the of the twenty-first century. It is projected that temperature will increase drastically in arid areas of Pakistan and India and western part of China [11]. Models predictions indicated that erratic rainfall with greater intensity would increase across the region, but higher intense rainfall will occur during summer monsoon season. Increase in aridity in South and Southeast Asia is projected due to decline in winter rainfall. Sea level will rise to 3–16 cm by 2030 and 7–50 cm by 2070 across the globe due to climatic abnormalities and in relation with regional sea level variability [11].

It is evident from the facts that lives of millions rural poorest people in Asia are highly vulnerable to climate change. There are evidences of prominent increase in intensity and frequency of many extreme events such as heat waves, erratic and uncertain rainfall patterns and more number of hot days, sustained dry spells, tropical cyclones and dust storms in the region. These countries accounted for 91% of the world's total death and 49% of the world's total damage due to natural disasters in the last century. South Asia is the most food insecure region with 262 million malnourished people in the world [6, 12]. Discussed facts showed (**Table 1**) that rural communities that already live in remote dry lands and deserts with inadequate natural resources are most prone to climate change. Agricultural systems being affected by abnormal climatic variables that disturbs the biological, physical and chemical processes of the systems. Number of hot days and warm nights are likely

Crops	Country/continent	Yield reduction (%)	References
Wheat	Australia	−32	[13]
	Iran	−37	[14]
	Worldwide	−5.5	[7]
	Mexico	+25	[15]
	China	−17.5	[16]
	Asia	−7.7	[17]
	India	−5.2	[19]
	Pakistan	−50	[20]
	Turkey	−20	[21]
Rice	India	−7	[22]
	Indonesia	−11	[23]
	India	−8	[24]
	Asia	−6.3	[25]
	Italy	−12	[26]
	Japan	−11.3	[27]
	Nepal	−24	[28]
Maize	Portugal	−17	[16]
	Ghana	+12	[29]
	Africa	−20	[30]
	USA	−50	[31]
	Ethiopia	−4.7	[32]
	China	−46	[33]
	Africa	−32	[34]
	Pakistan	−27	[4]
	China	−30	[35]
	USA	−27	[36]

Table 1.
Impact of climate change on cereal crop production.

to increase in the Asia from 1961 to 2003 and reduction in cool days and nights was observed especially in the years after the start of El Nino [37]. Tropical cyclones frequency and intensity has increased in Pacific from last few decades [38].

3. Impact of climate change on crop production

Global atmospheric concentrations of greenhouse gases have significantly increased relative to pre-industrial times [13, 39, 40]. As a result, greenhouse gas forcing is the main cause of the warming of the atmosphere during the past decades [14, 41, 42]. This warming is expected to substantially alter the climate system and change global food production, mainly because temperatures are predicted to increase which in turn will alter the precipitation pattern and increase the frequency of extreme events such as drought [15, 43–45]. Man-made greenhouse gas emissions as a result of industrialization and urbanization have made significant contributions to global warming and further changes in the global climate. As a result, global temperature rose by 0.83°C from 1906 to 2010 [10]. Global warming also causes changes in precipitation levels and patterns due to higher evapotranspiration and water vapor amounts in the atmosphere with several implications for the global hydrological cycle [16, 46]. As the major water consumer of the developing world and some developed countries, agriculture is one of the most vulnerable water sectors to climate change [17, 18]. Dramatic population growth, associated with reduction of productive land area and water resources, exerts extra pressure on the agricultural sector. To ensure sustainability of agriculture, studying the possible climate change impacts on this sector is essential [9, 19, 47].

Rate of plant growth and development is dependent upon the temperature surrounding the plant and each species has a specific temperature range represented by a minimum, maximum, and optimum [48–50]. The expected changes in temperature over the next 30–50 years are predicted to be in the range of 2–3°C Intergovernmental Panel for Climate Change [10]. Heat waves or extreme temperature events are projected to become more intense, more frequent, and last longer than what is being currently been observed in recent years [51, 52]. Extreme temperature events may have short-term durations of a few days with temperature increases of over 5°C above the normal temperatures [53]. Extreme events occurring during the summer period would have the most dramatic impact on plant productivity. A recent review by Barlow et al. [54] on the effect of temperature extremes, frost and heat, in wheat (*Triticum aestivum* L.) revealed that frost caused sterility and abortion of formed grains while excessive heat caused reduction in grain number and reduced duration of the grain filling period. Analysis by [55] revealed that daily minimum temperatures will increase more rapidly than daily maximum temperatures leading to the increase in the daily mean temperatures and a greater likelihood of extreme events and these changes could have detrimental effects on grain yield. If these changes in temperature are expected to occur over the next 30 years then understanding the potential impacts on plant growth and development will help develop adaptation strategies to offset these impacts [56, 57].

Previous studies of climate change impacts on agriculture, using crop yield simulation models [9, 58–60].or statistical models suggest that climate change will substantially affect productivity of major staple food crops such as maize, because growth and development of crops are mainly dependent on sunlight, temperature, and water [22, 23, 61]. Climate change may modify precipitation, soil water, runoff, and may reduce crop maturation period and increase yield variability and could reduce areas suitable for the production of many crops [24, 62, 63]. Climate change might limit crop production (the amount of a crop that is harvested in a farm, region, state, or country in kilograms or tons) in many areas [64–66].

Temperature increases affect most plants, leading to crop yield reduction and complex growth responses [25, 46, 67]. Nevertheless, the impact of increasing temperatures can vary widely between crops and regions. For example, a 1°C increase in the growing period temperature may reduce wheat production by about 3–10% [68], winter wheat productions may be decreased by 5–35%, respectively, under the future warmer and drier conditions [21, 26], and corn yield may be reduced by 2.4–45.6% due to higher temperatures [27, 69]. Even if precipitation is unchanged, the crop production may decrease by 15% on average due to the reduction in crop growth period and increased water stress as the result of higher temperature and evapotranspiration (Schlenker et al. [63]; Yang et al. [16]; Khanal et al. [28]) expected precipitation reductions in arid and semiarid regions of the world, where water is already limited, can have dramatic impacts on crop production [32–35]. For example, in northwestern Turkey, winter wheat yield may decline more than 20% under future climate change because the growth periods can be shortened as a result of increased temperature, exacerbated by a reduction in precipitation [21, 29–31]. Higher reduction in wheat yield of 50% was found in Pakistan as shown in **Table 2**. In some other areas, climatic change might have positive influences on agricultural crop yield, i.e., in dry areas rainfall enhances under wet climatic warming can lead to improved crop productions like in Mexico the wheat yield would be increase by 25% in future (**Table 2**). Maize, rice, winter wheat and potato crop yield can be enhanced with increasing air temperature and rainfall in the Plain of North China [73]. In Ghana maize yield would be increase by 12% in future (**Table 2**). The impact of climate change on sugarcane and cotton is shown in **Table 3**. Higher reduction in cotton yield of 17% and sugarcane yield of 40% was found at USA (**Table 3**). However, some positive impacts of climate change on sugarcane yield were found in few countries such as Brazil and Australia (**Table 3**). The impact of climate change on coarse grain, oilseed and other miner crops such as pearl millet, sorghum and sesame are shown in **Table 1**. Huge reduction in coarse grain and other

Crops	Country/continent	Yield reduction (%)	References
Cotton	China	−5.5	[48]
	USA	−17	[49]
	Africa	−7	[52]
	USA	−9	[51]
	Pakistan	−8	[53]
	Burkina Faso	−13	[56]
	Australia	−17	[59]
Sugarcane	Brazil	+15	[70]; [65]
	Switzerland	−9	[71]
	Australia	+20	[72]
	Africa	+11	[46]
	USA	−40	[66]
	Brazil	−27	[65]
	India	−30	[64]

Table 2.
Impact of climate change on other crop production.

Countries	Coarse grains	Oilseeds	Other crops
China	−22 to 2	−12 to 12	−22 to 2
Philippines	−17 to −3	−10 to 4	−17 to −3
Thailand	−17 to −3	−10 to 4	−17 to −3
Rest S+E Asia	−17 to −3	−10 to 4	−17 to −3
Bangladesh	−17 to −3	−10 to 4	−17 to −3
India	−17 to −3	−15 to 4	−17 to −3
Pakistan	−17 to −3	−15 to 4	−17 to −3
Rest S Asia	−17 to −3	−15 to 4	−17 to −3

Table 3.
Productivity shock due to climate change on rice, wheat, and coarse grains by 2030.

crops were found in china up to 2030. However, higher losses in oilseed crops were observed at India and Pakistan as shown in **Table 1**.

4. Adaptation strategies for agronomic crops

Climate change adaptation is the action to global warming, which helps to reduce the vulnerabilities in the social and biological system. The main objective of adaptation strategy is to build the resilient in societies against climate change [74].

Agriculture sector is highly vulnerable to changing climate. Extreme weather conditions and changing patterns of precipitation affects the crop development, growth and yield of crops. High temperature at critical growth stages could reduce the grain filling duration caused the grains sterility and consequently yields reduction. [4]. To avoid the risks in agriculture associated with climate change (CC), adaptation is the key factor that could help to mitigate the negative of climate change. Adaptation strategies provide an opportunity to address the CC challenges and to sustain the crop production [75].

In the recent year, climate change adaptation has been explored by the farmers in many ways. For example, in Pakistan and Brazil farmers has adapted the climate change variability by adjustment of planting tine and optimization of plant populations [9, 60]. Adjustment of planting date is important to explore the fully potential of crop. High temperature at grain filling stage, reduce the time for grain filling that lead to decrease the yield. Adjusting the planting time with the onset of rains and heat waves would decrease the yield losses. Number of plants per unit area plays a vital role for higher yield in crops especially wheat. The number of productive tillers dies or remains unproductive due to variation in temperature and moisture stress. The optimum plant population compensates the yield loss. The development of improved varieties such as early maturing, drought and heat tolerant are necessary to

Crop (s) name	Region/ country	Adaptation	References
Wheat	Pakistan Brazil	• Use of heat tolerant cultivars • Adjustment of planting dates • Optimum plant population	[9, 81]
Rice	Bangladesh Sri Lanka	• System of rice intensification with alternate wetting and draying • Direct planting	[77, 82]
Maize	Nepal Asia	• Raised bed planting • Early maturing cultivars • Precision nutrient management	[83, 84]
Cotton	Pakistan	• Heat and drought tolerant cultivars • Increase in plant population by 18%	[53]
Sugarcane	Swaziland India	• Ratoon management • Pit planting	[71, 72]
Chickpea	India	• Integrated weed control • Agro-forestry (Wind barrier) • Improved crop varieties (early maturity)	[85–87]

Table 4.
Climate change adaptations for agronomic crops.

sustain the productivity under changing climate. The new cultivars would increase the production per unit area under moisture stress and extreme temperatures [76].

Methane gas is produced form the flooded rice. Flood water in rice blocks the oxygen to penetrate in soil that creates the favorable condition for bacteria that emit the methane gas. So new methods of planting like direct seeded rice and system of rice intensification with Alternate wetting and draying reduce the methane emission and increase the water use efficiency [77].

Precision management of nutrients can increase the resilience in the crops by increasing the efficiency of fertilizers. Precision management of fertilizers in crops especially maize reduced the use of fertilizers that would enhance the production and soil health that lead to decrease the emission of greenhouse gases (GHGs) [78]. Ratoon crop of sugarcane is more adaptive to climatic vulnerabilities. Fuel consumption is less for tillage practices, and less soil is disturbed that lead to reduce the GHGs emission. Pit planting is new evolutionary method in sugarcane. In this methods sugarcane seedling are grown in a small pit under field condition. This method improved the aeration and solar radiation that lead to increase the quality of cane juice and number of canes for milling [79]. Weeds are serious issue in the chickpea cultivation. Weeds compete with the chickpea plants for water and nutrients that reduce the growth and yield of chickpea. So integrated weed control improves the yield. GHGs emissions are also reduced due to less use of synthetic weedicides [80] (**Table 4**).

5. Conclusion

Climate change and variability have negative effects on crop productivity. Change in precipitation pattern, increase in frequency, and intensity of extreme events such as heat waves and drought have detrimental effects on grain yield. Future projections showed that temperature would be increased by 2–3°C at the end of century. Number of hot days and warm night will be increased in Asia and high intense rainfall will occur in summer monsoon. This warming situation would cause a huge reduction in grain yield of crops by end of century. Wheat yield is expected to decrease by 50% in Pakistan, maize yield by 46% in China, cotton yield by 17% in USA and sugarcane yield would reduce by 30% at India. The negative effects of climate change can be mitigated by developing some adaptation measures. The development of heat tolerant cultivars, modification in current production technologies of crop can offset the negative effects of climate change. In future, climate change impacts should be studies by using low and high emission scenarios for early, mid and late century. The adaptation strategies should be quantified based on modeling approaches.

Author details

Ishfaq Ahmed[1*], Asmat Ullah[2], M. Habib ur Rahman[3], Burhan Ahmad[4], Syed Aftab Wajid[5], Ashfaq Ahmad[5] and Shakeel Ahmed[6]

1 Centre for Climate Research and Development (CCRD), COMSATS University Islamabad, Pakistan

2 Ayub Agricultural Research Institute, Faisalabad, Punjab, Pakistan

3 Department of Agronomy, MNS-University of Agriculture, Multan-Pakistan

4 Pakistan Meteorological Department, Islamabad, Pakistan

5 Department of Agronomy, University of Agriculture, Faisalabad, Pakistan

6 Department of Agronomy, Bahauddin Zakariya University, Multan-Pakistan

*Address all correspondence to: ishfaqkanjal@gmail.com

References

[1] Field CB. Managing the Risks of Extreme Events and Disasters to Advance Climate Change Adaptation: Special Report of the Intergovernmental Panel on Climate Change. United Kingdom: Cambridge University Press; 2012

[2] Rezaei EE, Webber H, Gaiser T, Naab J, Ewert F. Heat stress in cereals: Mechanisms and modelling. European Journal of Agronomy. 2015;**64**:98-113

[3] Ullah A, Ahmad A, Khaliq T, Akhtar J. Recognizing production options for pearl millet in Pakistan under changing climate scenarios. Journal of Integrative Agriculture. 2017;**16**(4):762-773

[4] Ahmed I, Rahman MH u, Ahmed S, Hussain J, Ullah A, Judge J. Assessing the impact of climate variability on maize using simulation modeling under semi-arid environment of Punjab, Pakistan. Environmental Science and Pollution Research. 2018;**25**(28):28413-28430

[5] Ullah A, Salehnia N, Kolsoumi S, Ahmad A, Khaliq T. Prediction of effective climate change indicators using statistical downscaling approach and impact assessment on pearl millet (*Pennisetum glaucum* L.) yield through genetic algorithm in Punjab, Pakistan. Ecological Indicators. 2018;**90**:569-576

[6] Lobell DB, Field CB. California perennial crops in a changing climate. Climatic Change. 2011;**109**(1):317-333

[7] Lobell DB, Schlenker W, Costa-Roberts J. Climate trends and global crop production since 1980. Science. 2011;**29**:616-620

[8] Lewis SC, King AD. Evolution of mean, variance and extremes in 21st century temperatures. Weather and Climate Extremes. 2017;**15**:1-10

[9] Ahmad I, Wajid SA, Ahmad A, Cheema MJM, Judge J. Assessing the impact of thermo-temporal changes on the productivity of spring maize under semi-arid environment. International Journal of Agriculture and Biology. 2018;**20**(10):2203-2210

[10] IPCC. Climate Change 2007—The Physical Science Basis: Working Group I Contribution to the Fourth Assessment Report of the IPCC. Vol. 4. United Kingdom: Cambridge University Press; 2007

[11] IPCC. IPCC, 2013: Climate Change 2013: The Physical Science Basis. Contribution of Working Group I to the Fifth Assessment Report of the Intergovernmental Panel on Climate Change, 1535 pp. Cambridge, UK, and New York: Cambridge University Press; 2013

[12] IPCC, "IPCC, 2012: Managing the Risks of Extreme Events and Disasters to Advance Climate Change Adaptation. A Special Report of Working Groups I and II of the Intergovernmental Panel on Climate Change," Cambridge University Press, Cambridge, UK, and New York, NY, USA; 2012. Vol. 30, no. 11, pp. 7575-7613

[13] Luo Q, Bellotti W, Williams M, Bryan B. Potential impact of climate change on wheat yield in South Australia. Agricultural and Forest Meteorology. 2005;**132**(3-4):273-285

[14] Valizadeh J, Ziaei SM, Mazloumzadeh SM. Assessing climate change impacts on wheat production (a case study). Journal of the Saudi Society of Agricultural Sciences. 2014;**13**(2):107-115

[15] Lobell DB, Ortiz-Monasterio JI, Asner GP, Matson PA, Naylor RL, Falcon WP. Analysis of wheat yield and climatic trends in Mexico. Field Crops Research. 2005;**94**(2-3):250-256

[16] Yang C, Fraga H, Van Ieperen W, Santos JA. Assessment of irrigated

maize yield response to climate change scenarios in Portugal. Agricultural Water Management. 2017;**184**:178-190

[17] Asseng S et al. Hot spots of wheat yield decline with rising temperatures. Global Change Biology. 2017;**23**(6):2464-2472

[18] Ahmad I, Wajid SA, Ahmad A, Cheema MJM, Judge J. Optimizing irrigation and nitrogen requirements for maize through empirical modeling in semi-arid environment. Environmental Science and Pollution Research. 2019;**26**:1227-1237

[19] Gupta R, Somanathan E, Dey S. Global warming and local air pollution have reduced wheat yields in India. Climatic Change. 2017;**140**(3-4):593-604

[20] Hussain J, Khaliq T, Ahmad A, Akhter J, Asseng S. Wheat responses to climate change and its adaptations: A focus on arid and semi-arid environment. International Journal of Environmental Research. 2018;**12**:1-10

[21] Özdoğan M. Modeling the impacts of climate change on wheat yields in Northwestern Turkey. Agriculture, Ecosystems and Environment. 2011;**141**(1-2):1-12

[22] Aggarwal PK, Mall RK. Climate change and rice yields in diverse agro-environments of India. II. Effect of uncertainties in scenarios and crop models on impact assessment. Climatic Change. 2002;**52**(3):331-343

[23] Naylor RL, Battisti DS, Vimont DJ, Falcon WP, Burke MB. Assessing risks of climate variability and climate change for Indonesian rice agriculture. Proceedings of the National Academy of Sciences. 2007;**104**(19):7752-7757

[24] Saseendran SA, Singh KK, Rathore LS, Singh SV, Sinha SK. Effects of climate change on rice production in the tropical humid climate of Kerala, India. Climatic Change. 2000;**44**(4):495-514

[25] Masutomi Y, Takahashi K, Harasawa H, Matsuoka Y. Impact assessment of climate change on rice production in Asia in comprehensive consideration of process/parameter uncertainty in general circulation models. Agriculture, Ecosystems and Environment. 2009;**131**(3-4):281-291

[26] Bregaglio S et al. Identifying trends and associated uncertainties in potential rice production under climate change in Mediterranean areas. Agricultural and Forest Meteorology. 2017;**237**:219-232

[27] Iizumi T, Yokozawa M, Nishimori M. Probabilistic evaluation of climate change impacts on paddy rice productivity in Japan. Climatic Change. 2011;**107**(3-4):391-415

[28] Khanal U, Wilson C, Hoang V-N, Lee B. Farmers' adaptation to climate change, its determinants and impacts on rice yield in Nepal. Ecological Economics. 2018;**144**:139-147

[29] Srivastava AK, Mboh CM, Zhao G, Gaiser T, Ewert F. Climate change impact under alternate realizations of climate scenarios on maize yield and biomass in Ghana. Agricultural Systems. 2018;**159**:157-174

[30] Rurinda J, Van Wijk MT, Mapfumo P, Descheemaeker K, Supit I, Giller KE. Climate change and maize yield in southern Africa: What can farm management do? Global Change Biology. 2015;**21**(12):4588-4601

[31] Xu H, Twine TE, Girvetz E. Climate change and maize yield in Iowa. PLoS ONE. 2016;**11**(5):e0156083

[32] Araya A, Girma A, Getachew F. Exploring impacts of climate change on maize yield in two contrasting agro-ecologies of Ethiopia. Asian Journal

of Applied Science and Engineering. 2015;**4**(1):26-36

[33] Tong DAI, Jing W, Di HE, Na W, meteorological Bureau J. Modelling the impacts of climate change on spring maize yield in Southwest China using the APSIM model. Resources Science. 2016;**1**:17

[34] Shi W, Tao F. Vulnerability of African maize yield to climate change and variability during 1961-2010. Food Security. 2014;**6**(4):471-481

[35] Xiao D, Tao F. Contributions of cultivar shift, management practice and climate change to maize yield in North China plain in 1981-2009. International Journal of Biometeorology. 2016;**60**(7):1111-1122

[36] Gunn KM et al. Modeled climate change impacts on subirrigated maize relative yield in northwest Ohio. Agricultural Water Management. 2018;**206**:56-66

[37] Cinco TA, de Guzman RG, Hilario FD, Wilson DM. Long-term trends and extremes in observed daily precipitation and near surface air temperature in the Philippines for the period 1951-2010. Atmospheric Research. 2014;**145**:12-26

[38] Emanuel K. Increasing destructiveness of tropical cyclones over the past 30 years. Nature. 2005;**436**(7051):686

[39] Belay A, Recha JW, Woldeamanuel T, Morton JF. Smallholder farmers' adaptation to climate change and determinants of their adaptation decisions in the Central Rift Valley of Ethiopia. Agriculture & Food Security. 2017;**6**(1):24

[40] Liang S et al. Response of crop yield and nitrogen use efficiency for wheat-maize cropping system to future climate change in northern China. Agricultural and Forest Meteorology. 2018;**262**:310-321

[41] Leng G, Huang M. Crop yield response to climate change varies with crop spatial distribution pattern. Scientific Reports. 2017;**7**(1):1463

[42] Clapp J, Newell P, Brent ZW. The global political economy of climate change, agriculture and food systems. Journal of Peasant Studies. 2018;**45**(1):80-88

[43] Fellmann T et al. Major challenges of integrating agriculture into climate change mitigation policy frameworks. Mitigation and Adaptation Strategies for Global Change. 2018;**23**(3):451-468

[44] Abbas G et al. Quantification the impacts of climate change and crop management on phenology of maize-based cropping system in Punjab, Pakistan. Agricultural and Forest Meteorology. 2017;**247**:42-55

[45] Ahmad S et al. Quantification of the effects of climate warming and crop management on sugarcane phenology. Climate Research. 2016;**71**(1):47-61

[46] Jones MR, Singels A, Ruane AC. Simulated impacts of climate change on water use and yield of irrigated sugarcane in South Africa. Agricultural Systems. 2015;**139**:260-270

[47] Lipper L et al. Climate-smart agriculture for food security. Nature Climate Change. 2014;**4**(12):1068

[48] Chen C, Pang Y, Pan X, Zhang L. Impacts of climate change on cotton yield in China from 1961 to 2010 based on provincial data. Journal of Meteorological Research. 2015;**29**(3):515-524

[49] Adhikari P et al. Simulating future climate change impacts on seed cotton yield in the Texas High Plains using the CSM-CROPGRO-cotton model. Agricultural Water Management. 2016;**164**:317-330

[50] Campbell BM et al. Reducing risks to food security from climate change. Global Food Security. 2016;**11**:34-43

[51] Reddy KR et al. Simulating the impacts of climate change on cotton production in the Mississippi delta. Climate Research. 2002;**22**(3):271-281

[52] Amouzou KA, Naab JB, Lamers JPA, Borgemeister C, Becker M, Vlek PLG. CROPGRO-cotton model for determining climate change impacts on yield, water-and N-use efficiencies of cotton in the dry savanna of West Africa. Agricultural Systems. 2018;**165**:85-96

[53] Rahman MH et al. Multi-model projections of future climate and climate change impacts uncertainty assessment for cotton production in Pakistan. Agricultural and Forest Meteorology. 2018;**253**:94-113

[54] Barlow KM, Christy BP, O'leary GJ, Riffkin PA, Nuttall JG. Simulating the impact of extreme heat and frost events on wheat crop production: A review. Field Crops Research. 2015;**171**:109-119

[55] Meehl Gerald A, et al. Global climate projections. Climate change 2007: The physical science basis. Contribution of Working Group I to the Fourth Assessment Report of the Intergovernmental Panel on Climate Change. 2007. pp. 747-845

[56] Diarra A, Barbier B, Zongo B, Yacouba H. Impact of climate change on cotton production in Burkina Faso. African Journal of Agricultural Research. 2017;**12**(7):494-501

[57] Lesk C, Rowhani P, Ramankutty N. Influence of extreme weather disasters on global crop production. Nature. 2016;**529**(7584):84

[58] Ewert F et al. Crop modelling for integrated assessment of risk to food production from climate change. Environmental Modelling & Software. 2015;**72**:287-303

[59] Williams A, White N, Mushtaq S, Cockfield G, Power B, Kouadio L. Quantifying the response of cotton production in eastern Australia to climate change. Climatic Change. 2015;**129**(1-2):183-196

[60] Ahmad A, Ashfaq M, Rasul G, Wajid SA, Khaliq T, Rasul F, et al. Impact of climate change on the cotton–wheat cropping system of Pakistan. In: Rosenzweig C, Hillel D, editors. Handbook of Climate Change and Agroecosystems: The Agricultural Model Intercomparison and Improvement Project (AgMIP). ICP Series on Climate Change Impacts, Adaptation, and Mitigation. London: Imperial College Press; 2018. pp. 219-258

[61] Deutsch CA et al. Increase in crop losses to insect pests in a warming climate. Science. 2018;**361**(6405):916-919

[62] Deryng D, Conway D, Ramankutty N, Price J, Warren R. Global crop yield response to extreme heat stress under multiple climate change futures. Environmental Research Letters. 2014;**9**(3):34011

[63] Schlenker W, Lobell DB. Robust negative impacts of climate change on African agriculture. Environmental Research Letters. 2010;**5**(1):14010

[64] Kumar A, Sharma P. Climate change and sugarcane productivity in India: An econometric analysis. Journal of Social and Development Sciences. 2014;**5**(2):111

[65] de Carvalho AL et al. Impact of climate changes on potential sugarcane yield in Pernambuco, northeastern region of Brazil. Renewable Energy. 2015;**78**:26-34

[66] Zhao D, Li Y-R. Climate change and sugarcane production: Potential impact and mitigation strategies. International Journal of Agronomy. 2015;**2015**

[67] Challinor AJ, Watson J, Lobell DB, Howden SM, Smith DR, Chhetri N. A meta-analysis of crop yield under climate change and adaptation. Nature Climate Change. 2014;**4**(4):287

[68] You L, Rosegrant MW, Wood S, Sun D. Impact of growing season temperature on wheat productivity in China. Agricultural and Forest Meteorology. 2009;**149**(6-7):1009-1014

[69] Tao F, Zhang Z. Adaptation of maize production to climate change in North China plain: Quantify the relative contributions of adaptation options. European Journal of Agronomy. 2010;**33**(2):103-116

[70] Marin FR et al. Climate change impacts on sugarcane attainable yield in southern Brazil. Climatic Change. 2013;**117**(1-2):227-239

[71] Knox JW, Díaz JAR, Nixon DJ, Mkhwanazi M. A preliminary assessment of climate change impacts on sugarcane in Swaziland. Agricultural Systems. 2010;**103**(2):63-72

[72] Singh J, Singh AK, Sharma MP, Singh PR, Srivastava AC. Mechanization of sugarcane cultivation in India. Sugar Tech. 2011;**13**(4):310-314

[73] Chavas DR, Izaurralde RC, Thomson AM, Gao X. Long-term climate change impacts on agricultural productivity in eastern China. Agricultural and Forest Meteorology. 2009;**149**(6-7):1118-1128

[74] Smit B, Wandel J. Adaptation, adaptive capacity and vulnerability. Global Environmental Change. 2006;**16**(3):282-292

[75] Fischer G, Shah M, van Velthuizen H. Climate change and agricultural vulnerability. A special report prepared as a contribution to the world summit on sustainable development. Laxenburg, Austria: International Institute for Applied Systems Analysis; 2002

[76] Deressa TT, Hassan RM, Ringler C, Alemu T, Yesuf M. Determinants of farmers' choice of adaptation methods to climate change in the Nile Basin of Ethiopia. Global Environmental Change. 2009;**19**(2):248-255

[77] Latif MA, Islam MR, Ali MY, Saleque MA. Validation of the system of rice intensification (SRI) in Bangladesh. Field Crops Research. 2005;**93**(2-3):281-292

[78] Srinivasan A. Handbook of Precision Agriculture: Principles and Applications. USA: CRC Press; 2006

[79] Yadav RL. Enhancing efficiency of fertilizer n use in sugarcane by ring-pit method of planting. Sugar Tech. 2004;**6**(3):169

[80] Pedde KC, Gore AK, Chavan AS. Integrated weed management in chickpea. Indian Journal of Weed Science. 2013;**45**(4):299

[81] Pimentel AJB, Rocha JR d AS, de Souza MA, Ribeiro G, Silva CR, Oliveira ICM. Characterization of heat tolerance in wheat cultivars and effects on production components. Revista Ceres. 2015;**62**(2):191-198

[82] Weerakoon WMW, Mutunayake MMP, Bandara C, Rao AN, Bhandari DC, Ladha JK. Direct-seeded rice culture in Sri Lanka: Lessons from farmers. Field Crops Research. 2011;**121**(1):53-63

[83] CIAT; World Bank; CCAFS and LI-BIRD. Climate-Smart Agriculture in Nepal. CSA Country Profiles for Asia Series. Washington, D.C: International Center for Tropical Agriculture (CIAT); The World

Bank; CGIAR Research Program on Climate Change, Agriculture and Food Security (CCAFS); Local Initiatives for Biodiversity Research and Development (LI-BIRD); 2017. 26 p

[84] Witt C, Pasuquin JM, Dobermann A. Towards a site-specific nutrient management approach for maize in Asia. Better Crops With Plant Food. 2006;**90**(2):28-31

[85] Ratnam M, Rao AS, Reddy TY. Integrated weed management in chickpea (*Cicer arietinum* L.). Indian Journal of Weed Science. 2011;**43**(1):70-72

[86] Thakur PS, Singh S. Effect of *Morus alba* canopy management on light transmission and performance of *Phaseolus mungo* and *Pisum sativum* under rainfed agroforestry. Indian Journal of Agroforestry. 2002;**4**(1):25-29

[87] Shiyani RL, Joshi PK, Asokan M, Bantilan MCS. Adoption of improved chickpea varieties: KRIBHCO experience in tribal region of Gujarat, India. Agricultural Economics. 2002;**27**(1):33-39

Chapter 5

Climate Change Impacts and Adaptation in Agricultural Sector: The Case of Local Responses in Punjab, Pakistan

Muhammad Mumtaz, Jose Antonio Puppim de Oliveira and Saleem H. Ali

Abstract

This study contributes to explore local responses to deal with the impacts of climate change on agriculture sector by looking the case of Punjab, Pakistan. Pakistan's agriculture is facing severe challenges due to the negative consequences of climate change. In this study, we investigate (a) What are the different initiatives taken at planned and autonomous level in Punjab province? (b) What are the drivers behind these initiatives? (c) How these initiatives are being transferred within farmer's community in Punjab and outside Punjab? and (d) What are the challenges for these farmers in adaptation to climate change and governance hurdles in the province? The government has launched massive level awareness campaign in the province. Other important initiatives are institutional capacity enhancement, promotion of climate change research, establishment of linkage with academics, enhancement of capacity building, and involvement of farmers' community in climate adaptation for agriculture sectors. The autonomous adaptation initiatives include changing planting dates, changing crops types, changing fertilizers, and planting shade trees. Planned level adaptation is primarily driven by coordination among the respective departments, engagement with academics, and availability of financial resources. Autonomous initiatives of the province are mainly driven by the previous experiences of farmers, sustainability in agriculture production, and knowledge sharing.

Keywords: adaptation, governance, climate change, agriculture, Punjab

1. Introduction

Climate change is a reality; according to Intergovernmental Panel on Climate Change (IPCC) fifth assessment report, humans are responsible for this unsustainable situation. From future climate change perspective, both human and natural systems are at risk [1]. It is a globally environmental challenge. Developing countries are more at risk due to a lower adaptive capacity and higher resource scarcity [2].

Climate change poses serious threats to Pakistan. Presently, climate change is considered one of the serious crises in Pakistan [3], as it is one of the most vulnerable countries, mostly due to its diverse geographical and climatic features [4]. As such, it has caused various disasters in the form of floods, droughts, and other natural calamities in the country [5]. In the aftermath of the 2010 floods, one fifth of the country's land area was submerged; it severely damaged the economy and infrastructure, impacted livelihoods of millions of people, and left 90 million people food insecure [6]. Pakistan is among those countries which were badly affected in 2012 due to climate change [7].

Climate change is a great challenge for the agrarian economy of Pakistan [8]. The agriculture sector of Pakistan is highly vulnerable to climate change [9]. Livelihoods of millions of Pakistanis are dependent on agriculture sector which is highly sensitive to climate change [10].

This sector contributes approximately 25% to the national Gross Domestic Product (GDP) and it absorbs about 42% of the labor force [11]. The sector generates over 75% of export revenue and it is the largest employer sector in the country [12]. Agriculture sector in Pakistan faces serious challenges from climate change-induced impacts, i.e., rising temperatures, floods, droughts and yield losses [13]. It is likely the variation in monsoons and increased temperature is a real challenge to the agriculture sector in Pakistan [14].

There are evidences that climate change will continuously pose threats throughout this century despite international efforts to curb greenhouse gas emissions [15]. In order to confront the challenge of climate change many efforts are made around the world. Various climate change policies are established at international, national, sub-national, and local levels to address the impacts of climate change. Traditionally, the focus of such policies remained on mitigation instead of adaptation measures despite urgent requirements for adaptation strategies being emphasized [16]. Adaptation actions are important response to climate change as these actions help to reduce the vulnerabilities in the social and biological system [17]. One of the major objectives of adaptation measures is to build the resilient in societies to face climate change [18].

The need for adaptation policies and actions is increasingly recognized [19]. Governments are being forced to rethink their ways to manage climatic impacts and to focus not only on mitigation but also adaptation [20, 21]. Due to increasing public interests, the adaptation policies are being recognized and gaining space on policy agenda [22]. In case of most vulnerable countries to climate change, adaptation is the focus of their strategies to tackle the negative consequences of climate change. Many countries including Pakistan recognize the need to focus on adaptation strategies to effectively address the challenge of climate change.

To manage the potentially fatal issue of climate change, Pakistan responded with various initiatives, mainly on adaptation measures. These initiatives are in the form of climate change policies, implementation frameworks and some other measures. The national climate change policy of Pakistan says, "Adaptation efforts are the focus of this document." Pakistan is ranked in the list of the countries that have the least adaptive capacity due to extreme poverty and lack of physical and financial resources [10].

The subnational governments are key institutions for the effective implementation of climate change related policies. The subnational governments are important to helping curb climate change due to their proximity to the consequences of climate change [22]. Subnational governments in Pakistan have taken multiple initiatives to manage climate change. For instance, in Punjab a massive level awareness campaigns has been launched, establishment of provincial climate change policy, and establishment of linkage among the related departments in the province are some of notable initiatives among others.

Pakistan recognizes the important role of subnational governments/provinces for effective response to climate change. After 18th constitutional amendment in 2010, the responsibility of implementing climate change policies rests with respective provinces/subnational governments in the country. This study is conducted to understand the adaptation governance initiatives for agriculture sector in the province of Punjab, Pakistan. This province is the major contributor of agriculture sector in Pakistan. The province accounts for 53% of the total GDP in the country. It is noted that agriculture adaptation actions reduce agricultural losses [23]. The subnational government of Punjab is taking adaptation steps to tackle climate change.

The objective of this study is to understand how adaptation to climate change for agriculture sector is happening in Punjab province. The study mainly focuses on the key factors: (a) What are the different initiatives taken at the planned and autonomous level in the province? (b) What are the drivers behind the initiatives (c) How these initiatives are being transferred within farmers' community in Punjab and outside Punjab? and (d) What are major challenges for these farmers in adaptation to the climate change and governance hurdles in the province?

The study contributes to the literature on climate adaptation governance and adaptation policies for the agriculture sector at subnational level by exploration of hidden adaptation measures at subnational level. It essentially contributes to explore the drivers behind adaptation measures in the province. Moreover, it contributes by identifying policy needs and research gaps for climate adaptation at subnational level. Additionally, the study provides some replicable lessons for other developing countries while devising adaptation policies and action plans at subnational levels.

2. An overview of adaptation strategies

The scientific community concluded with a strong consensus that climate is changing. According to the fifth report of IPCC [24], climate change is taking place in the world and developing countries are especially expected to suffer more, compared to the developed world.

To manage the consequences of climate change, two fundamental societal response options emerged in the form of mitigation and adaptation [25]. Füssel [26] maintained that mitigation and adaptation are complementary rather than mutually exclusive. Historically, the focus to address climate change remained on mitigation measures [27]. However, a shift emerged and it became widely accepted that mitigation alone is unlikely to be sufficient to cope with climate change [28].

Adaptation to climate change is getting much attention in the scientific and policy debate [29, 30]. Adaptation is defined as: "adjustment in ecological, social or economic systems in response to actual or expected climatic stimuli and their effects or impacts" [31]. According to Stern, "adaptation will be crucial in reducing vulnerability to climate change and is the only way to cope with the impacts that are inevitable over the next few decades" [32].

Many countries have set their strategies to cope with climatic event at national, provincial, state, district and local levels [33]. The first time that the dilemma of adaptation for developing countries was recognized, was in 2001 at the Seventh Conference of the Parties in Morocco [34].

The important role of climate change adaptation as a policy is well considered and recognized internationally [35]. For instance, Article 4.1b of the United Nations Framework Convention on Climate Change [36] states that parties are "committed to formulate and implement national and, where appropriate, regional programs containing measures to mitigate climate change and measures to facilitate adequate adaptation to climate change." Likewise, Article 10 of Kyoto Protocol also emphasized the promotion of adaptation and the incorporation of technological advancements for adaptation to overcome climate change [30]. Furthermore, in 2007 during the United Nations Conference on Climate Change in Bali, the need to enhance actions on adaptation by Parties to the Convention was emphasized (UNFCCC [37]).

Adaptation is essentially important to avoid impacts of climate change otherwise these impacts happen gradually [28]. It has been projected by climate models that there would be more frequent devastating floods, high rainfall events, and heat waves. Therefore, this scenario of climatic impacts necessitates the incorporation of adaptation strategies. Adaptation is considered a promising step to strengthen local capacity to tackle the forecasted and unexpected climatic conditions [38].

In Paris Agreement 2015, the adaptation to climate change is much emphasized. Dessai et al. [38] argued that mitigation alone is not sufficient to address climate change effectively. Therefore, adaptation attempts are made around the world to curb the challenge of climate change.

Pakistan recognizes the importance of adaptation to climate change. Change adaptation strategies are important to deal the issue of climate change in Pakistan (Ali and Erenstein [39]). Pakistan's policy response prioritizes adaptation measures. For instance, national climate change policy is much focused on climate adaptation initiatives. Mumtaz [40] argued that climate adaptation measures are unavoidable for Pakistan keeping in view its high vulnerability to climate change. It is reported that these adaptation measures are important for all major sector including agriculture sector.

3. Climatic impacts on agriculture sector

Agriculture is a sensitive sector to climate change and it is considered among the most vulnerable sectors to the impacts of climate change [35]. Extreme weather conditions and precipitation changes are affecting the crop development, growth and yield of crops [41]. Rise in temperature reduces the grain filling duration, caused the grains sterility and yield reduction [42].

For last few decades high temperature is reported in Asia and the Pacific regions [43]. The agriculture sector in these regions is more vulnerable considering that Asia and the Pacific are responsible for 37% of the total world emissions from agriculture production. Most vulnerable countries in these regions are: Bhutan, Indonesia, Pakistan, Papua New Guinea, Sri Lanka, Thailand, Timor-Leste, Uzbekistan, and Vietnam [44]. It is also reported that agriculture sector may disturb the climate [45]. It is indicated that 14% of nitric oxide and methane is coming from the agriculture sector and 18% is due to deforestation for agriculture use [46].

Agriculture sector is the backbone of the economy of Pakistan. This sector in Pakistan faces serious challenges due to climate change which impacts in the form of rising temperatures, floods, droughts, and yield losses [13]. The continuously occurrence of floods in Pakistan and other climate change impacts is costing the country 14 billion dollars per year, which is around five% of gross domestic product (GDP) to its economy [47].

Agriculture sector feeds food to the fast growing population of Pakistan according to Economic Survey of 2010–2011. Climate change is a great challenge for Pakistan's agrarian economy [8]. Agriculture productivity is affected by various factors including rainfall pattern, variation in temperature, and variation in dates of harvesting and sowing, availability of water, and evaporation along with suitability of land [9]. It is projected with 1°C rise in temperature will cause (6–9%) decline in wheat productivity [48].

To face the risks in agriculture associated with climate change, adaptation is the key factor to address the negative impacts of climate change. Adaptation strategies are important opportunities to tackle climate change effectively and to sustain the crop production [49]. Adaptation is an important policy response to climate change in agriculture sector [31, 35]. The IPCC emphasizes that it is very fundamental for the agricultural sector to adapt to climate change.

During the last two decades, the role of subnational states in the realm of global climate governance has grown significantly [50]. They further pointed out that the role of subnational governments has been extended as influential actors in international climate change policies. It is normal to find subnational governments playing a leading role in climate change policies [51].

The existing literature highlights the role of subnational policies for climate governance. As such, Jörgensen [52] argued that subnational state policies are a key aspect of climate governance and function as laboratories of experimentation which could promote policy change through policy-learning. Jänicke [53] considered that provincial (subnational) and state level activities for climate change have increased significantly in the recent years.

Climate change adaptation poses many complex governance questions and has therefore been called a "wicked problem par excellence" [54, 55]. The "wicked problem" needs comprehensive and proper solutions. Adaptation governance faces many difficulties, hindrances and opportunities involved in dealing with the "wicked problems" [56]. They further argued that due to the novelty and complexity of adaptation governance, a number of fundamental governance dilemmas have to be (re)addressed in developing the governance of adaptation to climate change. For instance, which ministry or agency is responsible for climate policy in general and climate adaptation policy in particular? Are existing divisions of responsibilities adequate for tackling climate adaptation issues?

It is evident that nation states, multilateral and bilateral development organizations, citizen's groups and communities are expected to respond to the negative impacts of a changing climate. There is a consensus in adaptation/adaptive capacity literature that there is a need to build adaptive capacity in the form of free flow of ideas, knowledge and technology, capable institution and government schemes, and other policies for an effective response to climate change [57]. They further highlighted that it is significantly unclear that how this adaptive capacity is actually built or enhanced.

To deal with the complex issue of climate change, equally complex solutions are required which involve several fields of human activity and different stakeholders. Multiple stakeholders, such as civil society, research institutions, universities, private sectors, etc., play an important role in the production of responses to the climate crisis together with governmental representatives [58]. The linkage of subnational/local governments with international networks provides a great potential for the development of effective policies and actions as responses to climate change [59].

Presently, the subnational or local level governments have an important governance role to tackle climate change in general and climate change adaptation in particular. However, there has been relatively little research in the area of adaptation policies especially in developing countries. There are studies in Pakistan which highlight the need for adaptation action but actual field-based studies for adaptation responses to climate change are rare [10]. The need for field-based studies is increasingly recognized as important for better understanding of the local level vulnerability and adaptation responses to climate change [60]. Therefore, this study is important to highlight the responses of subnational governments for climate adaptation.

4. Climate adaptation in Punjab

This study was conducted for the period of November 2016–April 2017. Apart from desk research, 30 in-depth semi-structured interviews were conducted with relevant stakeholders. These stakeholders include policy experts, government officials, think tanks working in the area, related nongovernmental organizations, academics, ministry of climate change, provincial environmental protection agency, farmer community, civil society, and climate change activists in the province of Punjab.

Punjab is geographically located approximately at 30,000 N, 70,000 E in the semi-arid lowlands zone [61]. It is the most populous and second largest province of Pakistan. Punjab is a fertile agricultural region which holds an extensive irrigation network and plays a leading role in the development of the economy [62]. The province accounts for 56.2% of the total cultivated area, 53% of the total agricultural gross domestic product and 74% of the total cereal production in the country [63, 64]. Punjab mainly contributes for agricultural sector in the percentage of land (57.2%) in agricultural sector and the percentage share (53%) of Pakistan's agricultural gross domestic product [65]. Agriculture sector in Punjab is facing the impacts of climate change. Below is the map of Pakistan, highlighting light green portion as our case of study that is Punjab province (**Figure 1**).

During the last decade or so, climate change adaptation gained a space on policy agenda. Subnational governments, being closer to the locale of climatic impacts, play a key role in the effective implementation of climate change policies. In the case of Pakistan, the implementation of climate change and other related policies rests with subnational governments. In this section, we analyze the adaptation strategies in Punjab Province. **Table 1** shows the adaptation initiatives and drivers behind these initiatives in Punjab.

4.1 Awareness campaign

The Punjab government has launched an awareness campaign about climate change and agriculture. They have set up a radio station, which gives information to farmers about weather conditions. The station broadcasts multiple programs to increase farmers'

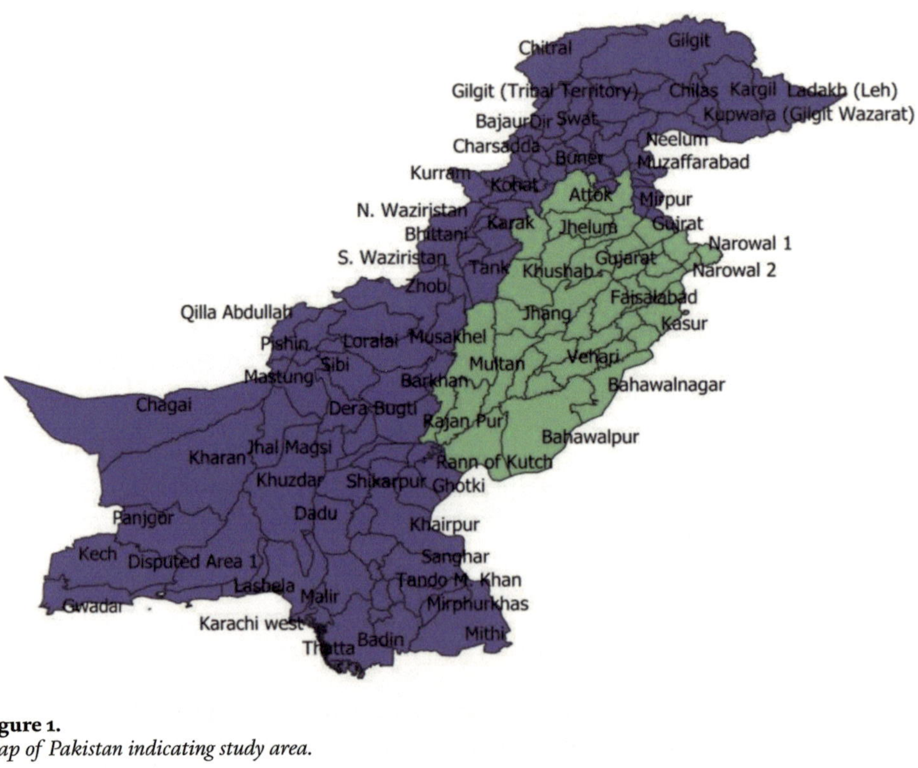

Figure 1.
Map of Pakistan indicating study area.

Numbers	Adaptation initiatives	Drivers behind the initiatives
1	Awareness campaigns	To educate farmers, control damages, improve productivity, linkage with NGOs
2	Training programs for farmers	To expedite implementation, enhance understanding of farmers, to prepare the farmer's community for climate change
3	Research and innovation	Understanding of climate change dynamics, to produce novel techniques
4	Institutional capacity	Effective implementation of climate policies, to address the challenge of lack of professionalism
5	Role of academics	Better research environment, promote climate research, establishment of climate policies with sound scientific backing
6	Departmental coordination	To expedite implementation mechanism, comprehend the situation well in time
7	Autonomous adaptation	Past experiences, knowledge sharing, avoid damages and enhance productivity

Table 1.
Source: The table is created by the authors.

awareness about climate change, its impacts on agriculture sectors and possible strategies to confront the challenge. The impact of this radio station has been positively observed in the province. For example, many farmers are regularly tuning into the radio for information about weather conditions, planning dates, and to know how they will be affected due to climate change. Some of interviewees told us that they are regular listeners of the radio because they get valuable information, such as, climatic conditions, advice about fertilizer use and seeds, and government subsidy schemes for the farmer community. In some programs, climate experts and agro experts among others are invited to discuss climate change, its impacts on agriculture sector and possible solutions. One farmer told us that he regularly follows such programs on Radio because it is helpful to get novel ideas and best practices from agricultural experts.

4.2 Training programs for farmers

The government has set up a formal mechanism to give practical training to the farmers. They arrange sessions with the farmers to teach them based on scientific data how climate change is threatening agriculture sector. Agriculture extension departments which work closely with the farmers at local levels collect data from the fields and gives training to the farmers at local levels. Based on these initiatives, the farmers are able to execute what they learned in their farming practices. For instance, they are advised that they should plant seeds which have been tested in scientific labs and shown to have the capabilities to survive severe weather conditions. It is noted that many trained farmers approached the agriculture extension departments to obtain suitable seeds with respect to weather conditions. We were told by the agriculture extension department in Faisalabad that after attending the training, the farmers' community is encouraged to approach the department for more information about climate change, suitable seeds, solutions for damages due to pests, etc.

The provision of financial help to farmers can be helpful when they are facing financial constraints. On the part of the government, it is providing certain subsidies to the farming community. For example, it provides the farmers with the best seeds and best quality fertilizers keeping in view the exposure of fields to climate change at nominal prices. The farmers' community is also contributing by handling climate change through their adaptation practices.

4.3 Research and innovation

Research and innovation can play a key role in promoting adaptation. Research and innovative techniques are already in place for agriculture adaptation in the province, where experiments have been conducted to find the best varieties of seed that can survive extreme weathers. The government is providing the best varieties of seeds which can survive in hot seasons and produce good results. For instance, the Punjab Seed Corporation is established to provide quality seeds to the farmers according to the conditions of climatic zones in various parts of the province. The subnational government is focusing on research and innovative strategies to address the impacts of climate change. At the institutional level, they are giving training to government officials so that they can comprehend the situation more amicably and address the situation scientifically.

4.4 Institutional capacity and role of academics

Institutional capacity is important for the implementation of any policy, programs or plans. They arrange proper training for the people working in the area of climate change in order to understand the actual scenario, especially the impact assessment of climate change on the agriculture sector in the province. For example, over 150 individuals working in relevant departments in the province are trained in national and international institutions. Many others are encouraged to go abroad and conduct research in area of climate and agriculture sector. It is very likely that well trained staff will play a key role in bringing positive results for effective handling of climate change. Engagement of other stakeholders, especially academics, is another core agenda of government.

Academics in the province are contributing to and conducting studies on climate change and agriculture sector. For example, Agriculture University of Faisalabad has published some work on climate change adaptation and highlighted the importance of adaptation in the province. The university has linkages with international institutions on climate change research. The linkages with international institutions provide opportunities for the professors and researchers at the university to learn innovative adaptation techniques from other parts of the world. They can put into practice in Pakistan the relevant activities they have learned for climate adaptation towards agriculture sector.

4.5 Departmental coordination

Coordination among relevant line departments is essentially important for the implementation of any policy. The subnational government of Punjab has established a link among the 26 agriculture institutes throughout the province in order to set up comprehension strategies for climate change and the agriculture sector. They regularly arrange meetings among these institutes to discuss the new challenges and the existing strategies to manage the negative impacts of climate change. For instance, the Ayub Research Centre (ARC), which manages climate change related activities, is well familiar with what is happening in the agriculture extension departments at various levels and vice versa. By being aware of the activities of agriculture extension departments and others, the ARC can disseminate the positive activities among other institutions and set new targets accordingly.

4.6 Autonomous adaptation and transference of adaptation initiatives

It is noted that the farmers in the province are adjusting to climate change. They are adjusting their agricultural activities with changing climate. In the recent year, adaptation strategies are being explored by the farmers in many ways. For instance, in Pakistan and Brazil farmers' community is adapting climate change variability by adjustment of planting time and optimization of plant populations [66]. These adjustments are very important adaptation strategies to get the maximum potential of the crops and secure expected productivity. Likewise, the improved and heat tolerant seeds are another important strategy which is being used in Punjab province by the farmers. The development of improved varieties such as early maturing, drought and heat tolerant are necessary to sustain the productivity under changing climate. It is very likely by incorporating such adaptation techniques, the production can increase under moisture stress and extreme temperatures [67]. The autonomous strategies are identified: changing seeds types, changing sowing dates, looking for new fertilizers and planning shade trees. It is the experience; enhance productivity, and knowledge sharing in the farmers' community which encourages them to opt for these adaptation actions.

It is observed that the adaptation strategies are being transferred from one place to another place. They successful strategies are being shared among the farmers' community within the Punjab province and beyond. The successful strategies are applied by other farmers. For instance, a farmer's production was suffering due to rise in temperature. He brought the wheat seed from another place where the temperature was already high; probably these seeds are more heat tolerant. By applying this strategy his production was increased. He shared his practice with other farmers in the area so they also started the same practice and that was quite successful. These strategies are not only limited to the Punjab province but it is also reported that the successful adaptation initiatives either they are planned or autonomous are replicated in other province. For instance, the framers in the Khyber Pakhtunkhwa (KPK) with bordering area of Punjab province are learning the successful stories from Punjab farmers and replicated in their areas in the KPK. This is again shared and transferred within the farmers' community in the KPK. Therefore, it is concluded that the adaptation initiatives in Punjab province is not only limited and transferred within Punjab province but also transferred and implemented in other areas and provinces. Despite all these efforts, there are certain challenges as well.

4.7 Key challenges

Despite these promising initiatives the province and local farmers are facing some key challenges for the effective adaptation to climate change. The data indicates that the major hurdles are in the form of a lack of institutional and human capacity, scarcity of financial resources, a lack of technological advancement, lack of research and innovation, and a weak integration of adaptation policy with other related policies. On the other hand the local farmers face lack the awareness about climate change, weak capacity building, financial constraints, and technicality issues to opt with adaptation measures.

5. Conclusion

Climate change adaptation in the agriculture sector is considered a striking strategy to manage the impacts of climate change. Theoretically, climate change adaptation is a new field and it creates a space for experimentation and new forms of governance. In recent years, subnational governments have shown that they have an effective role in dealing with climate change. The subnational government Punjab is firmly committed to addressing the challenge of climate change.

This study has found that the province is actively dealing with the consequences of climate change. They have been taking initiatives for climate adaptation for the agriculture sector in the province. The government has launched a major awareness campaign in the province. Moreover, institutional capacity enhancement, promotion of climate change research, establishment of linkage with academics, enhancement of capacity building, and involvement of farmers' community in climate adaptation for agriculture sectors are some of other important steps taken by Punjab province. It is pointed out that autonomous adaptation is also taking place in the province. The identified autonomous adaptation practices include changing planting dates, changing crops types, changing fertilizers, and planting shade trees. Our study identified the drivers behind planned and autonomous level adaptation. These differences at planned level adaptation are primarily driven by coordination among the respective departments, engagement with academics, and availability of financial resources. On the other hand, autonomous initiatives of the province are mainly driven by the previous experiences of farmers, sustainability in agriculture production, and knowledge sharing. The study found that the adaptation strategies are being transformed from one place to another place within the province and outside the province. This transformation is happening by successful experience sharing among the farmer community. Moreover, the study identified key challenges for adaptation to climate change in the province. These challenges are in the form of lack of institutional and human capacity, scarcity of financial resources, lack of research and innovation, and integration of adaptation policy with other related policies.

Conflicts of interest

The authors declared no conflict of interest in this manuscript.

Author details

Muhammad Mumtaz[1*], Jose Antonio Puppim de Oliveira[1] and Saleem H. Ali[2]

1 São Paulo School of Management, Fundação Getulio Vargas, Brazil

2 Energy and the Environment, University of Delaware, USA

*Address all correspondence to: mumtaz86@hotmail.com

References

[1] Walker BJA, Adger WN, Russel D. Institutional barriers to climate change adaptation in decentralised governance structures: Transport planning in England. Urban Studies. 2014;**52**:1-16

[2] Lemos MC, Agrawal A, Eakin H, Nelson DR, Engle NL, Johns O. Building adaptive capacity to climate change in less developed countries. In: Asrar GR, Hurrell JW, editors. Climate science for serving society: Research, modeling and prediction priorities. The Netherlands: Springer, Dordrecht; 2013. pp. 437-457. DOI: 10.1007/978-94-007-6692-1_16

[3] Salik KM, Ishfaq S, Saeed F, Noel E, Syed Q. Pakistan: Country situation assessment. Islamabad: Sustainable Development Policy Institute; 2015

[4] Rasul G, Ahmad B. Climate Change in Pakistan. Focused on Sindh Province, Pakistan Meteorof Logical Department Report. 2012

[5] Banoori W. Pakistan: Climate Report. 2012. Available from: http://www.thefrontierpost.com/article/163613

[6] Rasmussen B, Allen S, Rasmussen ZA, Bajwa R. Learning from disasters to save lives every day in Pakistan. Lancet. 2013. DOI: 10.1016/S0140-6736(13)60652-2. Available from: https://www.ncbi.nlm.nih.gov/pubmed/23684253

[7] Kreft S, Eckstein D. Global Climate Risk Index 2014. Bonn: Germanwatch; 2013

[8] Shakoor U, Saboor A, Ali I, Mohsin AQ. Impact of climate change on agriculture: Empirical evidence from arid region. Pakistan Journal of Agricultural Sciences. 2011;**48**:327-333

[9] Janjua PZ, Samad G, Khan NU, Nasir M. Impact of climate change on wheat production: A case study of Pakistan. The Pakistan Development Review. 2010;**49**:799-822

[10] Abid M, Scheffran J, Schneider UA, Ashfaq M. Farmers' perceptions of and adaptation strategies to climate change and their determinants: The case of Punjab province. Pakistan Earth System Dynamics. 2015;**6**:225

[11] Go P. Pakistan Economic Survey 2015-2016. Islamabad: Ministry of Finance, Government of Pakistan. 2016. Available from: http://www.finance.gov.pk

[12] TDAP. Statistics for 2015-2016. Islamabad: Trade Development Authority of Pakistan (TDAP). 2016. Available from: http://www.tdap.gov.pk/tdap-statistics.php

[13] Noman MA, Schmitz M. Economic assessment of the impact of climate change on the agriculture of Pakistan, business econ. Horizons. 2011;**4**:1-12

[14] Abid M, Schneider UA, Scheffran J. Adaptation to climate change and its impacts on food productivity and crop income: Perspectives of farmers in rural Pakistan. Journal of Rural Studies Part A. 2016;**47**:254-266

[15] Wilson E. Adapting to climate change at the local level: The spatial planning response. Local Environment: The International Journal of Justice and Sustainability. 2006;**11**(6):609-662

[16] Biesbroek GR, Swart RJ, van der Knaap WGM. The mitigation-adaptation dichotomy and the role of spatial planning. Habitat International. 2009;**33**(3):230-237

[17] Adger WN et al. Are there social limits to adaptation to climate change? Climatic Change. 2009;**93**:335-354

[18] Smit B, Wandel J. Adaptation, adaptive capacity and vulnerability.

Global Environmental Change-Human and Policy Dimensions. 2006;**16**:282-292. DOI: 10.1016/j.gloenvcha.2006.03.008

[19] Nyong A, Adesina F, Elasha BO. The value of indigenous knowledge in climate change mitigation and adaptation strategies in the African Sahel. Mitigation and Adaptation Strategies in Global Change. 2007;**12**:787-797

[20] Porter J, Demeritt D, Dessai S. The right stuff. Informing adaptation to climate change in British local government. SRI Papers. 2014;**76**:1-30

[21] Rijke J, Brown R, Zevenbergen C, Ashley R, Farrelly M, Morison P, et al. Fit-for-purpose governance: A framework to make adaptive governance operational. Environmental Science and Policy. 2012;**22**:73-84

[22] Puppim de Oliveira JA. The implementation of climate change related policies at the subnational level: An analysis of three countries. Habitat International. 2009;**33**:253-259

[23] Di Falco S. Adaptation to climate change in sub-Saharan agriculture: Assessing the evidence and rethinking the drivers. European Review of Agricultural Economics. 2014;**41**(3):405-430

[24] IPCC. Fifth Assessment Report. Cambridge: Cambridge University Press; 2014

[25] Kpadonou RAB, Adégbola PY, Tovignan SD. Local knowledge and adaptation to climate change in Ouémé valley, Benin. African Crop Science Journal. 2012;**20**(2):181-192

[26] Füssel H-M. Adaptation planning for climate change: Concepts, assessment approaches, and key lessons. Sustainability Science. 2007;**2**:265-275

[27] Burton I, Diringer E, Smith J. Adaptation to Climate Change: International Policy Options. Arlington, VA: Pew Center on Global Climate Change; 2006

[28] Kurukulasuriya P, Rosenthal S. Climate Change and Agriculture. World Bank Environment Department Paper 91. 2003

[29] McCarthy JJ, Canziani OF, Leary NA, Dokken DJ, White KS, editors. Climate Change 2001: Impacts, Adaptation, and Vulnerability. Cambridge: Cambridge University Press 2001. pp. 877-912

[30] United Nations Framework Convention on Climate Change (UNFCCC). The Kyoto protocol to the UNFCCC. In: UNFCCC, Report of the Conference of the Parties Third Session. Kyoto: UNFCCC; 1998. pp. 4-29

[31] Mizina SV, Smith JB, Gossen E, Spiecker KF, Witkowski SL. An evaluation of adaptation options for climate change impacts on agriculture in Kazakhstan. Mitigation and Adaptation Strategies for Global Change. 1999;**4**:25-41

[32] Stern N. The economics of climate change. In: The Stern Review. 2006. https://www.brown.edu/Departments/Economics/Faculty/Matthew_Turner/ec1340/readings/Sternreview_full.pdf

[33] Selvaraju R, Subbiah AR, Baas S, Juergens I. Livelihood Adaptation to Climate Variability and Change in Drought-Prone Areas of Bangladesh. Developing Institutions and Options; 2006. http://www.fao.org/3/a-a0820e.pdf

[34] Adger WN, Huq S, Brown K, Conway D, Hulme M. Adaptation to climate change in the developing world. Progress in Development Studies. 2003;**3**:179-195

[35] Smit B, Skinner MW. Adaptation options in agriculture to climate change:

A typology. Mitigation and Adaptation Strategies for Global Change. 2002;7(1):85-114

[36] United Nations Framework Convention on Climate Change (UNFCCC). United Nations Framework Convention on Climate Change. Text, Geneva: World Meteorological Organization and United Nations Environment Program; 1992

[37] UNFCCC Climate Change Country Parties Discuss Approaches to Reduce Emissions From Deforestation in Developing Countries. 2007. unfccc.int/press/news_room/newsletter/in_focus/items/3944.php

[38] Dessai S, Lu X, Risbey JS. On the role of climate scenarios for adaptation planning. Global Environmental Change. 2005;**15**(2):87-97

[39] Ali A. Erenstein O. Assessing farmer use of climate change adaptation practices and impacts on food security and poverty in Pakistan. Climate Risk Management. 2017;**16**:183-194

[40] Mumtaz M. The National Climate Change Policy of Pakistan: An evaluation of its impact on institutional change. Earth Systems and Environment. 2018;**2**:525-535

[41] Leng G, Huang M. Crop yield response to climate change varies with crop spatial distribution pattern. Scientific Reports. 2017;7:1463

[42] Hatfield JL, Boote KJ, Kimball BA, Ziska LH, Izaurralde RC, Ort D, et al. Climate impacts on agriculture: Implications for crop production. Agronomy Journal. 2011;**103**:351-370. DOI: 10.2134/agronj2010.0303

[43] Preston BL, Suppiah R, Macadam I, Bathols J. Climate change in the Asia/Pacific region. In: A Consultancy Report Prepared for the Climate Change and Development Roundtable. Aspendale Australia: CSIRO Marine and Atmospheric Research; 2006

[44] Asian Development Bank (ADB). The Economics of Climate Change in Southeast Asia: A Regional Review. Manila: Asian Development Bank; 2009. https://www.adb.org/sites/default/files/publication/29657/economics-climate-change-se-asia.pdf

[45] Paul E, Ernsting A, Semino S, Gura S, Lorch A, Agriculture and climate change: Real problems, false solutions. In: Grupo de Reflexión Rural, BiofuelsWatch, EcoNexus. Oxford, UK: The Development Fund Norway and NOAH; 2009

[46] Siddiqui R, Samad G, Nasir M, Jalil HH. The impact of climate change on major agricultural crops: Evidence from Punjab, Pakistan. The Pakistan Development Review. 2012;**51**:261-276

[47] TFCC—Task Force on Climate Change: Planning Commission, Government of Pakistan. Available from: http://pc.gov.pk/usefull_links/Taskforces/TFCC_Final_Report.pdf

[48] Ali S, Liu Y, Ishaq M, Shah T, Ilyas A, Din IU. Climate change and its impact on the yield of major food crops: Evidence from Pakistan. Food. 2017;**6**(6):39

[49] Tubiello FN, Fischer G. Reducing climate change impacts on agriculture: Global and regional effects of mitigation, 2000-2080. Technological Forecasting and Social Change. 2007;**74**:1030-1056

[50] Jörgensen K, Jogesh A, Mishra A. Multi-level climate governance and the € role of the subnational level. Journal of Integrative Environmental Sciences. 2015;**12**:235-245. DOI: 10.1080/1943815X.2015.1096797

[51] Roppongi H, Suwa A, Puppim De Oliveira JA. Innovating in sub-national climate policy: The mandatory emissions reduction scheme in Tokyo. Climate Policy. 2017;**17**(4):516-532

[52] Jörgensen K. Climate Initiatives at the Subnational Level of the Indian States and their Interplay with Federal Policies. 2011. https://www.polsoz.fu-berlin.de/polwiss/forschung/systeme/ffu/publikationen/2011/11_joergensen_montreal/isa11_joergensen_draft.pdf

[53] Jänicke M. The multi-level system of global climate governance—The model and its current state. Environmental Policy and Governance. 2017;**27**(2):108-121

[54] Davoudi S, Crawford J, Mehmood A. Planning for Climate Change: Strategies for Mitigation and Adaptation for Spatial Planners. London: Earthscan/James and James; 2009

[55] Jordan R, Birkin M, Evans A. Agent-Based Modelling of Residential Mobility, Housing Choice and Regeneration. Netherlands: Springer; 2012

[56] Termeer CJAM, Dewulf ARPJ, Rijswick HFMW, van Buuren A, van Huitema D, Meijerink S, et al. The regional governance of climate adaptation: A framework for developing legitimate, effective and resilient governance arrangements. Climate Law. 2011;**2**(2):159-179

[57] Eakin H, Lemos MC. Adaptation and the state: Latin America and the challenge of capacity-building under globalization. Global Environmental Change. 2006;**16**:7-18

[58] Bulkeley H, Newell P. Governing Climate Change. New York: Routledge; 2010

[59] Lindseth G. The cities for climate protection campaign (CCPC) and the framing of local climate policy. Local Environment. 2004;**9**(4):325-336

[60] Moser SC, Luers AL. Managing climate risks in California: The need to engage resource managers for successful adaptation to change. 2008. https://link.springer.com/article/10.1007/s10584-007-9384-7

[61] Ahmed H, Khan MR, Panadero-Fontan R, Lopez-Sández C, Iqbal MF, Naqvi SMS, et al. Geographical distribution of hypodermosis (*Hypoderma* sp.) in northern Punjab, Pakistan. Kafkas Universitesi Veteriner Fakultesi Dergisi. 2012;**18**:A215-A219

[62] Abid M, Ashfaq M, Hassan S, Fatima N. A resource use efficiency analysis of small of cotton farmers in Punjab, Pakistan. Pakistan Journal of Agricultural Sciences. 2011;**48**:65-71

[63] Badar H, Ghafoor A, Adil SA. Factors affecting agricultural production of Punjab (Pakistan). Pakistan Journal of Agricultural Sciences. 2007;**44**:3

[64] PBS—Pakistan Bureau of Statistics: Agricultural Statistics of Pakistan. Government of Pakistan, Statistics Division, Pakistan Bureau of Statistics. 2011. Available from: http://www.pbs.gov.pk/content/agricultural-statistics-pakistan-2010-11 [Accessed: 1 August 2014]

[65] Hanif U, Syed SH, Ahmad R, Malik KA, Nasir M. Economic impact of climate change on the agricultural sector of Punjab [with comments]. The Pakistan Development Review. 2010:771-779. http://www.pide.org.pk/pdf/PDR/2010/Volume4/771-798.pdf

[66] Ali A, Erenstein O. Assessing farmer use of climate change adaptation practices and impacts on food security and poverty in Pakistan. Clinical Risk Management. 2017;**16**:183-194

[67] Deressa TT, Hassan RM, Ringler C, Alemu T, Yesuf M. Determinants of farmers' choice of adaptation methods to climate change in the Nile Basin of Ethiopia. Global Environmental Change. 2009;**19**:248-255

Chapter 6

Environment, Agriculture, and Land Use Pattern

Saifullah Khan, Mehmood Ul Hassan and Aslam Khan

Abstract

This study aimed at the environment, agriculture, and land use pattern and in the arid region of Pakistan. Physiography and location of the study area with respect to coastal region are the key factors that control the climate. There are a number of factors that have their influence on the cropping pattern in the area apart from climate. They include the type of soils, availability of irrigation water, government policies, socioeconomic condition, advance technologies, market value, human demand, etc. The soil of irrigated plain in lower Punjab and Sind is more suitable for the agriculture than other parts, where the water is insufficient for cultivation.

Keywords: environment, agriculture zones, arid region, land use pattern, soil, natural vegetation

1. Introduction

The chapter covers the environment of the arid region in relation to agriculture and land use pattern. The carrying capacity of any area in the biosphere to produce enough food and other crops for the human population and domesticated animals depends a lot on environmental conditions and human technical capabilities. The work elaborates the environmental conditions in the arid region of Pakistan and how the farming patterns have been adjusted to these conditions. It has been divided into four sections. The first section on physical setting explores the landform, soil, climate, and natural vegetation and their role in agriculture of the arid region. The second section on human environment discusses the distribution of population, human settlements, and the socioeconomic setup and their relationship to agriculture sector. Furthermore, the third section discusses the agroecological zones and its characteristics with reference to agricultural. The concluding section highlights the findings of the chapter.

The arid region extends northeast to southwest from latitude 37°N into $23^1/_2$°N and longitude 60°E to 75°E. The northeastern and northwestern part of the arid region consists of high mountain ranges like the Himalayas [1], Hindu Kush, and Karakorum with highest peaks like K-2 (8475 m), Rakaposhi (7665 m), and Tirich Mir (7569 m). The study area covers about 63 districts and total area of 676,400 km^2 (77.1%) on which approximately 78.7 million population resides (**Figure 1**).

It extends from the Arabian Sea to the interior of Pakistan, while in the extreme north, it covers the whole Gilgit-Baltistan province and a part of the Chitral and Dera Ismail Khan district in Khyber Pakhtunkhwa. The arid region of Pakistan includes southern and central Baluchistan, southern Punjab, southern and northern Khyber Pakhtunkhwa, Sind, and the Gilgit-Baltistan province. The well-known deserts of the arid region comprises of Cholistan, Thar, Thal, and Kharan Desert.

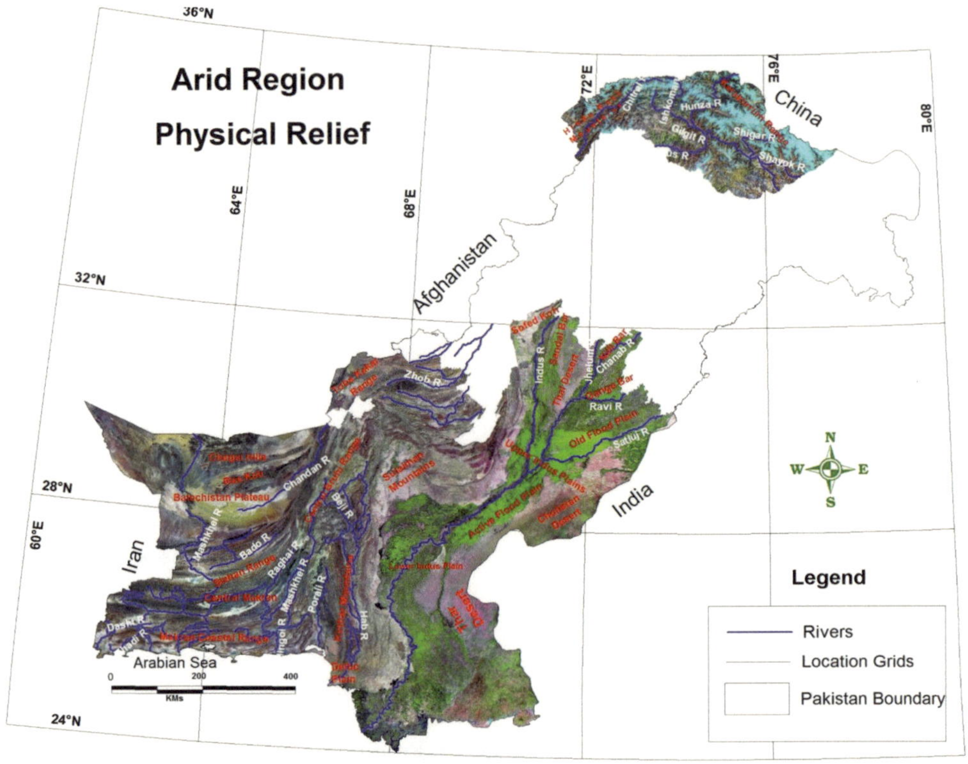

Figure 1.
Arid Region Location, GoUSA [2], online, ftp://ftp.glcf.umiacs.umd.edu.

2. Physical setting

The physical features of the earth play a vital role in controlling of temperature, precipitation, natural disasters, evapotranspiration, etc. The temperature decreases from low to high altitude, while it is converse for precipitation. The moisture contents of the soil decrease from high to low altitudes that influence the crop productivity at low lands. Besides, the height of trees and crops and the length of cropping seasons also somehow depended on the topography of the area. In most cases, the crops in arid and semiarid areas are confined to isolated patches, where the soil is fertile and the water is available. There are two arid regions identified in Pakistan, that is, the southern and northern arid region, separated by a wide zone of semiarid, subhumid, and humid region of central and upper Punjab and Khyber Pakhtunkhwa provinces. Therefore, it is necessary to discuss the physiography of the arid regions in Pakistan.

The southern arid region covers Sind, Baluchistan, lower Punjab, and Dera Ismail (DI) Khan as well as Chitral district of the Khyber Pakhtunkhwa province. The landforms of the southern arid region have three main units—the northwestern highland, Baluchistan plateau, and the Indus plain (**Figure 1**). The northwestern highland of Baluchistan consists of two parallel regions of Toba Kakar extending from northeast to southwest in Baluchistan and another parallel range of Sulaiman-Dera Bugti mountains along the Indus plain. The second major physiographic unit is the Baluchistan plateau that lies between the Toba Kakar ranges and the Sulaiman-Kirthar mountains and is also a dry rugged area with harsh environment. Mainly, pastoral agriculture is practiced on the Baluchistan plateau. Crops are grown only in those areas where water is available [3].

The third major landform unit is the Indus plain, which lies south of the Himalayas and the Salt Range stretching southward to the Arabian Sea. The Sind province and lower Punjab of the Indus plain fall in the arid region. This is the main agricultural area, where the canal irrigation has developed during the nineteenth century. A deltaic plain that has developed at the mouth of the Indus River is also an important agricultural area. The rolling sand plain covers an extensive area to the southeast of the canal-irrigated arid region covering Cholistan and Thar Deserts in Sind and lower Punjab. To the west of Kalat Plateau is the rolling sand plain of Kharan and Chaghi, while in the southwestern Baluchistan, there are dry and arid mountain ranges of the Makran coastal region. They are not very conducive to agriculture because of their dryness and rough topography, but agriculture is practiced to some extent in the Dasht River valley around Turbat in Kech district.

The northern arid region covers the extreme north of Pakistan, consisting of the Transhimalayas or the Karakoram and Hindu Kush mountains. The slopes are very steep, and natural hazards particularly the landslide are frequent, and the agriculture land is confined to the narrow belts along the rivers and streams in the valley. Therefore, in this rugged mountainous belt, agriculture is practiced mainly in the valleys primarily in the form of terraced cultivation (**Figure 1**).

3. Weather and climates

The term climatology is a Greek word, which literally means "inclination," i.e., inclination of the sunrays to the ground, to denote the mean weather condition. The word climate refers to the mean or normal condition over a long period, such as 20, 30, and 100 years [4]. According to Petterson [5], *"climatology or statistical meteorology determines the statistical relations, mean values, normal, frequencies, variation distribution, etc., of the meteorological elements, such as temperature, pressure, precipitation, wind speed and direction, humidity, sunshine, cloudiness and number of rainy days etc."*

Blair [6] has defined the climate as "*The summation of weather conditions in historical time*" or "*Climate is the summary of all the manifold weather influences.*" Miller [7] is of the opinion that the science that discusses the weather condition of the earth surface is known as climate. Thus, the wind, temperature, humidity, precipitation, vegetation, sunshine, etc. are subjected to continuous variations, which are more or less invariant at a given place. According to Oliver [8], "*climate is the aggregate of weather at a given area for a given time period*." The term weather refers to the more or less instantaneous conditions of these elements over a relatively short time period. The generalized picture of weather is called climate [9]. Climatology deals with atmospheric conditions over a longer time period and, as a result, is often defined inadequately as "*average weather*."

3.1 Seasonal division

This section covers the temporal and areal climate divisions and climate regions. The temporal division is based on the division of the year into various seasons and the description of each season. The areal division is based on classifying the climate of region only and thus dividing arid region of Pakistan into areas of similar climate/physical features.

The distinct period into which the year may be divided, in terms of duration of daylight and climate conditions, as a result of changes in duration and intensity of sunshine and rainfall, is termed as season [10]. According to Moore [11], season is defined as, "Those periods of the year, which are characterized by special climate conditions, mainly caused by the inclination of the earth's Axis to the plane of the Ecliptic and the revolution of the earth about the sun." In order to study the seasonal variation of weather elements, the year has been divided into two main seasons, that is, summer and winter. The interrelation of factors affecting climate of the arid region in Pakistan shows that the summer month in coastal areas may not be the summer month inland and a summer month in plain may not be that of the mountains. Therefore, months of the year having mean temperature of above 22°C, maximum temperature of above 34°C, and minimum temperature of above 10°C are suggested as summer months, otherwise winter (**Figure 2**). In spite of this, months having positive deviation from the mean temperature are considered as summer months, whereas months having negative deviation from the mean temperature have been placed in winter season. Generally, in plain the summer lasts from April to October and winter from November to March. In highland, the summer has a span from May to September and winter from October to April. These two main seasons are further condensed into four sub-seasons that are cold, hot, warm, and monsoon seasons. The cold season of the country is from mid-November to mid-April with all months' mean monthly temperature below 20°C, rainfall below 6 cm (2.4 inches), moderate humidity, high pressure, minimum sunshine period, and low evapotranspiration. The hot season varies from mid-April to June with each month's mean monthly temperature of above 30°C; low rainfall, humidity, and pressure; maximum sunshine period; and high evapotranspiration. The monsoon season ranges from July to mid-September with all months' mean monthly temperature below 30°C, rainfall above 6 cm (2.4 inches), moderate humidity, low pressure, and high sunshine period and evapotranspiration. Nonetheless, the warm season remains from mid-September to mid-November, with mean monthly temperature below 25°C, low rainfall and humidity, moderate pressure and sunshine period, and high evapotranspiration. However, November is completely the winter month in hilly areas, while April is the summer month in plain. These months are thus divided into two parts, and the average of 15 days has been added to each season.

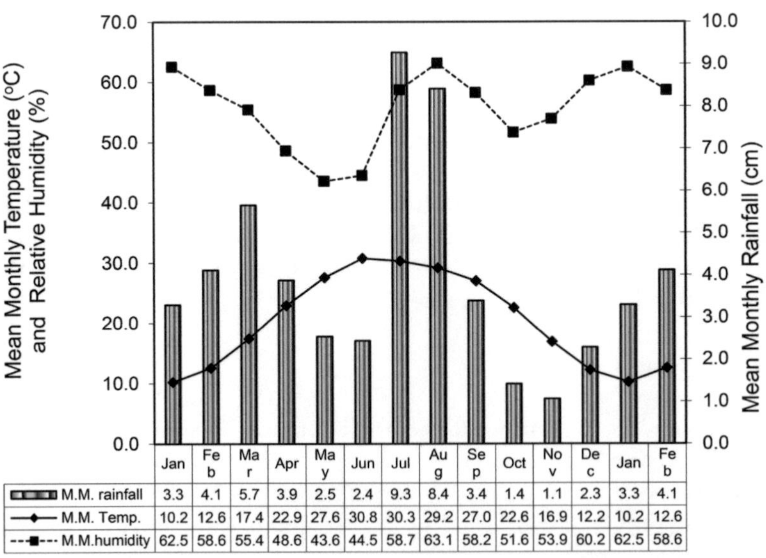

	Jan	Feb	Mar	Apr	May	Jun	Jul	Aug	Sep	Oct	Nov	Dec	Jan	Feb
M.M. rainfall	3.3	4.1	5.7	3.9	2.5	2.4	9.3	8.4	3.4	1.4	1.1	2.3	3.3	4.1
M.M. Temp.	10.2	12.6	17.4	22.9	27.6	30.8	30.3	29.2	27.0	22.6	16.9	12.2	10.2	12.6
M.M.humidity	62.5	58.6	55.4	48.6	43.6	44.5	58.7	63.1	58.2	51.6	53.9	60.2	62.5	58.6

Figure 2.
Pakistan arid region temperature (°C), rainfall (cm) and relative humidity (%) 1931–2017.

The average rainfall indicates that it increases with a decrease in temperature from December to March, while at the rise of temperature, a decrease occurs from March to June. In July it exceeds 9 cm (3.6 inches) and then decreases till September and onward up to November (below 3 cm or 1.2 inches). The annual variation of rainfall and humidity also shows two-time positive and negative deviation from the mean condition (**Figure 2**). From September to November, the arid region of Pakistan is covered by anticyclone and reversible monsoon lows. These months, with moderate temperature and low rainfall, constitute as warm season. Generally, the local thunderstorms give high rainfall in the Northern Areas and constitute as a rainy season of the northern arid region in Pakistan. The month of the year having a mean monthly rainfall below 3 cm and temperature above 17°C has been suggested as a dry month, otherwise moist. The characteristics of each season are as follows [12].

3.1.1 Cold season

The cold season lasts from mid-November to mid-April, which is completely winter months in the hilly areas. It is the season when, because of the prevalence of anticyclone, air subsides over Pakistan and the weather is feeblest. A cross section of the atmosphere about 78° east meridian shows the southern branch of the jet stream over northern Pakistan and India just south of the Himalayas, with the middle latitude westerly reaching down to the surface or, nearly so, north of about 25° [13]. The modest winter rainfall over northern Indo-Pakistan is associated with disturbances, which enter the area in the extreme northwest, after passing through Iraq, Iran, and Afghanistan. These disturbances reach their maximum development in winter when the jet stream lies south of the highlands, but they also occur though less frequently, in fall and spring. On the surface synoptic charts, the western disturbances usually first appear in the northwest in the vicinity of the surface polar front [7]. A cold continental air from eastern Europe and Western Asia break through the lower highland and spills out on the Indus lowlands. The front form between the continental polar air and the dry but warmer air of the Pakistani lowlands may not in the beginning be very active weather-wise. If the resulting depression acts to pull in a vigorous inflow of humid tropical maritime air mass from the Arabian Sea, the convergence is likely to produce extensive light rains. These are mid-latitude disturbances, a good proportion of that are not well-developed cold or warm fronts either at the surface or aloft.

The rainfall generated by these western disturbances of the cooler season is usually fairly widespread and light to moderate. It is locally heavy where thunderstorms are associated with the disturbances particularly in hot dry season. This fall in the cool season, when losses from evaporation are low, is highly effective for the growth of winter crops. On the plain, the total fall for the three winter months is only 1–3 inches, and yet this is of vital importance. Such storms also provide a much larger amount of winter precipitation in the form of snow and ice in the highland, whose water melts in hot season (summer) and furnishes the indispensable irrigation water for the Indo-Gengetic plain. These disturbances give high rains over the northwestern parts of Pakistan, which progressively decrease toward southeast and west. This variation, in general, leads to its long trajectory over land areas [12].

The season is usually characterized by cold weather with low mean monthly temperature, moderate rains, high humidity and pressure, low sunshine period, and gentle breeze. In cold season, most of the rains are caused by low-pressure depressions called western disturbances. These waves travel from the Mediterranean Sea and enter to Pakistan at the western margin after passing through Iraq, Afghanistan, and Iran. These low-pressure waves give 3.83 cm (1.5 inches) average rainfall in the arid region of Pakistan. Areas located at the western border of the country record high rains in cold season as compared to other seasons. The highest rainfall from these winds is recorded at Dir (24.22 cm or 9.7 inches) in March, which is the moistest month of the season. The highest mean monthly total precipitation of the season is 83.5 cm or 33.4 inches at Dir, and lowest is 0.8 cm or 0.32 inches at Mohenjo Daro with highest relative humidity recorded at Ormara and lowest at Khuzdar. The lower Indus plain has almost recorded high temperature, while it decreases up to −7.8°C at Astore. Most of the Northern Areas receive rain from these winds as compared to other sources. The rainfall from western depression progressively decreases from the northern mountainous region, toward northwest and southeast due to long trajectory of these lows over continental areas. Most of the highland receives rain in the form of snow, which is a guarantee for the agriculture activities in the hot season. The temperature of the country decreases from southwest toward northeast, but sometime cold waves are caused by heavy snowfall in mountains and decrease temperature after sunset below freezing point in Baluchistan plateau and the lowlands, influencing plant growth and human activities.

3.1.2 Hot season

The season ranges from mid-April to June and is characterized by high temperature and aridity. It is sometime called as the hot dry period to distinguish it from the cool dry season of winter and the hot wet season of summer and early fall. From April to June, the anticyclone subsidence and clear skies, characteristic of the winter months, still prevail, and this in combination with a much stronger solar radiation sets the weather pattern for the season. Temperatures are high, and a heavy, dry haze envelops in the interior, but drought still grips most of the country particularly upper and lower Indus plain.

The areal rainfall distribution in hot season is not fundamentally different from that of winter. The upper and lower Indus plain as well as Baluchistan, where maximum subsidence prevails, is still the driest parts of the country, and the Northern Areas with mountainous north are the wettest. The most perceptible areas of rainfall increase are the northern mountains and Azad Jammu and Kashmir, while most of the plains have less than 2 cm rains in these particular months.

According to Trewartha [13], in the far south, the added rainfall reflects the creeping northward with the sun of the ITC and the equatorial westerly. The increased rainfall in the north is furnished by the western disturbances. These perturbations are able to produce much precipitation in the more humid air of the north, they yield much more abundant rainfall, and some of it is associated with strong convective systems.

The hot season, in general, is characterized by violent weather, in the form of thunder squalls in the northern part of the country. In the drier land of Pakistan, the rainfall accompanying this vigorous convective system is low, but occasionally well-developed cumulonimbus clouds are generated with strong squall wind and violent dust storms.

In hot season, the mean monthly temperature exceeds 30°C with rare rains, low humidity and pressure, and high sunshine duration, evapotranspiration, and wind speed. These are the specific determinants, which cause parching of leaves in plants and evaporation of sweats from human bodies. In hot season, the western depression continues to travel along the northern latitude of Pakistan, which caused thunderstorms (with some rains) over the mountains and dust storms or dust-raising winds over the plain and Baluchistan plateau. The cold waves are rare in April, but hailstorms are frequent in April and rare in May, which caused widespread damages to fruit trees in the western and Northern Areas of Pakistan. June is the driest month of the season with mean monthly rainfall of 2.42 cm or 0.96 inches and also the hottest with mean monthly temperature of above 30°C. The highest mean monthly total precipitation of the season is 35.1 cm or 14.04 inches at Garhi Dupatta, and the lowest is 0.1 cm (0.04 inches) at Mohenjo Daro (arid region). The highest mean monthly humidity is recorded at Ormara, and the lowest is at Chilas and Dalbandin (arid region). The highest mean monthly temperature of the season is recorded at Sibi, and the lowest is at Astore. In hot season, the Northern Areas of the country receive rainfall above 10 cm or 4 inches, which decreases toward south. This high temperature in the southern part of Pakistan produces a trough of low pressure, which attracts monsoon depression in the hot moist season. The variation in temperature, generally, shows the same pattern as in cold season. As the hot season progresses, the belt of highest day temperature takes over Pakistan from south to north. Moreover, in this season the average rainfall of Pakistan is 2.9 cm or 0.99 inches, which is insufficient for vegetation growth and human activities [12].

3.1.3 Monsoon season

According to Moore [11], the term monsoon is derived from the Arabic word "mausim" which literally means "season." Originally, it was applied to the regular winds of the Arabian Sea, blowing for 6 months from the northeast and for the remaining 6 months from the southwest. Now generally, the term is applied to those and some other winds that blow with considerable regularity in different seasons of the year, due to the seasonal reversal of pressure over land masses and their neighboring oceans. In the typical area of the Indian subcontinent and southeast Asia, it is the seasonal inflowing moist winds that bring rains; hence, the monsoon season is considered as synonymous with the rainy season, and the term monsoon is applied to the rain without reference to the winds. The monsoon season of Pakistan ranges from July to mid-September, while in some areas it continues up to October. The deflected monsoon currents, generally, travel westward along the foothills of the Himalayas and reach Pakistan in July and are well established by the middle of that month [10]. The Arabian Sea branch of monsoon reaches to Sind-Makran coast by the end of June. However, it is of low vertical extant and, generally, produces stratus clouds in the coastal areas. The monsoon currents remain steady till it begins retreating toward the beginning of September. The second monsoon current enters Pakistan at upper Indus plain and gives more rains in Punjab, upper Sind, and northeastern Baluchistan. The third branch of the monsoon lows arrives at Kashmir Vale and the northern mountainous region of the country. Due to mountain trigger and high moisture index, these areas record high rains from these lows as compared to other regions of Pakistan. The rainfall from these winds generally decreases from northeast to southwest. This variation in rainfall intensity from monsoon is due to its long trajectory decreasing the moisture index of these depressions as they travel over continental areas. These winds are the only source, saving the lower Indus plain of Pakistan from aridity. In monsoon season, the mean monthly temperature of Pakistan drops below 30°C with high rainfall of 3–9 cm (1.2–3.6 inches) and humidity, low pressure, maximum sunshine period, and high evapotranspiration and wind speed. These winds give torrential rains with showers and caused damages to residential areas, crops, and vegetables. It gives above 7 cm or 2.8 inches of rainfall in Pakistan, which is higher than the other seasons. The maximum rainfall from these winds has been recorded at Balakot in July being the moistest month of the season with mean monthly rainfall of above 9 cm or 3.6 inches. However, in some areas especially in the mountainous north, August is the moistest month of the season. The highest mean monthly total precipitation of the season is recorded at Murree and lowest at Nok Kundi (arid region), with highest mean monthly humidity at Jiwani and lowest at Chilas. The lower Indus plain as well as Baluchistan (arid region) has recorded high temperatures, while the lowest is at Murree and the adjoining areas of Malakand and Mansehra divisions (humid region). The rainfall of Pakistan, in monsoon season, generally, decreases from northeast to southwest [12].

In Pakistan, the transition from the relatively dry and weather less spring to the cloudy, rainy season of summer, with its numerous perturbations, is abrupt and is usually associated with strongly disturbed weather. The so-called monsoon rains begin over Burma in May or even late April [13], but in Pakistan, they arrive from 15th of June in coastal areas, and the first week of July in mountainous north, and are well-established up to the mid-July in the whole country. Usually, the advance of the monsoon currents toward north over India and Pakistan is accompanied by turbulent weather in the form of thunderstorms, but its frequency decreases after the summer circulation is established. The monsoon begins to retreat from northern Pakistan in late August, and the withdrawal continues southward through September and October [12].

The later arrival of monsoon currents in Pakistan may result from the fact that during the winter and spring, there is an orographically determined upper trough, oriented Northsouth at about 85°E, over the western Bay of Bengal. It acts to accelerate the southwesterly monsoon flow over Burma located east of this trough while at the same time retarding it over Indo-Pakistan to the west [14]. As the subcontinent is heated intensively in April and May, the zonal westerly's over northern India and Pakistan begin to move northward, and it changes its direction under the influence of the mountain ranges, toward southwest. As a result, the jet stream, which had been south of the highland at about 30°N during winter and spring, tends to disappear and then reappear alternately south of the mountains. Disappearance becomes more frequent as the season advances, and each disappearance is associated with a northward surge of the summer monsoon. Finally, in late May or early June, the jet disappears completely over northern Pakistan and takes up a position at about 40°N of the Himalayas and Tibet. Simultaneously, there occurs a shift of the low-latitude trough and ridge positions, and the upper trough which previously was located at about 85°E quickly moves westward some 10° and takes up a position over western Indo-Pakistan subcontinent at approximately 75°E. With the disappearance of the jet over northern India and a westward shift of the upper trough, the equatorial westerly, or summer monsoon, surges northward over India accompanied by unsettled weather. The heating of the subcontinent and the development of a surface pressure trough are unable to produce a northward advance of the ITC until large-scale dynamic features of the circulation aloft become favorable [13]. When the jet stream reappears south of the Himalayas, again in fall, the summer monsoon again retreats southward and is called reversible monsoon.

3.1.4 Warm season

The season varies from mid-September to mid-November and is characterized by pleasant weather with moderate temperature and low rainfall. As the thermal trough over northern Pakistan weakens, paralleling a decline in insulation, the flow of southwesterly equatorial air across Indo-Pakistan subcontinent and up to the Bay of Bengal likewise weakens; the ITC, as well as the paths of the monsoon depressions, retreats slowly southward; and rainfall declines in the north [7]. With this retreat of the equatorial westerly, a greater intrusion of the north Pacific trades into the Bay of Bengal and over Indo-Pakistan. By the late fall, the trough of low pressure, separating the easterly and westerly air currents, becomes established over the southern part of the Bay of Bengal and adjacent to southern Indo-Pakistan subcontinent. Along the discontinuity between the equatorial westerly and the zonal easterly, various kinds of perturbations develop ranging all the way from weak monsoon depressions to hurricanes. The depressions follow less well definite tracks than in summer, but in general, their progress is westward so that their rainfall effects are concentrated in coastal region of Pakistan, which lies in close juxtaposition to the earth's most active region of cool season tropical cyclogenesis, located in the southwestern part of the Bay of Bengal.

Within Pakistan and India, severe hurricane storms are more numerous in this season particularly in the coastal region of Pakistan and Bay of Bengal. Some of these storms, in weakened form, appear to have the western north of the Arabian Sea regenerating over coastal areas of Indo-Pakistan.

During the fall months, the dynamic features of the circulation aloft, including jet stream and the orographically imposed troughs and ridges, begin to approach their cool season positions, with the reappearance, south of the mountainous north of Indo-Pakistan subcontinent in October or November of the middle latitude westerly and the jet stream, and the re-establishment of the polar front in the extreme northwest of Pakistan. The western disturbances once more become an important control of weather in northernmost India and Pakistan [13].

In warm season, the temperature of the country falls below 30°C with low rainfall and humidity, moderate pressure and sunshine, high evapotranspiration, and gentle breeze. In these months the monsoon lows give way to those of winter currents (western depression), and most of the rains are caused by thunderstorms, which develop due to local low pressure, especially in mountainous areas. The retreat of monsoon from the north Arabian Sea is marked by disappearance of the stratus clouds with a gradual increase in daytime temperature over Sind-Makran coast [10]. The highest mean monthly total precipitation of the season is recorded at Sialkot, while the lowest is at Nok Kundi with highest relative humidity at Ormara and lowest at Chilas (Arid region). The highest temperature of the season marked at Las Bela with lowest at Astore. In warm season, the rainfall of Pakistan decreases from northeast to south, while the temperature decreases from south toward north.

3.2 Climates of the arid region

The arid climate is characterized by low rainfall and covers a vast region in the southern as well as northern parts of Pakistan, where the annual total rainfall is between 0–10 inches. According to Hasan and Khan [12], the arid region of Pakistan is classified into the following subregions.

3.2.1 Hot long summer and mild short winters (HsMw')

The region is designated by hot and arid climates with mean temperature of the hottest month, June, above 32°C and warmest month, November, 10–21°C with average summer temperature more than 25°C. This climate type is divided into the following micro-regions.

3.2.1.1 Winter dry and summer rainfall with average day humidity of 55–60% (wx')

The region consists of Hyderabad, Mirpur Khas, and parts of Dadu District. The annual total rainfall of the area is 5–10 inches with summer concentration and annual day relative humidity of 55–60%. The mean monthly evapotranspiration is 7.1 mm or 0.28 inches at Hyderabad with annual sunshine of 8.2 h/day. The subregion has long, hot, and moist summers and short, mild, and dry winters. The mean monthly summer temperature exceeds 25°C, and the winter temperature is less than 15°C. June with mean temperature of above 32°C is the hottest month of the region, and January with mean temperature of less than 20°C is the coldest month. The mean monthly temperature of the area varies between 25°C and 30°C, with 35°C–40°C maxima and 10°C–15°C minima. The mean daily range is above 10°C, with 30–35°C daily maximum and 20°C–25°C daily minimum temperature.

The region is characterized by severe summers and moderate winters having 50°C ever recorded maximum temperature at Hyderabad in June, while the lowest minimum temperature of 1°C recorded in January. This subregion has 8.9 numbers of rainy days, while the average wind speed is 5.1 knots.

High-loam and clayey soil characterizes the zone with some pediment plain near the foothill of Kirthar mountains. High water table has caused waterlogging and salinization problems, depressing crop growth or making land unproductive for agriculture. However, areas having well-developed canal system and less salinization are characterized by intensive agriculture. Agriculture is the main activity in this zone, but there are also areas of rough grazing, riverine forests, and tracts of flooded backswamps. Wood resources come from riverine and thorn forests and trees of farmlands, much of which are used for firewood.

3.2.1.2 Winter dry and summer rainfall with relative humidity of 60–65% (wv)

The subregion includes Badin and northeastern part of Thatta. June is the hottest month with mean monthly temperature above 32°C, while January is the coldest month with mean temperature less than 20°C. The summers are hot, long, and moist, while the winters are short, mild, and dry. The area receives much of rainfall from monsoon lows particularly from the Arabian Sea branch. The annual total rainfall of the area is 5–10 inches with summer concentration, and average day relative humidity is 55–60%. The mean temperature of the region is 25°C–30°C, with 35°C–40°C maxima and 15°C–20°C minima. The mean daily range of temperature varies between 10°C and 15°C, with 30°C–35°C daily maximum and 15°C–20°C daily minimum temperature. The extreme maximum temperature of the subregion is 49.4°C recorded in June, and the lowest minimum is −2°C recorded in January. The annual evapotranspiration is 7 mm or 0.28 inches with annual sunshine of 8.2 h/day. The number of rainy days is 9.2 with average wind speed of 6.7 knots.

The region is characterized by sandy clayey soil, which is not suitable for agriculture, and therefore, the area has very little agriculture activities excluding those parts where water is easily available for irrigation. The water logging and salinization are the dominant factors of low agriculture production of the region. Vegetation resembles that of arid desert, with the addition of thorn scrub and light scrub forests. The land is mostly used for grazing of cattle, sheep, and goats. Moreover, the groundwater is acidic and is not suitable for drinking as well as vegetation growth.

3.2.1.3 Uniform rains with summer concentration and relative humidity of 55–60% (Usx')

Thal Desert and Dera Ismail Khan districts belong to this subregion, where the mean temperature of the coldest month that is January is above 10°C and hottest month that is June exceeds 32°C. The area receives both summer and winter rains, but the contribution of summer rains particularly from monsoon lows is high as compared to winter rainfall. The annual total rainfall of the region is 5–10 inches with average day relative humidity of 55–60%. The evapotranspiration of the region is 4.3 mm or 0.17 inches with annual sunshine of 8.4 h/day, which makes the summer of the region hottest as compared to the adjoining regions. The mean temperature of the subregion particularly at Dera Ismail Khan is 20°C–25°C, with 36.7°C maxima and 13°C minima. The area has high range of temperature both in summer and in winter seasons. The annual range of temperature is 10°C–15°C, with 30°C–35°C as the mean daily maximum and 15°C–20°C as the daily minimum temperature. The extreme maximum temperature of the region is 50°C, recorded in May, the hottest month of the region, while the lowest minimum temperature is −3°C, recorded in January. The number of rainy days is 18.2 with average wind speed of 2 knots.

The area is characterized by low agriculture practices due to nonavailability of water and poor soil. In some areas of Dera Ismail Khan, soil is fertile and having good agriculture, while most of the Thal Desert has sandy soil with some loams. The summer's "loo" (hot wind) is the important characteristic of the region, which not only increases the temperature of the region but also affects plant growth and its production, due to high evapotranspiration. The area has subtropical thorn forests with some irrigated plantation near canal banks and is used, in general, as a grazing land for livestock.

3.2.1.4 Summer dry and winter rainfall with average relative humidity of 45–50% (sy')

Panjgur and Turbat (Baluchistan) are included in this subregion, where the summers are long and dry, while the winters are short and moist. The annual total rainfall is less than 5 inches with winter concentration particularly from western disturbances. The area has severe summers and mild winters. The mean temperature varies between 20°C and 25°C, with 30°C–35°C maxima and 5°C–10°C minima. The mean daily range exceeds 15°C, with 25°C–30°C daily maximum and 10°C–15°C daily minimum temperature. The ever-recorded maximum temperature of the region is 45°C, recorded in June, the hottest month, while the lowest minimum temperature is −7.8°C, recorded in December, the coldest month of the year. The number of rainy days of the area is 10.2, while the wind speed is 4.5 knots with average day relative humidity of 45–50%.

Hot arid desert covers the area in the border with Iran and areas near the Makran coast. The soil of the area is sandy with sand dunes, badly eroded by running water. Vegetation is closely related to soil moisture, with grass along "nullah" (seasonal rivers) and on piedmont plain, associated with small trees and shrubs. The main species are *Pistacia* spp. and *Quercus ilex* with evergreen oaks occurring above 1200 m. Woodlands here provide fuel wood, nuts, fruits, and brows for camels. Wetter areas have relatively good grazing for livestock.

The land is used mainly for grazing. Pockets of cultivation are fed by water from karez system or spate irrigation along rivers. Vegetation is extremely variable, from completely barren ground to medium-density shrub and tree cover. Any well-watered land is cultivated, and all the vegetation are grazed, browsed, or cut for domestic requirements, particularly near urban centers.

3.2.2 Hot long summer and cool short winters (HsCw')

It is generally characterized by warm desert, where the June temperature exceeds 32°C, while the mean January temperature is between 0°C and 10°C with mean summer temperature of above 25°C. It is further divided into the following subregions [12].

3.2.2.1 Summer dry and winter rainfall with average day relative humidity of 30–35% (sz)

Parts of Chaghi and Kharan districts (Nok Kundi) belong to this region, where the annual total rainfall is less than 5 inches. The lowest rainfall of

Pakistan that is 1.4 inches has been recorded in this region, and the average day relative humidity varies between 30 and 35%. The area receives more rains in winter season, while the summer season is dry, with two dry spells in a year: the first is from mid-April to June, while the second is from mid-September to mid-December. The summers are long and hot, while the winters are moist and cool. The annual evapotranspiration is 5.4 mm or 0.22 inches with mean sunshine duration of 8.5 h/day. The mean temperature of the region is 20°C–25°C, with 35°C–40°C maxima and 10°C–15°C minima. The mean daily range is above 15°C with mean daily maximum temperature of 30°C–35°C and with a daily minimum above 15°C. The number of rainy days of the subregion is 4 with annual wind speed of 7.6 knots. The ever-recorded maximum temperature exceeds 45°C in June, while the lowest minimum temperature is less than −10°C in December, due to cold Siberian currents.

The soil of the region is sandy with arid piedmont plain and barren Chaghi hills. It constitutes as the western Baluchistan rangelands. The land is used, in general, for grazing, but some patches of cultivated land are also located in the karez-fed areas. The land is badly eroded by geomorphic agents due to sparse vegetation. Vegetation varies from barren ground and mountains to medium-density scrubs and trees, with acacia and jajoba as the main species. The forests are mostly used for domestic purposes and browsed for camels.

3.2.2.2 Summer dry and winter rainfall with average day relative humidity of 35–40% (sz')

Parts of Kharan and Chaghi districts (Dalbandin) are the specific areas of the region. The summers are long and dry, while the winters are short and moist. The area receives most of the rains in winter from western disturbances with annual total rainfall below 5 inches. The evapotranspiration of the area is 5.1 mm or 0.2 inches with annual sunshine duration of 8.6 h/day and the average day relative humidity of 35–40%. The mean monthly temperature of the area is 20°C–25°C, with mean maximum temperature of 35°C–40°C and minimum temperature of 5°C–10°C. The annual daily range varies between 15 and 20°C with mean daily maxima of 30°C–35°C and mean daily minima of 10°C–20°C. The highest maximum temperature of the area is above 50°C, recorded in June, and the lowest is below 10°C, recorded in December, due to Siberian cold winds. The number of rainy days of the area is 8.7, and the average wind speed is 3.2 knots.

The physiography and soil of the land are the same pattern as given in 2.1. subregion. However, the area is badly eroded and having barren mountains, particularly the Chaghi hills. Most of the areas near Dalbandin are piedmont plain, while toward south, the area is sandy, and sandy dunes cover a vast area with scrubs and bushes. The land is mostly used for grazing and browsed for camels. Due to sandy soil, agriculture practices are confined to isolated patches in the area. The barren Chaghi hills increase the daytime temperature while decrease the nighttime temperature of the region due to high absorption, deflection, reflection, and releasing of solar energy.

3.2.3 Hot long summer and short warm winters (HsWw')

The region is a hot desert, where mean temperature of the hottest month June is above 35°C and that of coldest month January is between 21°C and 32°C. The mean summer temperature of the area exceeds 25°C. It is further divided into the following sub-climatic regions [12].

3.2.3.1 Winter dry and summer rainfall, with average day relative humidity of 55–60% (wx')

Bahawalpur, Bahawalnagar, Mianwali, and Multan districts are included in this region. The area is characterized by hot long moist summers and short warm dry winters. The annual total rainfall varies between 5 and 10 inches with average day relative humidity of 55–60%. The mean evapotranspiration of the region is 5.5 mm or 0.22 inches, with annual sunshine of 8–10 h/day. The number of rainy days is 10–15, while the wind speed is 2–4 knots. The mean monthly temperature ranges between 25°C and 30°C, with 35°C–40°C maxima and 10°C–20°C minima. The mean daily range of temperature is 10°C–15°C, with 30–35°C daily maximum and 15–20°C daily minimum. The extreme maximum temperature of the region varies between 45°C and 51°C, recorded in May and June, while the lowest minimum temperature is 0°C to −5°C, recorded in January. May and June are the hottest months, whereas January is the coldest month of the region.

It is the extension of the Rajasthan Desert, but due to well-developed canal system, intensive agriculture is practiced in parts of the region. The area has sandy soil, but alluvial soil has also developed near the banks of Indus and its tributaries. The eastern part of the region has sandy loams with terraces of the "Hakra River" (India), sand ridges, inter-dune valleys, and saline lakes and flats.

Natural vegetation comprises xerophytic trees, shrubs, and grasses, but some irrigated plantations have also been grown along the river–/canal banks. The pressure on these limited tracts of shrubs vegetation from fuel woodcutters is intense. Land, otherwise, is used only for livestock production, hunting, and agriculture. Due to rapid deforestation, temperature of the region increases with passage of time and the area continuously going toward severe aridity.

3.2.3.2 Winter dry and summer rainfall with relative humidity 40–45% (wy)

Kach-Sibi areas fall in this region. It has hot long moist summers and warm short dry winters. The mean annual total rainfall of the area varies between 5 and 10 inches, with average day relative humidity of 40–45%. The mean annual evapotranspiration of the region is 4 mm or 0.16 inches, with mean sunshine duration of 8.2 h/day, number of rainy days 12.1, and wind speed of 2.9 knots. The mean temperature of the area is 25°C–30°C, with 30°C–35°C maxima and 10°C–15°C minima. The mean daily range of temperature varies between 15 and 20°C, with 35–40°C daily maximum and 15°C–20°C daily minimum. In Sibi, more than 50°C maximum temperature has been recorded for three times in June, making it the hottest region of the country. The lowest minimum temperature of the area is −3.3°C, recorded in January, the coldest month of the region.

Aridity prevail the whole region, with less fertile sandy soil. Kachhi plain is a desert, with no vegetation. However, in the irrigated areas, the natural vegetation in general is xerophytes with shrubs and grasslands, with medium size trees. The area is mostly used for grazing, where there are karezes and there are some agriculture patches of land.

3.2.3.3 Winter dry and summer rainfall with relative humidity of 50–55% (wx)

The region consists of Khanpur and the upper irrigated Sind. The annual total rainfall of the region is less than 5 inches with annual day relative humidity of 55–60%. The summers are long hot and moist, while the winters are warm short and dry. The annual evapotranspiration is 6.7 mm or 0.27 inches (Jacobabad), and sunshine is 8.7 h/day. The number of rainy days is 5–10, with a wind speed of 2–4 knots. The mean monthly temperature varies between 25°C and 30°C, with 35°C–40°C maxima and 10°C–20°C minima. The mean daily range is 10–15°C, with 30–35°C daily maximum and 15–20°C daily minimum temperature. The highest maximum temperature of the region is 45°C–52°C recorded in May and June, while the lowest minimum temperature is 0°C to −5°C, recorded in January and December.

The soil of the region used to be subjected to annual flooding before the river embankments were constructed, and because of this, the soils are rich loamy and clayey. High water tables have caused water logging and salinization problems, and the land is unproductive for agriculture growth.

Agriculture is the main activity in this zone, but there are also areas of rough grazing, riverine forest, and tracts of flooded backswamps. Wood resources come from riverine forests and trees on farmlands, much of which is used for firewood or send to Baluchistan as a mining timber.

3.2.3.4 Winter dry and summer rainfall with average day relative humidity of 60–65% (wv)

The region consists of Nawabshah and Padidan in central Sind. The total rainfall of the region varies from 5 to 10 inches with summer concentration, while the winters are short and dry. The average day relative humidity varies from 60–65%, with evapotranspiration of 6.9 mm or 0.28 inches and sunshine of 8.8 h/day. The mean monthly temperature varies between 25 and 30°C, with 35–40°C maxima and 10–15°C minima. The mean daily range of temperature is 15–20°C, with 35–40°C daily maximum and 15–20°C daily minimum temperature. The maximum temperature exceeds 50°C in June, while in January, it falls below freezing. The number of rainy days is 5–10 per year, with average wind speed of 0–5 knots. The soil and natural vegetation as well as physical relief of this climatic zone are the same as given in 3.3 climate type.

3.2.3.5 Winter dry and summer rainfall with relative humidity of 45–50% (wy')

Thar Desert in Sind and Cholistan in Punjab constitute in this micro-region. It has long hot moist summers and short warm dry winters. The annual total rainfall of the area varies from 5 to 10 inches, but in Cholistan, it is less than 5 inches. The average day relative humidity of the region is 45–50%, with annual evapotranspiration of above 5 mm or 0.2 inches and sunshine duration of more than 8 h/day. The mean monthly temperature varies between 25–30°C, with 35–40°C maximum and 10–15°C minimum temperature. The mean daily range is 15–20°C, with 30–35°C daily maximum and 15–20°C daily minimum temperature. The extreme maximum temperature of the region is above 50°C, recorded in June, and the lowest minimum temperature is 0–5°C, in January.

Arid desert covers the eastern part of Sind and lower Punjab. These major landforms of the region are a series of parallel and linear sand dunes "bets" oriented northwest to southeast. Valleys between these dunes cover about 30% of the area. There are also salt lakes fed by water. Natural vegetation is of two types, *Cenchrus-Panicum* dune type and *Eleusine-Cenchrus* inter-dune type. Livestock is the mainstay of the region, with some secondary dryland farming in inter-dune valleys. Cholistan has sandy soil and is the extension of the Great Indian Desert that includes terraces of Hakra River, sand ridges, inter-dune valleys, and saline lakes and flats. Natural vegetation is xerophytic trees, shrubs, and grasslands. The region, in general, is used for hunting and grazing, with some patches of rainfed agriculture.

3.2.4 Warm long summers and mild short winters (WsMw')

This sub-climatic type is characterized by mean temperature of the hottest month June which is 21–32°C and the warmest month November which is 10–21°C, with mean summer temperature of above 25°C. It is further divided into the following sub-types.

3.2.4.1 Winter dry and summer rainfall with relative humidity of 55–60% (wx')

The region characterizes by Marine continental climates with an annual total rainfall of 5–10 inches and average day relative humidity of 55–60%. It includes Las Bela and Dadu. This sub-type is characterized by short warm dry winters and long hot moist summers. The annual evapotranspiration of the area is 4.4 mm or 0.18 inches, with sunshine duration of 8.1 h/day, number of rainy days 14.5 per year, and average wind speed of 3 knots. The mean monthly temperature varies between 25–30°C, with 35–40°C maxima and 10–15°C minima. The mean daily range is 10–15°C, with 30–35°C daily maximum and 15–20°C daily minimum temperature. The ever-recorded maximum temperature of the area is 51°C at Las Bela, recorded in May, the hottest month of the region, while the lowest minimum temperature is below freezing recorded in January.

Vegetation of the region is closely related to soil moisture, with grasses along streams and seasonal rivers. Woodlands here provide fuel wood, nuts and fruits, and brows for camels. Wetter areas have relatively good grazing patches for livestock. The land is mainly used for grazing. Pockets of cultivation are fed by water from karez systems or from spate irrigation along streams and seasonal rivers. Vegetation is extremely variable, from completely barren ground to medium-density shrubs and tree cover. Any well-watered land is cultivated, while the vegetation is grazed, browsed, or cut for domestic purposes, particularly near Karachi and Dadu.

3.2.4.2 Uniform rains with winter concentration and relative humidity of 65–70% (Uwv')

Karachi west belongs to this region, where the mean annual total rainfall is 5–10 inches, with average day relative humidity of 65–70%. The area receives both summer and winter rains, but the contribution of winter rains is more than summer season. It has long warm summer and short mild winters with sea and land breezes throughout the year. June is the hottest month, with maximum temperature of above 32°C, and January is the coldest month, with minimum temperature of less than 5°C. The mean evapotranspiration of the area is less than 4 mm or 0.16 inches, with sunshine duration of more than 8 h/day, number of rainy days 5–10 per year, and wind speed of 6.6 knots. The mean monthly temperature of the region is 25–30°C, with 35–40°C maximum and 15–20°C minimum temperature. The mean daily range varies between 10 and 15°C with 30–35°C daily maximum and 20–25°C daily minimum temperature. The highest maximum temperature of the region is 45°C, recorded in June, and the lowest is 0°C, recorded in January. The physiography, soil, and natural vegetation are the same as given in A.4.5. sub-type.

3.2.4.3 Summer dry and winter rainfall with average relative humidity of 70–75% (su)

The region is marine coastal has climates with annual total rainfall of less than 5 inches and with average day relative humidity of 60–75%. Pasni and parts of Jiwani belong to this region. It has warm long summers and mild short winters with winter rainfalls, but summers are also not dry. The annual evapotranspiration is above

4 mm or 0.16 inches with annual sunshine duration of 8.3 h/day, number of rainy days 6.5 per year, and wind speed of 6.9 knots. The mean monthly temperature varies between 25°C and 30°C, with 35°C–40°C maxima and 15°C–20°C minima temperature. The annual daily range of temperature is 10°C–15°C, with 30°C–35°C daily maximum and 20°C–25°C daily minimum temperature. The highest maximum temperature of the area is 47°C, recorded in June, the hottest month of the year, whereas the lowest minimum is 2°C, recorded in December, the coldest month of the region.

The region is also characterized by land and sea breezes with saline sandy soil, sparsely covered by scrub trees particularly of Bela forest (medium size trees). Due to low soil fertility, agriculture is not very important, and most of the area is used for grazing and woodlands.

3.2.4.4 Summer dry and winter rainfall with relative humidity of 75–80% (su')

The region consists of Ormara and Jiwani with warm long dry summers and short mild moist winters. Sometime, Ormara receives both summer and winter rains, while the intensity of winter rains is high at Jiwani as compared to summer. The annual total rainfall of the area varies between 5 and 10 inches, with mean evapotranspiration of less than 5 mm or 0.2 inches and sunshine duration of 9 h/day. The average day humidity is 75–80%, with number of rainy days between 5 and 10 per year and wind speed of 5–8 knots. The mean monthly temperature of the area varies between 25°C and 30°C, with 30°C–40°C maxima and 15°C–20°C minima. The mean daily range of temperature is 5°C–10°C, with 30°C–35°C daily maximum and 20°C–25°C daily minimum temperature. The highest maximum temperature of the region is 48°C, recorded in June, and the lowest minimum temperature is 0°C–5°C, recorded in January. June is the hottest month of the area, while January is the coldest month. The general characteristics of the region particularly of soil and vegetation are the same as stated in A.4.3. subregion.

3.2.4.5 Uniform rains with summer concentration with relative humidity of 65–70% (Usv')

Marine coastal, Indus Delta (Thata), and eastern Karachi are included in this region. Both summer and winter rains are received in the area, but the contribution of summer rains is higher than winter. The annual total rainfall of the area varies from 5 to 10 inches with annual evapotranspiration of 4.9 mm or 0.2 inches and sunshine duration of 8.1 h/day. The average day relative humidity varies from 60–65%, with number of rainy days less than 10 per year and wind speed of 6.6 knots. The mean monthly temperature of the area is 25°C–30°C, with 35°C–40°C maxima and 15°C–20°C minima. The mean daily range of the area varies from 10°C–15°C, with 30°C–35°C daily maximum and 20°C–25°C daily minimum temperature. In May, the maximum temperature reaches to 48°C and falls to 0°C till January.

The Indus Delta has its apex some distance northeast of Thata, where the distributaries fan-out to form the deltaic plain near Thata. Two of the large distributaries are Ochito and the Gungro. Many of the channels perform the dual function of distributaries and estuaries. The channel beds and their levees are higher than the adjacent lands, and the shallow troughs between them are often filled with water, resulting in swamps. The tidal delta is submerged at high tide and has mangrove swamps and tamarisk groves in its western section. The eastern section is Rann of Kutch, a saline marshy land. The coast is low except between Karachi and Cape Monze, where the Pab hills approach the shore.

The Karachi plain has a thin mantle of soil over weathered bedrock. A few low hills rise above 15 m. Shallow depressions are known as Dhand. One of these, Heleji Dhand, is used as water reservoir for Karachi city. The vegetation of the region is mostly Bela, with some irrigated plantation on Indus banks and roadsides. The area is characterized by sea and land breeze with poor sandy soil, due to which very rare cultivation is practiced in the area. The forests, in general, are used for grazing and domestic purposes.

3.2.5 Warm short summers and cool long winters (Ws'Cw)

The region includes areas having mean temperature of the hottest month June between 21°C and 32°C and mean temperature of the coldest month January as 0°C–10°C, with mean winter temperature of above/equal to 15°C. The region has further been divided into the following sub-types.

3.2.5.1 Summer dry and winter rains, with relative humidity of 40–45% (sy)

The region includes Kalat in the south and Chaman and Loralai in the north. These areas have short warm dry summers and cool long moist winters. The mean annual total rainfall of the area varies between 5 and 10 inches, with average day humidity of 40–45%. The annual evapotranspiration of the sub-region is 5 mm or 0.2 inches, with sunshine duration of 8.7 h/day, number of rainy days 15–20 per year, and wind speed of 3 knots. The mean temperature of the area varies between 15°C and 20°C, with 25°C–30°C maxima and 0°C–5°C minima. The mean daily range is 15°C–20°C, with 20°C–25°C daily maximum and 5°C–10°C daily minimum temperature. The highest maximum temperature recorded at Kalat is 38°C, in June, while the lowest is 0°C to −18°C, recorded in January.

The region is characterized by extensive woodlands of *Juniperus*, *Pinus gerardiana* (*chilgoza*), etc., with many shrubs including a well-known medicinal plant, e.g., *Ephedra nebrodensis*. Excessive timber and fuel woodcutting has completely degraded these woodlands, leaving many former forest areas as shrubby and grassy steppes. The relatively higher productivity of forest in these areas has encouraged denser populations, and consequently greater human exploitation of scrub and woodlands has declined the stocks of wood. The soils, in general, are regosols and lithosols with some fertile alluvial soils in the valleys. Terrace agriculture is practiced in the valleys; otherwise, the area is used for grazing. The topography of the region is mostly rough with steep slopes and narrow valleys, intensively eroded by running water.

3.2.5.2 Summer dry and winter rainfall with relative humidity of 30–35% (sz)

Nushki and its surrounding in the Chaghi district falls in this sub-type, where the annual total rainfall is between 0 and 5 inches, with average day humidity of 30–35%. This low relative humidity of the region causes high evapotranspiration from plants as well as the human skin. The summers are, in general, hot long and dry, while the winters are mild cool and moist. The cold Siberian winds in winters decrease the temperature of the area up to freezing. The mean annual evapotranspiration is less than 5 mm or 0.2 inches with sunshine duration of 8 h/day. The mean temperature of the region is 20–25°C, with 30°C–35°C maxima and 5°C–10°C minima. The mean daily range is between 15°C and 20°C, with 25°C–30°C daily maximum and 15°C–20°C daily minimum temperature. The soils consist of shingles, pebbles, and sands (regosols) and are very poor for cultivation. However, some agriculture patches are located in karez-feeding areas. The forests are mostly scrubs with grasslands and also having barren lands and mountains.

3.2.5.3 Uniform rains with summer concentration and relative humidity of 35–40% (Usz')

The climatic region includes Khuzdar, which receives both summer and winter rains with summer concentration. The summers are long and warm, while the winters are cool and short. The annual total rainfall of the region is 5–10 inches, with average day relative humidity of 35–40%. The annual evapotranspiration is less than 5 mm or 0.2 inches with mean sunshine duration of less than 8 h/day. The number of rainy days is 19.3 per year, with wind speed of 3 knots. The mean temperature of the region is 20°C–25°C, with 30°C–35°C maxima and 5°C–10°C minima. The daily range of temperature is 10°C–15°C, with 25°C–30°C daily maximum and 10°C–15°C daily minimum temperature. The extreme maximum temperature of the area is 43°C, recorded in July, the hottest month of the year, whereas the lowest minimum temperature is −5 to −10°C, recorded in January, the coldest month of the year.

The region characterized by steep slopes having some grasslands and sparse shrubs but have mainly barren. The mainstay of the land is grazing, with pockets of cultivation fed by tube well irrigation. Farmers grow mainly fruits such as apples, apricots, grapes, peaches, and plums. The vegetation is mainly used for domestic purposes and livestock. Due to rough topography, some patches of terrace agriculture are at the tube-well-fed areas.

*3.2.6 Warm short summers and cold long winters (Ws'C*w)*

In this region the mean temperature of the hottest month June is between 21 and 32°C, with the coldest month January having temperature below freezing and the mean winter temperature above 15°C. It is further divided into the following sub-climate types.

3.2.6.1 Winter dry and summer rainfall particularly from local thunderstorms with relative humidity of 40–45% (wty)

The region consists of Gupis in the Northern Areas, where the annual total rainfall is less than 5 inches with cold long dry winters and short moist summers. In winters low precipitation mostly in the form of snow and ice is received, whereas the summer rains are high, particularly from the local thunderstorms. The winters are very cold, while the summers are warm. The evapotranspiration of the region is 2.8 mm or 0.11 inches, with annual sunshine of 6.6 h/day, while the average day relative humidity is 40–45%. The mountain slopes are mostly covered by snow particularly in Ishkoman valley, while the ground and lower mountain slopes are barren. The physiography is generally rough with sparse vegetation and cultivation. The mean temperature of the region is 10–15°C, with 20°C–25°C maxima and 5°C–10°C minima. The daily range of temperature is between 10°C and 15°C, with 15°C–20°C daily maximum and 5°C–10°C daily minimum temperature. The highest maximum temperature of the region is 41.1°C, recorded in July, the hottest month of the area, while the lowest minimum temperature is −10 to −15°C, recorded in January, the coldest month of the year. The number of rainy days of the region is 13.8, while the wind speed is 2.6 knots.

The zone is characterized, in general, by glacier feed areas, deep narrow valleys, fans, and terraces. Upper mountain slopes have poor soils, but valley bottoms have deep, clay-rich soils overlying the colluviums on lower slopes, fans, and terraces. Crops can be cultivated on those fertile fans and terraces, irrigated by the streams.

Slopes rely on rainfall to produce maize, wheat, and orchards. Energy demands are high because of the cold winters. The forests are, in general, orchards with medium size trees, which are mostly used for domestic purposes.

3.2.6.2 Summer dry and winter rainfall with relative humidity of 50–55% (sx)

The region consists of Skardu, Ladakh, in occupied Kashmir and the adjoining areas of China (Tibetan Plateau). The annual total precipitation varies from 5 to 10 inches, particularly in the form of snow and ice in winter season. However, the summers are not dry, and the region receives more than 1 cm or 0.18 inches rainfall in April and May from local thunderstorms. The annual evapotranspiration is 4.4 mm or 0.18 inches, with annual sunshine of 6.8 h/day, while the average day relative humidity is between 50 and 55%. The mean temperature varies from 10–15°C, with 20–25°C maxima and 0°C minima. The mean daily range of temperature is between 15 and 20°C, with 15–20°C daily maximum and 5–10°C daily minimum temperature. The highest maximum temperature of the region is 41°C, recorded in August, and the lowest minimum is −20 to −25°C, recorded in January. The mean temperature in January as well as December reaches below the freezing point. Solar energy is needed for the plant protection particularly in winter season. The number of rainy days of the region is 21.5 per year, with wind speed of 1.1 knots.

The area is characterized by barren mountains and glaciers. The soils are poor at higher mountain slopes and are fertile near foothills. The agriculture is practiced only on terraces in well-watered areas. The mountains are, in general, barren due to glaciations; however, some deciduous forests have grown at the foothills and riverbanks.

3.2.7 Hot long summers and cool short winters (HsCw')

The region is characterized by hot climates, with June temperature above 32°C and with January temperature between 0 and 10°C, while the mean summer temperature exceeds 25°C. The climatic zone is further divided into the following sub-types.

3.2.7.1 Winter dry and summer rainfall particularly from local thunderstorms with relative humidity of 40–60% (wtY)

Gilgit district and Bunji, located close to Hunza and Indus Rivers, fall in this region. The annual total rainfall of the area varies from 5 to 10 inches. This sub-type receives both summer and winter rains with summer concentration particularly from local thunderstorms. The evapotranspiration of the area is 2.7 mm or 0.11 inches (Gilgit), with annual sunshine of 6.5 h/day and average day humidity of 40–60%. The summers are hot and moist, while the winters are cool and dry. The mean temperature of the region varies between 15 and 20°C, with 25–30°C maxima and 0–10°C minima. The mean daily range is 10–15°C, with 20–25°C daily maximum and 5–15°C daily minimum temperature. The highest maximum temperature of the region is 45–50°C, recorded in July and August, while the extreme minimum temperature is –5 to −15°C, recorded in December and January. The number of rainy days varies between 14 and 20 per year, with annual wind speed of 0–3 knots.

The region consists of major valleys and high permanent snowcapped mountains. Snowfall received is between 2 and 6 m/year. Vegetation at 3800-4000 meters is alpine scrub of small deciduous and evergreen forests. Between 3500 m and 3800 m, vegetation is mixed coniferous forests. High alpine scrub is important for summer grazing, but livestock also enter the forests, and large tracts of forests have been cleared. Rainfed cultivation from the valleys and lower slopes is expanded upward, when fertile land is no more available in the valleys. The soil at foothills is usually fertile, but due to the lack of water, agriculture is confined to well-watered areas.

3.2.7.2 Uniform rains with summer concentration from local thunderstorms and relative humidity of 35–40% (Ustz')

Chilas in the Northern Areas and parts of Kohistan district are included in this region, where the annual total rainfall varies between 5 and 10 inches. Rainfall is received both in summers and winters, with summer concentration especially from local thunderstorms. Physiography is rough with high barren mountains and glacial topography. The evapotranspiration of the region is 3.3 mm or 0.13 inches, with annual sunshine of 6.5 h/day and average day relative humidity of 35–40%. The mean temperature of the region is between 20 and 25°C, with 30 and 35°C maxima and 5 and 10°C minima. The mean daily range varies between 10 and 15°C, with 25–30°C daily maximum and 10–15°C daily minimum temperature. The highest maximum temperature of the region is 48.1°C, in August, and the lowest minimum is −4.4°C, recorded in January. The number of rainy days is 17.3 per year, with average wind speed of 1 knot.

The upper mountain slopes have poor soils, but the valley bottom has deep clay-rich soils overlying the colluviums on lower slopes, fans, and terraces. Crops can be cultivated on these fertile fans and terraces, obviously irrigated by the streams. Slopes rely on rainfall to produce maize, wheat, and orchards of walnuts, apricots, and plums. The forests are usually deciduous with most of the barren mountains and lands. Agriculture is mostly confined to rainfed areas.

4. Soil

Agriculture and climate of the arid region have also been greatly influenced by the soil composition and structure. "The soil has three distinct constituents, which are solid particle, air, and water." These three major components have also great effects on the climate and crop productivity of a region. For example, salts, minerals, and organic matters help in the reflection, deflection, and absorption of solar radiation; consequently, the temperature of the earth surface and atmosphere fluctuates [3].

The arid region of Pakistan has peculiar types of soil due to its topography, climate, vegetation cover, parental materials, and time period of formation (main factors of soil formation). Based on structure and composition, the soil of the arid region can be classified into "soil of the flood plain, loamy soils of bar uplands, loamy and clayey, partly saline soils of the piedmont plains, rolling to hilly sandy soils of Aeolian desert, highlands, Baluchistan plateau." The soil of the floodplain is subclassified into loamy and some sandy stratified soil of recent river plains, mainly loamy and clayey soils of subrecent river plains, mainly loamy saline soils of estuary plains, and mainly salty and clayey wet saline soils of tidal plains.

The loamy and clayey soils are rich in soil contents and suitable for agriculture activities. Such soil is located at the bar upland in lower Punjab and also on both sides of the Indus River in Sind (arid region), and both of the areas produced a major part of the agriculture economy on national level. Most of the arid region of Baluchistan, eastern Sind, and lower eastern Punjab is characterized by the rolling to hilly sandy soils, which are poor for crop cultivation, while the remaining areas are converted to cultivated lands after establishment of better irrigation system that originated from Indus and its tributaries (**Figure 3**).

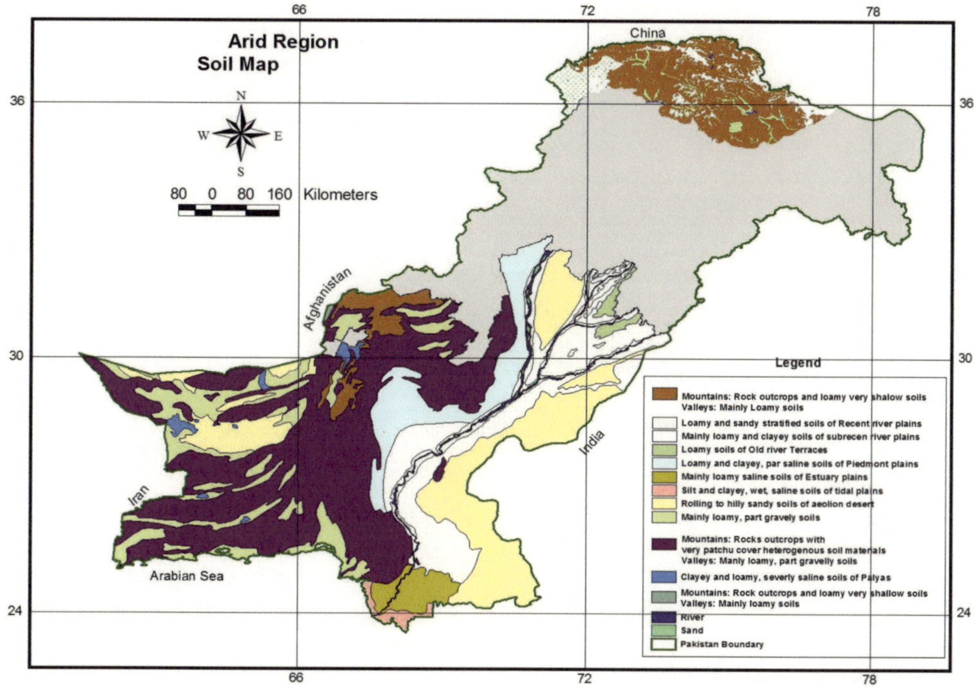

Figure 3.
Soil map of the arid region, GoP [15].

5. Natural vegetation

It is one of earth's most vital natural resources that provide food, fiber, and fuel for the sustainability of human life, while vegetation preserves biodiversity and soil and plays an outstanding role in the hydrological cycle. The inhabitants of the arid region of Pakistan have cleared the land surface by cutting the natural vegetation cover for the cultivation of crops. While the vegetation cover controls the soil erosion, storage and filtration of water, and wildlife habitat, playing a key role in the transfer of gasses between the atmosphere, soil, and water bodies is basic to the crop cultivation, production, and growth.

Dense forests can influence the local climate to some extent. Temperature is lower in a region of dense forests, light is reduced greatly, and the soil is several degrees colder. Humidity is greater, and dew and fog are formed readily over the adjoining fields. Evapotranspiration from the soil and wind velocities under a forests covered area are greatly reduced and also increase the capacity of water storage of the soil. The barren land and mountains have inverse climate condition.

In arid region, there is a variety of natural vegetation due to variation in climates, topography, and soil condition. "Natural vegetation covers about 4.8% of the total area of Pakistan, while about 20–25% of forests are needed for the balanced economy of a country" [3]. The province-based distribution shows that Khyber Pukhtunkhwa has the leading percentage (15.6%), followed by Sind (2.1%), Gilgit-Baltistan (13.5%), and Azad Jammu and Kashmir (13.6%) and the remaining 55.2% in Punjab and Baluchistan province. The arid region has two distinct types of forests, that is, natural vegetation and the plantation forests (**Figure 4**).

The natural vegetation consists of different type of forests including alpine forests, coniferous forests or subtropical pine forests, subtropical dry forests, tropical thorn forests, and coastal mangrove forests. Plantation forests are divided into riverine and irrigated forests that are located at the roadsides, canals, and rivers.

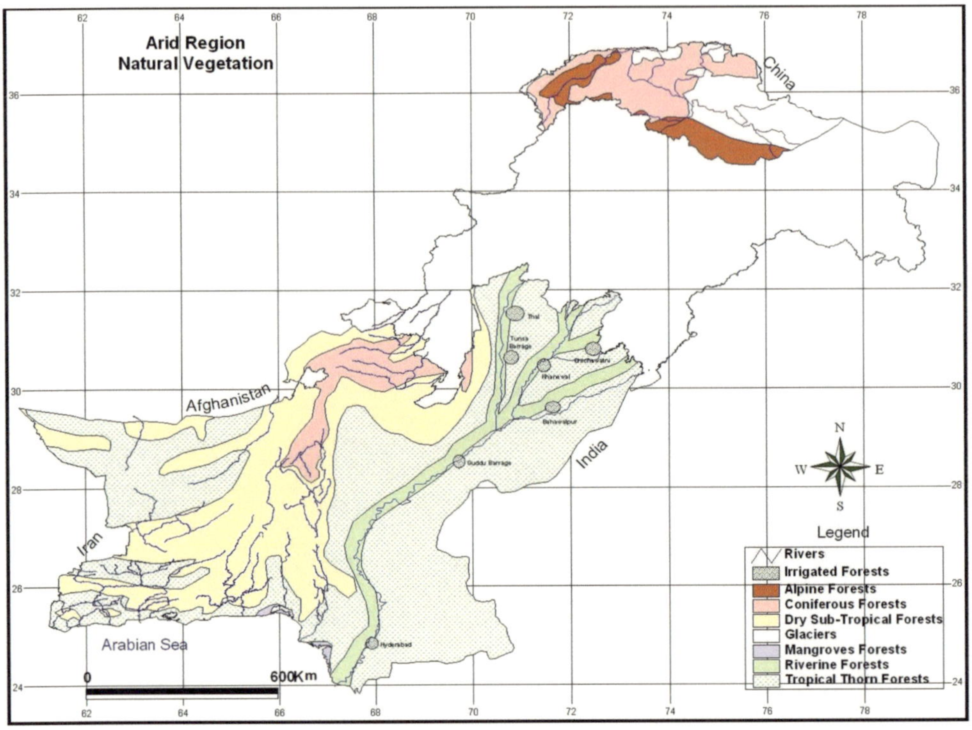

Figure 4.
Natural vegetation of the arid region, Pakistan, Khan [3].

6. Human environment

Agriculture is practiced in diverse economic, social, and environmental conditions. Hence, great diversity has developed in agriculture practices and products. In some areas of the arid region, farming is carried out to meet the local demand; in others it is primarily cultivated for commercial purposes. "The different types of farming in the arid region of Pakistan are intensive subsistence farming, dairy farming, plantation farming, truck farming, and extensive herding" [16]. Among the factors that have influenced the type of farming include distribution of population and human settlements, as well as socioeconomic conditions.

6.1 Population

Pakistan has sustained significant population growth in the past century. In early 1891, the population of the country was 18.4 million, which dropped to 16.6 million in 1901. Then the population was estimated at 19.4 million in 1911, which increased to 21.2 million in 1921 and 23.6 million in 1931, while in 1941 it again increased up to 28.3 million. According to the first well-organized census in 1951, Pakistani population was 33.7 million having 6 million urban and 27.8 million rural population. In 1961, this figure was increased to 42.9 with 9.6 million urban and 33.3 million rural inhabitants. Onward, in 1971, it i up to 65.4 million with 7.6 million urban and 48.8 million rural population. The increase in population continues in 1981, where population of the country reached to 84.3 million having 23.9 million urban and 60.5 million rural population. As per 1998 census, the population had quadrupled to 132 million inhabitants and almost double (207.8 million in 2017) (**Figure 5, Table 1**).

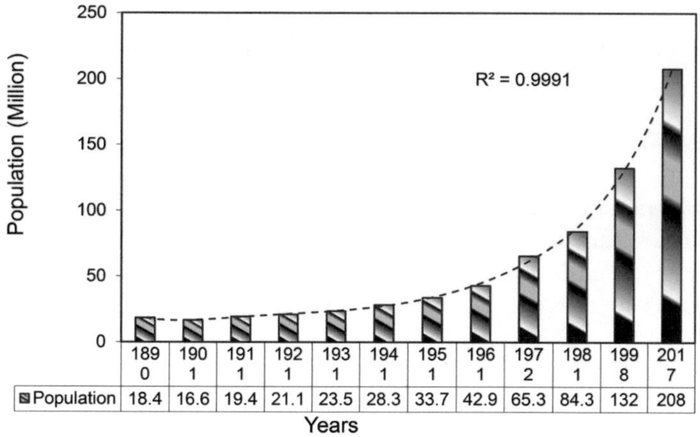

Figure 5.
Pakistan population 1890–2017.

Year	Population	Year	Population density	Year	Annual growth	Year	Urban	Rural
1890	18.4	1890	NA	1890	NA	1890	NA	NA
1901	16.6	1901	20.8	1901	1.6	1901	9.8	90.2
1911	19.4	1911	24.4	1911	0.8	1911	8.7	91.3
1921	21.2	1921	26.5	1921	1.1	1921	9.8	90.2
1931	23.6	1931	29.6	1931	1.9	1931	11.8	88.2
1941	28.3	1941	35.9	1941	1.8	1941	14.2	85.8
1951	33.8	1951	42.4	1951	2.4	1951	17.7	82.3
1961	42.9	1961	53.9	1961	2.3	1961	22.5	78.0
1972	65.4	1972	82.1	1972	3.3	1972	28.3	71.7
1981	84.3	1981	105.8	1981	3.3	1981	28.3	71.7
1998	132.4	1998	166.3	1998	2.7	1998	32.5	67.5
2017	207.8	2017	256.0	2017	2.4	2017	36.4	63.6

Source: GoP [17].

Table 1.
Arid region of Pakistan demographic characteristics.

In early 1994, the population of Pakistan was estimated to be 126 million, making it the ninth most populous country in the world. Its land area, however, ranks 32nd among nations. Thus, Pakistan has about 2% of the world's population living on less than 0.7% of the world's land. The population growth rate is among the world's highest, officially estimated at 3.1% per year but privately thought to be closer to 3.3% per year by many planners involved in population programs (**Table 1**).

The total population of the arid region in 1998 was 60.2 million comprising 31.5 million male and 28.7 million female. It is estimated that the projected population of the arid zone will be 78.7 million including 41.2 million male and 37.5 million female populations by the year 2010.

The population density of the arid region is 137 persons/km^2 area in 1998, while it is expected that the projected population will be 179 persons/km^2 by the year 2010.

6.2 Population distribution

The distribution of rural population brings out the man-land relationship effectively in an agriculture area like Pakistan. The pressure of population on land in arid region has been increasing with passage of time. The population density of the arid region has been categorized into the following major classes.

6.2.1 Thinly populated region

The thinly populated areas of the arid region mostly cover the entire area of Baluchistan and Gilgit-Baltistan provinces. These areas comprise of hills, mountains, and barren sandy lands. Most of the land consists of rainfed areas having scared agriculture activities (**Figure 6**).

6.2.2 Moderately populated region

Moderate population varies from 101 to 200 persons/km^2. The moderate population is mostly located in the northwestern part of Baluchistan (Pishin district), parts of Sind, and lower Punjab provinces. Most of the agriculture is based on rainfall or canal system with low productivity (**Figure 6**).

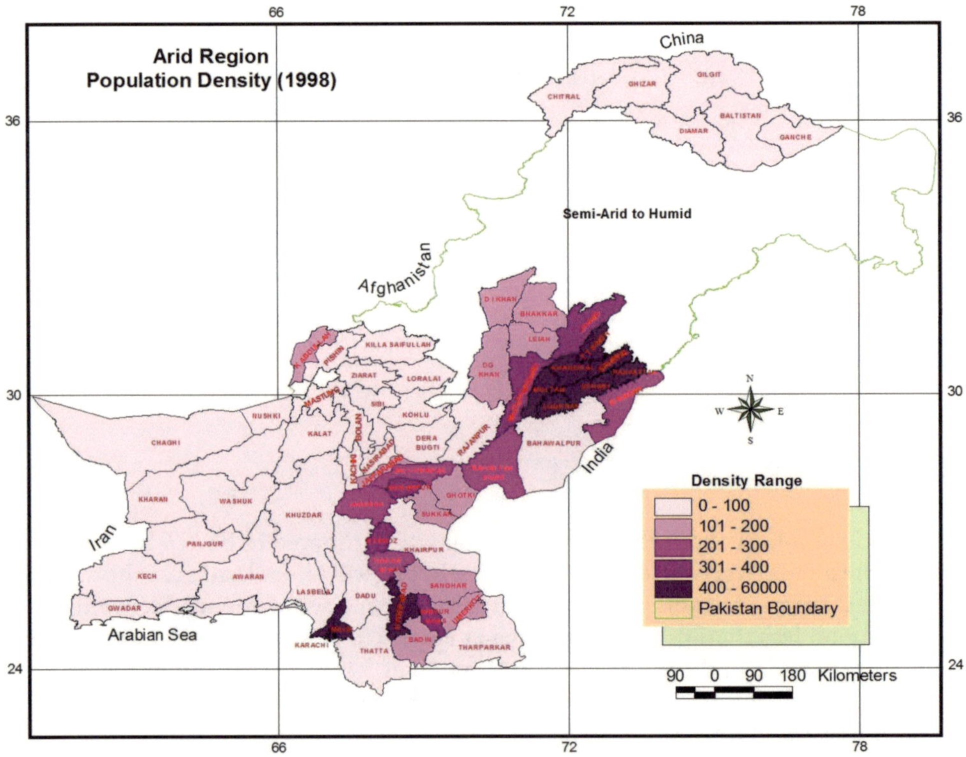

Figure 6.
Arid region projected population (2011–2012), GoP [18].

6.2.3 Thickly populated region

The range for this class has been considered between 201 and 300 persons/km^2 area. The arid areas having thick population mostly cover the fertile land of Sind and Punjab. These are the areas, where intensive agriculture is carried out with a high productivity (**Figure 6**).

6.2.4 Very thickly populated region

The very thickly populated region varies from 301 to 400 and comprises parts of Punjab and northwestern Sind. Most of these areas are characterized by fertile soils and located close to the major cities. These are the most important agriculture lands forming source of food for the entire region (**Figure 6**).

6.2.5 Most thickly populated region

There are 13 most thickly populated districts in the arid region where the population density is above 400 persons/km^2. These are located around the major cities and fall in the old floodplains of the Punjab and active floodplain of Sind excluding Karachi (**Figure 6**).

7. The agroecological zones and land use pattern

The agriculture sector continues to play a basic role in the arid region economy. "It is the second largest sector, accounting for over 22% of GDP, and remains by far the largest employer, absorbing 45% of the country's total labor force. Nearly 62% of the country's population resides in rural areas, and is directly or indirectly linked with agriculture for their livelihood" [18].

About nine different agroecological zones have been identified in the arid region of Pakistan by GoP [19], and the map produced has been updated on the basis of the recent data. The general description, physiography, land use, and environmental condition of each agroecological zone have been given below.

7.1 Indus Delta

The region comprises of the Indus Delta and the surrounding areas of Thatta, Badin districts, and parts of Hyderabad that are formed through successive advancement of land into the sea. The tall grasses and salt bushes had hampered the spread of agriculture in large areas. The changing nature of the Indus distributaries, floods, and marshes makes the reclamation of the land a major challenge (**Figure 7**). The region is characterized by arid tropical marine climate with moderate hot summers and very mild winters. However, due to prevalence of sea breeze during summer, the day temperatures are not very high.

There are two types of soils in the region comprising of clayey and salty soils. The clayey soils cover about one half of the area (**Figure 8**). Generally, the saline and clayey soils of the region are barren. Parts of clayey soils are under irrigated cultivation to grow mainly rice and some sugarcane and pulses. Berseem fodder has high potential in the rotation of the crops. Locally, banana and orchards are also located on clayey soils.

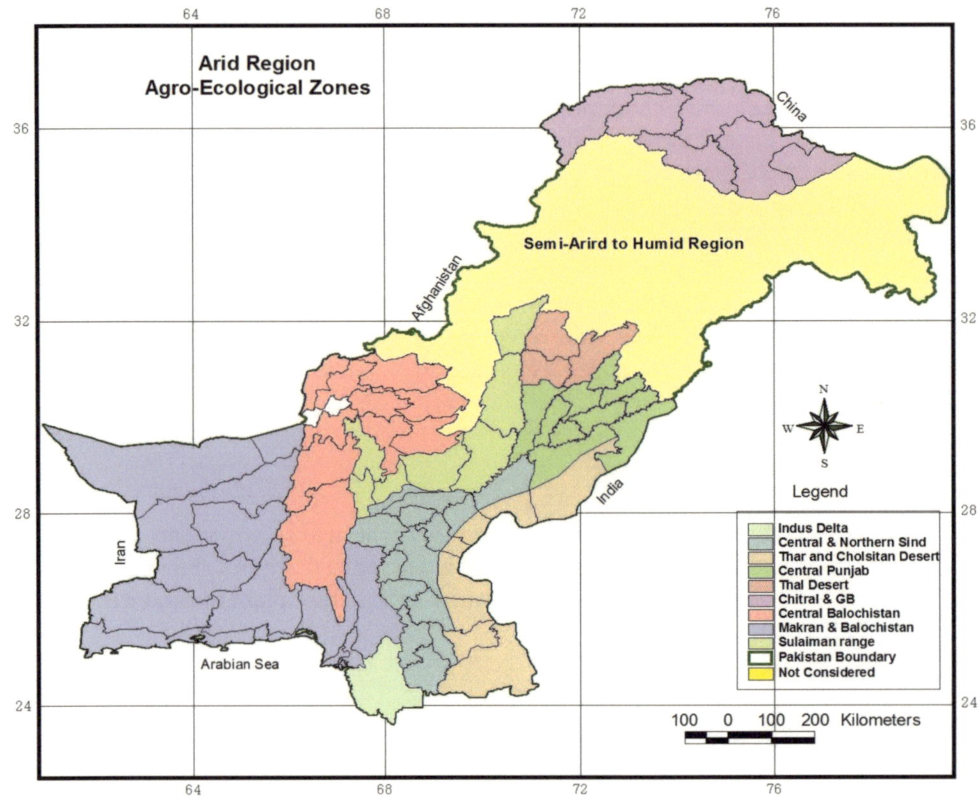

Figure 7.
Agro-ecological zones of Pakistan, GoP [19].

7.2 Central and northern Sind

The region consists of Hyderabad, Badin, Tharparkar, Sanghar, Dadu, Khairpur, Larkana, Nawabshah, Jacobabad, Sukkur, and Shikarpur in Sind, Nasirabad in Baluchistan, and Rahim Yar Khan of Punjab (**Figure** 7).

The climate is arid subtropical continental with hot summer and mild winter. The zone could be subdivided into northern and southern regions. The northern region is extremely hot in summer with mean daily maximum temperature of above 40°C. The precipitation of the region varies from 16 to 20 mm. The minimum temperature in winter remains below 10°C with lowest mean monthly of 2.4°C.

The region is characterized by three main soil types with loam as a common feature. Along the river one meets the silt and sandy loam soils associated with the active floodplain. Outside the range of active floodplain in the upper part of the region, calcareous, loamy, and clayey soils cover vast stretches. In the rest of the region, soil texture is almost the same, but there are saline patches (**Figure 8**).

Most of this area comprises of arable irrigated land, with small patches of unused land under rough grazing. There are distinct cropping patterns that emerged with varying availability of irrigation water.

Canal-irrigated agriculture is the predominant land use of the region. Cotton and wheat, colza (mustard), sugarcane, and berseem fodder are the main crops in the area on the left bank of the Indus. Rice, wheat, gram, and berseem fodder are the main crops in the area on the right bank. *Sorghum bicolor* is the main crop in southern part of Dadu district because of water shortage.

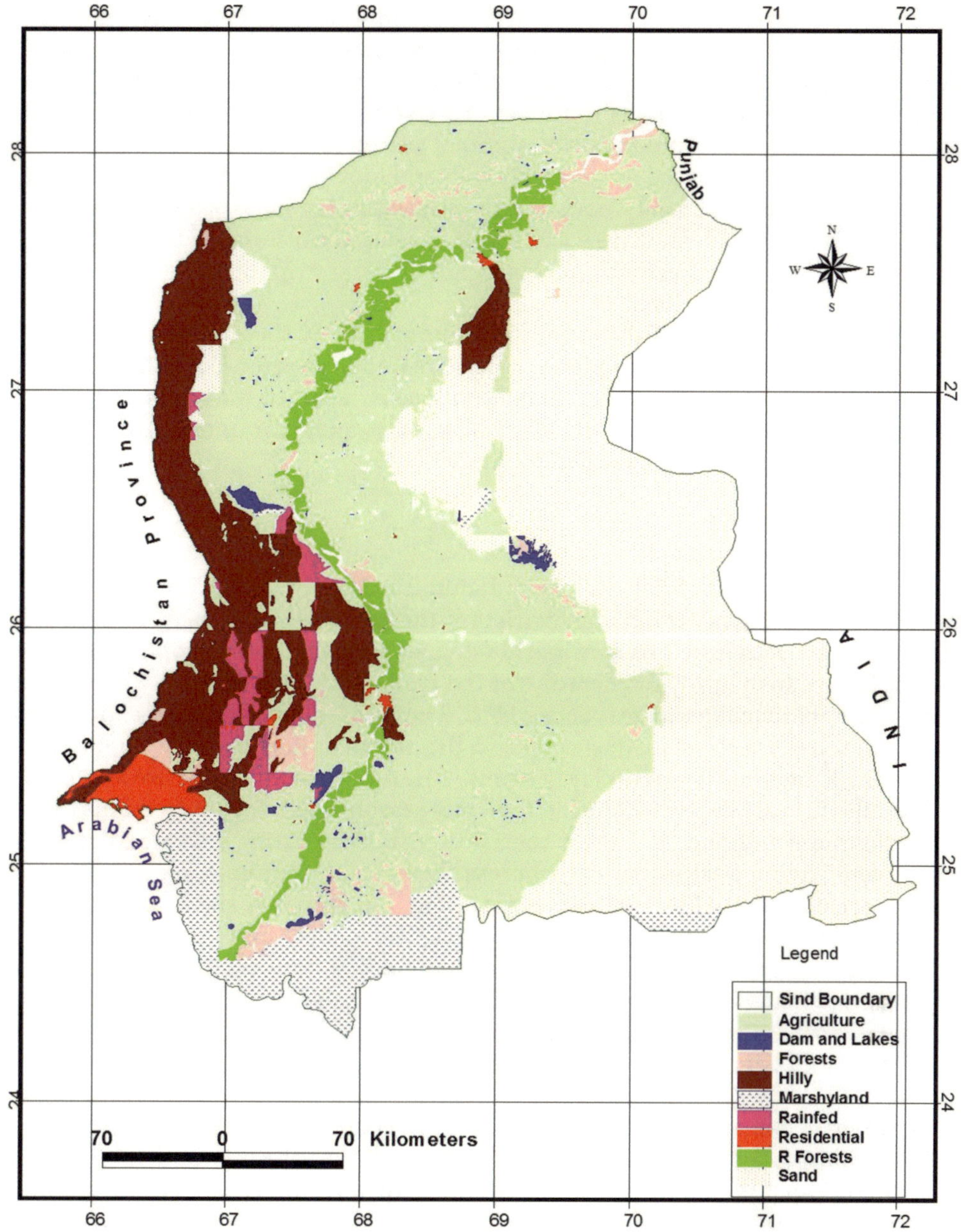

Figure 8.
Sind province land use.

7.3 Thar and Cholistan Desert

Parts of Tharparkar, Khairpur, Nawabshah, and Sanghar in Sind and Rahim Yar Khan and Bahawalpur in Punjab cover this region, which is a part of the Great Indian Desert. The area is higher than the adjoining Indus plain, characterized by elongated sand ridges formed by wind process. The desert is covered by thorny bushes.

The inter-dune areas are devoid of shifting sand. During the rainy season, the runoff from the adjoining dunes is collected in the central part providing enough moisture to support some scanty agriculture in the south eastern Sind. Most part of the desert is rainfed. The depth of the underground water is at several 100 m and difficult to support the fauna, agriculture, and human needs. The climate of the desert is tropical with hot long summers and short warm winters. The climate is desert type with very hot days in summer and cold nights in winter with frost and fogs. Dust storms are common during summer season (**Figure** 7).

The area is characterized by the sandy soils and moving sand dunes. The valleys between the dunes have sandy loam, but these cover a very small proportion of the area. However, in the southern part, sandy loam soil covers considerably large patches of land. The western part of this region has long strips of clayey soils formed in the deposits of Hakra River.

The region comprises of the pastures for the grazing of sheep, goats, camels, and cattle. In the southern part jowar millets are important crops, which are grown in years of favorable precipitation. In the southeastern part of the region, where rainfall is about 300 mm (Tharparkar), wheat is also an important crop on loamy soils and castor on sandy loam soils (**Figure 8**).

7.4 Central Punjab

The region consists of Bahawalnagar, Rahim Yar Khan, Multan, Vehari, Muzaffargarh, Sahiwal, and Jhang districts of the south central Punjab, famous for the agriculture activities. The area is arid (steppe) with subtropical continental climates. The mean monthly temperature of the area remains above 35°C, whereas the winter minimum temperature is above 10°C. The area is characterized by hot summers and warm winters with dust storms in the month of June and July (**Figure** 7).

The soils are sandy loam to clayey loams [10]. Along the rivers, narrow strips of recent alluvium are deposited during the rainy season when the flow remains high. In the southern and central parts of central Punjab, dominating soils are calcareous salty loams having weak structure, whereas clayey soils occur in minor patches.

The area is covered by canal system for irrigation. Both *rabi* and *kharif* crops are cultivated in the region with high productivity. The main crops of the region are cotton, sugarcane, maize, and wheat. The area is also famous for the fruit orchards of citrus and mangoes throughout the arid region (**Figure 9**).

7.5 Chitral and Gilgit-Baltistan province

The region comprises of Gilgit-Baltistan province and part of Chitral district in Khyber Pakhtunkhwa. Enclosed by high mountains, the valleys located in Chitral, Gilgit-Baltistan, are characterized by extreme aridity. However, a large number of glacier-fed streams bring abundant water, which is used for irrigation of terraced agriculture. The terraces formed of fluvioglacial materials are highly fertile. The higher slopes of these valleys and the mountains in the neighborhood of snow line are covered with a narrow belt of pine forests (**Figure** 7).

The climate of the region varies from arid climate to undifferentiated highlands. The tops of high mountains are covered with snow generally for the greater part of the year. The summers are mild and the winters are cold.

"The soils are generally deep, clayey and are formed in colluvial material accumulated on lower parts of mountain slopes and in alluvial deposits in narrow valleys." [10]. Most of the area is used for grazing, and a part is under the scrub forest. Deep soils of valleys and lower parts of the mountains are used for growing maize and wheat under rainfed cultivation. Locally, in favorable condition rice is also grown with irrigation. Fruit orchards are confined to flanks of streams where irrigation water is available (**Figures 10** and **11**).

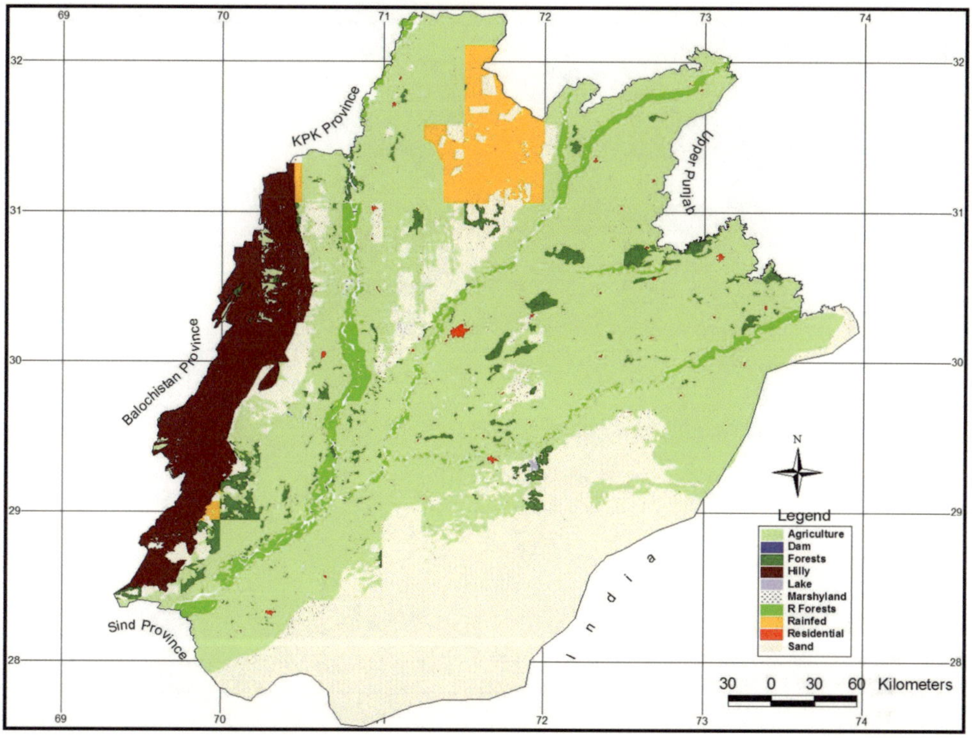

Figure 9.
Lower Punjab land use.

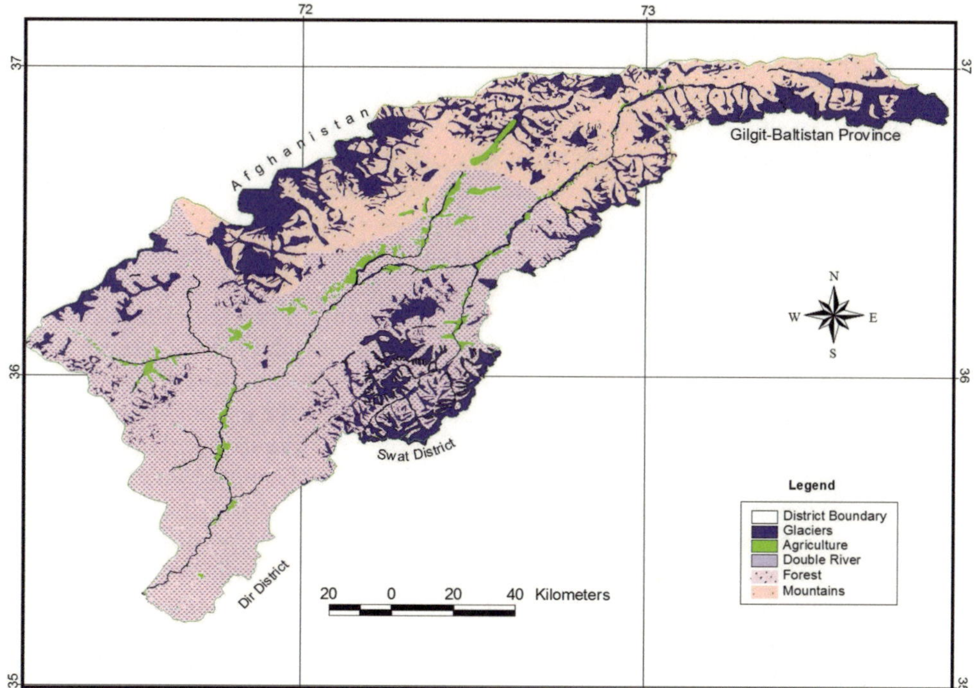

Figure 10.
Chitral district land use.

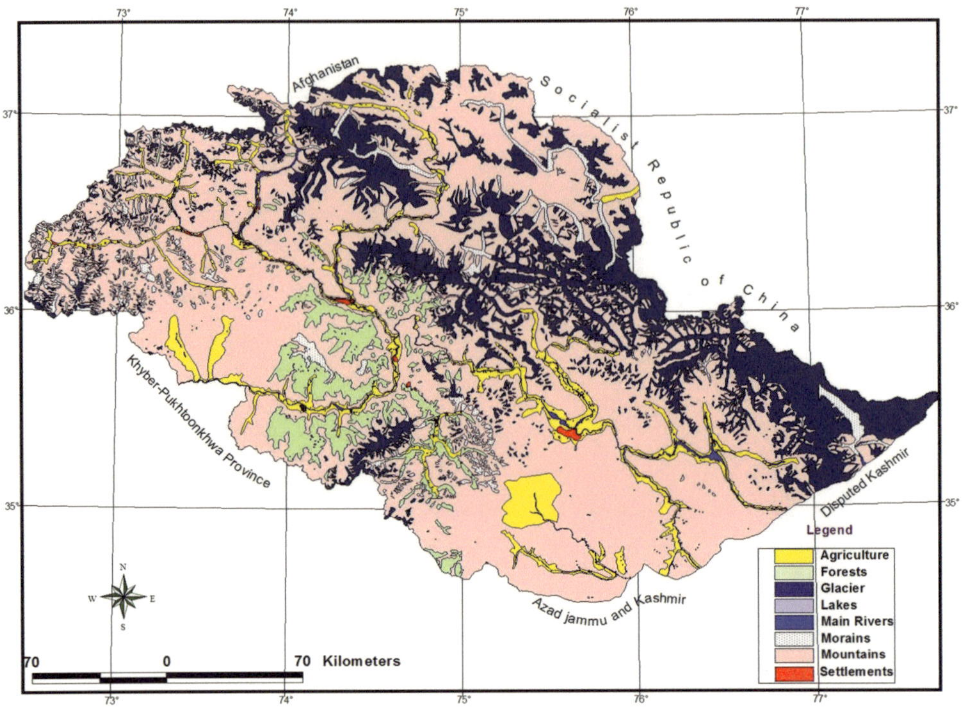

Figure 11.
Gilgit-Baltistan province land use.

7.6 Central Baluchistan

The region consists of Northern and central Baluchistan having barren hills with steep slopes, intervened by valleys filled with Pleistocene deposits and gravels, alluvium, and loessic materials in Kachhi plain.

Due to steep slopes, soil erosion is severe and extensive. Most of the land is almost devoid of vegetation except some xerophyte bushes and poor grasses, which are being worn down by human and animal exploitation. There are numerous hill torrents that carry flashfloods and occasionally cause considerable damage to land and population. The area is mountainous having an arid climate with mild short summers and cold long winters (**Figure** 7).

The area is mountainous with deep valleys. The main soils of the valleys are strongly calcareous but deep and loamy, with weak subangular blocky structure and low organic matter content of 0.3–0.5%. Only at high altitudes, the virgin soils have 1–1.5% organic matter content. The higher parts of valleys have gravelly soils; the lowest parts are occupied by plains containing strongly saline soils and in between are loamy. The mountains have either very shallow soils or bare rocks with soil materials only in crevices (**Figure 12**).

Predominant land use of the area is grazing. Parts of the loamy soils of the valleys are, however, used for growing wheat under a system of spate irrigation by diverting torrent floods into fields having high embankments. A very small proportion of the area is under irrigated fruit orchards, mainly apples, peaches, plums, apricots, and grapes. Also grown under the irrigation are crops of wheat, maize, and alfalfa.

7.7 Makran coast, southwestern Baluchistan, and Sind

This agroecological zone covers Karachi and Dadu districts in Sind and Gwadar, Kharan, Chaghi, and Las Bela districts in Baluchistan. The climate of the region

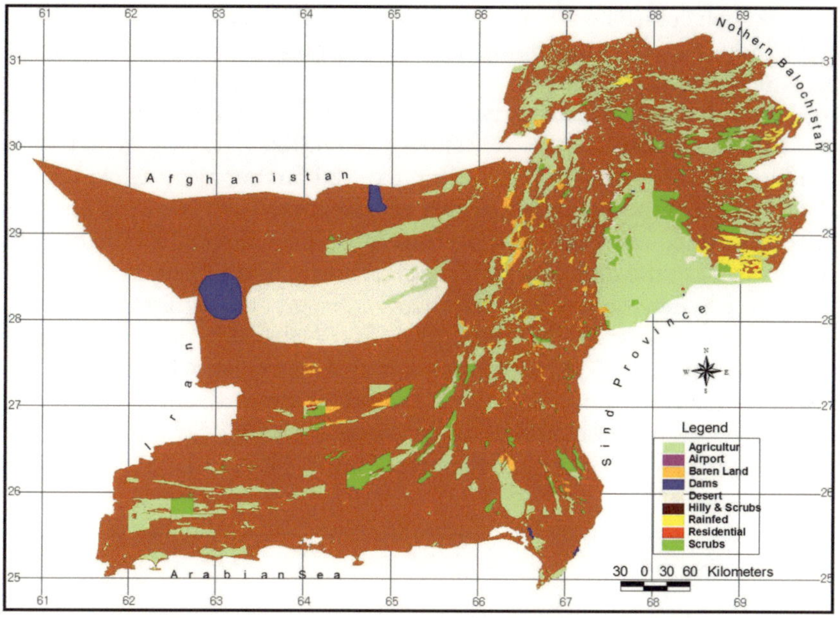

Figure 12.
Baluchistan province land use.

comprises of long hot summers and mild short winters with dust storms in June and July. The winter precipitation is higher than the summer monsoon. The area is characterized by the mountains with intermountain basins, plateaux, and desert. It is the watershed between the Indus drainage system and the rivers that drains directly into the Arabian Sea including Porali, Nall, Hingol, and Dasht Rivers of Baluchistan. In the north and the west, it has inland drainage system (**Figure 7**) with isolated patches of agriculture scattered in the area.

The valleys generally have their floors covered with recent and subrecent deposits of alluvium derived from the adjoining hill ranges. Most of the non-perennial streams from these valleys in the area converge on Hamun-e-Mashkel and conspicuous salt pan in western Kharan Desert.

The soils of the plain area in this region are deep, strongly calcareous silt loams with weak structure. The slopes of the mountains and hills comprise either bare rocks or have very shallow soils. The lower parts of hills and higher parts of the plain have gravelly soils with a strong zone of lime accumulation at 30–40 cm of depth. Vegetation is xerophytes and is characterized by thorny scrubs and poor grasses in the lower regions. On higher altitudes there are forests of junipers and wild olive, which merge with barren lands with scanty bushes and grasses in low-lying areas (**Figure 12**).

The most important land use of the area is grazing. Cultivation on deep valley soils depends mainly on spate irrigation practiced by diverting torrent water into fields, which have high embankments for pending of water. The soil is soaked with 30–50 cm of water to grow the crop. In the north wheat is the main crop but some melons are also grown. In the south *Sorghum bicolor* and millets are the important crops along the coast; castor bean is grown quite extensible. Fruits, vegetable, and wheat are grown wherever water is available from springs or karez.

7.8 Piedmont zone of the Sulaiman range

The region contains Dera Ismail Khan in Khyber Pakhtunkhwa, Dera Ghazi Khan and Rajanpur of the Punjab, and Dera Bugti, Nasirabad, Jhal Magsi, and Bolan districts in Baluchistan. The piedmont plains of the Sulaiman range are sloping toward the Indus River. A large number of alluvial fans have been built by the streams, which slope from the hills on to the piedmont slope. Barkhan, Dera Bugti, Nasirabad, and Jhal Magsi districts lie at the foot of the Sulaiman range to the west.

The piedmont deposits in this area show a generalized grading in texture, with gravel and boulders near the hills and silty material farther away. However, this grading of material is not in evidence everywhere, because it has been obscured by subsequent fine or course materials, producing geohydrological inequalities laterally and vertically in the deposits. Severe soil erosion, heterogeneity of soils, shifting of stream channels, lack of vegetation and meager, and highly variable rainfall have made it into one of the most desolate areas of the arid region. The weather is generally hot and arid with hot long summers and short cool winters.

The soils are loamy on gentle slope near mountains but clayey in level areas. All these soils are strongly calcareous and low in organic matter content. Strong salinity occurs only in narrow strip, which is the junction of piedmont plain and the river floodplain.

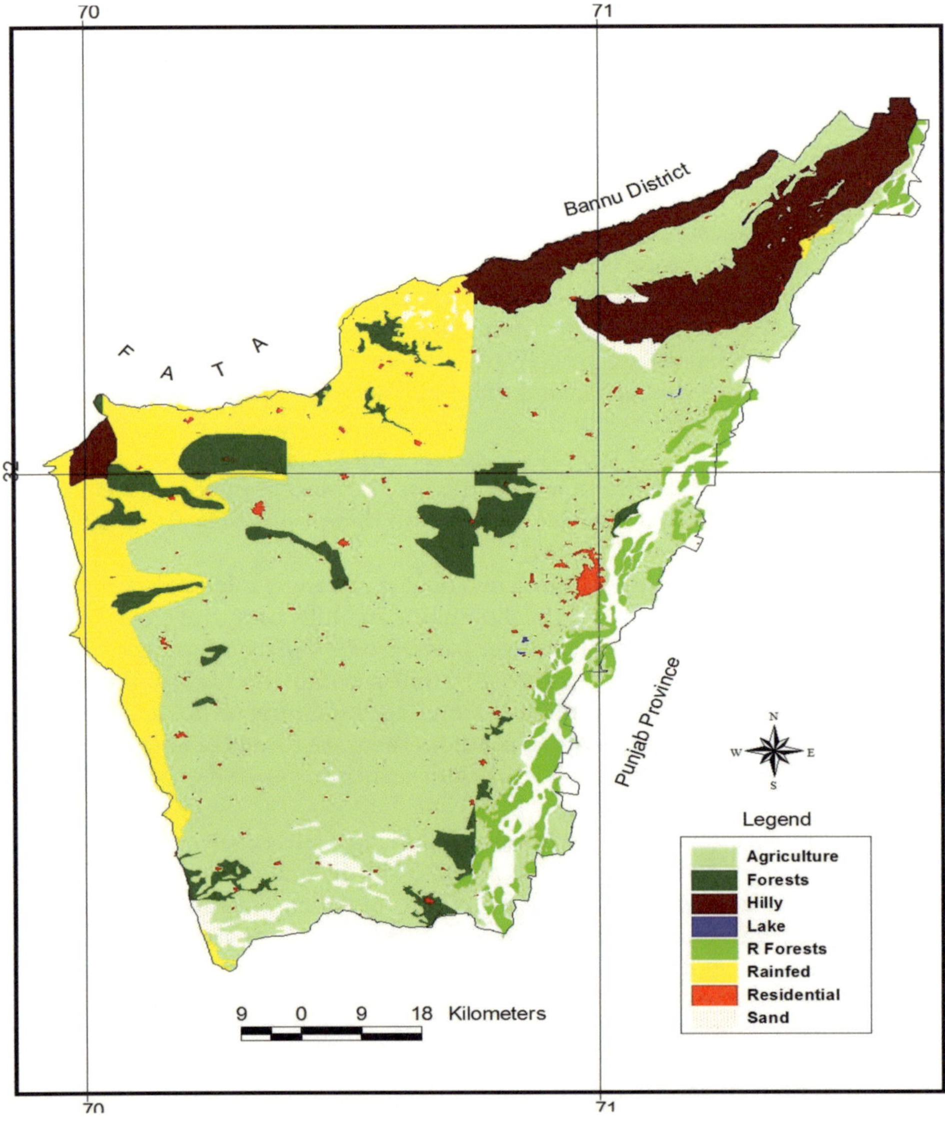

Figure 13.
Dera Ismail Khan land use.

Torrent-watered cultivation is the main land use, under which wheat, *Sorghum bicolor*, millets, and some gram are grown. Some parts of the clayey soils in the central part of the region are under canal irrigation, and *Sorghum bicolor* and colza are the main crops. Rice is grown in a narrow strip forming the junction of the piedmont and river plains. Extensive grazing is carried out in some parts, especially in torrent beds where coarse tall grasses grow (**Figure 13**).

7.9 Thal Desert

The region consists of Layyah, Bhakkar, and Khushab districts in western Punjab. This is a typical desert area with desert landforms including sand ridges, dunes, and sand sheets. However, it places silty and clayey deposits occurring in narrow strips. *The sand ridges are 5–15 m high* [19]. Between the sand ridges, there are depressions that turn into ponds after the rain. In the central parts of the desert, there are large elongated channels showing the palaeochannels the Indus. The desert is quite profusely dotted with vegetation comprising short trees and canals for irrigation. The climate is arid with hot long summers and warm short winters.

"This is an area of the stable sand ridges, which have sand and loamy fine sand soils" [20]. The hollows between the sand ridges have sandy loams and loams, which account for about 10% of the area. However, in the southwestern part of the area, the proportion of loamy soils increases. All the soils are moderately calcareous and have low organic matter content. In addition, there are some narrow strips of silty and clayey soils, which are moderately to strongly calcareous and locally saline. Predominant land use of the area is grazing of livestock especially goats, sheep, camels, and cattle (**Figures 9** and **10**).

A part of the area is used for dry farming of mainly gram and wheat, while cotton, sugarcane sorghum, millets, and wheat are also grown by canal irrigation.

Author details

Saifullah Khan[1]*, Mehmood Ul Hassan[2] and Aslam Khan[2]

1 Institute of Social Sciences, Bahauddin Zakariya University, Multan, Punjab, Pakistan

2 Department of Geography, University of Peshawar, Khyber Pakhtunkhwa, Pakistan

*Address all correspondence to: saifullahkhan33@gmail.com

References

[1] Kureshy KU. Geography of Pakistan. Lahore: National Book Service; 1968-1988. pp. 13-62

[2] GoUSA, Geodast Space Flight Center, Global Change Master Dictionary, The Global Land Cover Facility (GLCF) Data Products and Satellite Imagery, NASA, USA. 2005. Available from: https://gcmd.nasa.gov/records/01-global_land_cov_facility-00.html or ftp://ftp.glcf.umiacs.umd.edu [Accessed: 12 December, 2018]

[3] Khan FK. A Geography of Pakistan. Environment, People Economy. Pakistan: Oxford University Press Karachi; 1991-1993. p. 245

[4] Trewartha GT. An Introduction to Climate. 3rd ed. Mc Grawhill Book Company. Inc.; 1937, 1943, 1954. p. 395

[5] Petterson S. Introduction to Meteorology. New York, London: McGraw-Hills Book Company, Inc.; 1941. p. 260

[6] Blair TA. Climatology: General and Regional. New York: Prentice-Hall, Inc.; 1942. p. 478

[7] Miller AA. Climatology. New York: Mehtewn London and E.P. Dulton and Co. Inc.; 1959. p. 313

[8] Oliver JE. Climatology: Selected Applications. Bungay, Suffolk: Richard Clay Ltd; 1981. p. 260

[9] Anwar MM. Geography Pakistan. Islamabad and Lahore: Book-World; 1993. pp. 20-60

[10] Khan JA. The Climate of Pakistan. Karachi: Rahber Publishers; 1993. p. 79

[11] Moore WG. The Penguin Dictionary of geography. In: Definitions and Explanations of Terms Used in Geography. 6th ed. England: Penguin Book Ltd.; 1991. p. 370

[12] Hasan M, Khan S. Climate classification of Pakistan. In: BALWOIS Conference Proceedings, Ohrid, Macedonia. 2010. p. 42

[13] Trewartha GT. The Earth's Problem is Climates. Baltimore, Maryland: Waverly Press, Inc.; 1961. p. 334

[14] Yim TM. A synoptic aeorology study of the summer monsoon over India and Burma. Journal of Meteorology, Japan. 1949;**vi**:23

[15] GoP. Soil Distribution of Pakistan. Lahore: An Atlas of Soil Survey of Pakistan; 1998. p. 70

[16] Khan FK. Economic Geography. 9th ed. Karachi: Saleem Publishing House; 1989. p. 268

[17] GoP. Population Census of Pakistan. Islamabad: Population Census Organization, Statistical Division, Govt of Pakistan; 1998. p. 300

[18] GoP. Statistical Supplement of the Economic Survey. Chapter-2. Ministry of Finance, Govt of Pakistan, Islamabad. 2008-2009. pp. 21-60. Available from: http://www.finance.gov.pk/s_survey_0708.html

[19] GoP. Agro-Ecological Zones of Pakistan. Islamabad: Pakistan Agriculture Research Council; 1972. p. 70

[20] GoP. Climate Change Vulnerabilities in Agriculture in Pakistan. A Joint Report of the Ministry of Environment and IUCN, Islamabad. 2004. p. 6

Chapter 7

Resiliency of Prairie Agriculture to Climate Change

Surendra Kulshreshtha

Abstract

The Prairie Region, consisting of the provinces of Manitoba, Saskatchewan, and Alberta, is a major agricultural region of Canada. Climate change will have a significant impact on its agriculture and through these changes on related industries in the region, as well as on other parts of Canada and rest of the world. This study is a synthesis of literature on various aspects of climate change—impacts, adaptation, and knowledge gaps for the Prairie Region. These impacts include potential to increase crop production through higher yields, improved livestock productivity, and higher income for producers through higher export sales. Agriculture may also expand to higher latitude areas that are currently not capable of sustaining such an activity. However, the dampening effect on the region would be through higher frequency of extreme events—droughts and floods. In spite of positive effect of climate change on the region, adaptation to climate change by producers will be virtually necessary. Producers are willing to adopt such measures, although there are major knowledge gaps in how climate change would affect and how one can adapt to it. Major uncertainties are carbon fertilization, ability of northern areas to sustain production with shifting ecozones, and impact of new pest and diseases, among others, which may affect the degree of resiliency of the prairie agriculture to climate change. This review concludes that the prairie agriculture is resilient except under prolong and intense extreme events and lack of adaptation to them.

Keywords: climate change, Canadian Prairies, economic impacts, adaptation, extreme events

1. Introduction

In its fifth assessment report [1], the Intergovernmental Panel on Climate Change (IPCC) concluded, "Warming of the climate system is unequivocal, and since the 1950s, many of the observed changes are unprecedented over decades to millennia. The atmosphere and ocean have warmed, the amounts of snow and ice have diminished, sea level has risen, and the concentrations of greenhouse gases have increased." This warming of the climate is commonly referred to as the climate change.

Climate change is a change in the long-term weather conditions of a region and is often used to refer specifically to anthropogenic climate change (also known as global warming). Weather records from across Canada show that every year since 1998—that is 20 years ago now—has been warmer than the twentieth century average [2]. This means that a whole generation of Canadians has never experienced what most of modern history considered a "normal" Canadian climate.

Climate change presents opportunities as well as risks for agricultural production world over and Canada is no exception to it. A warmer climate and a longer growing season could benefit many aspects of Canadian agriculture [3, 4] and provide new opportunities. However, the key here may lie in the manner in which producers adapt to the new climate and take advantage of them. Appropriate adaptations would allow agriculture to minimize the losses by reducing negative impacts and maximize profits through capitalizing on the benefits. Adoption of proper policies may also play a critical role in not only minimizing the losses from climate change but also providing proper incentives for adaptation to it, thereby becoming more resilient to climate change.

1.1 Adaptation and resilience

Under a changing climate, society must make adjustment, called adaptations. If such adaptations are done properly, the industry may become more resilient to climate change. The term resilience has Latin roots, *resilire* (meaning 'to jump back'), and can be defined as a measure of the resistance of systems and of their ability to absorb change and disturbance and still maintain the same relationship between populations or state variables [5]. Hammond et al. [6] have defined it as the amount of change a system can undergo, while retaining similar function and structure. Resilience is closely related to the notions of sustainability but emphasizes unpredictable dynamic environments [7]. Achieving resilience is through improving adaptive capacity, the ability to react effectively to change over time in order to maintain a desirable system state.

1.2 Objectives of the study

This study is based on a review of available literature for the Prairie Region and other similar parts of the world. Its major objective is to review impacts of climate change on agriculture in the Prairie Region in the context of resiliency. Both direct and indirect impacts are included in the review along with adaptation measures that could be adopted to reduce the negative impacts of climate change and to create resiliency.

2. Description of the study region

Agriculture (including crop and livestock enterprises) is an important industry in the Prairie Region, which includes three provinces: Manitoba, Saskatchewan, and Alberta (**Figure 1**). The region is land-locked, located in the western part of Canada. In 2017, over half of the agricultural activities (measured in terms of farm cash receipts) originated in the region (**Figure 2**). In fact, these three provinces could represent the epicenter of Canada's agricultural production, as over half of the Canadian farm income is generated in this region. However, in terms of land base devoted to agricultural pursuits, the region is the most important one within Canada, as over two-fifth (84.4%) of crop and pasturelands is in the region (**Table 1**).

Climate is an important factor that not only determines the geographical limits to agricultural production in a region but also produces year to year variability in crop yields, productivity of livestock enterprises, and through these ultimately in economic returns to producers. Because of climate- and soil-related limitations, not all land area in the region can be devoted to agriculture. For example, in Saskatchewan, of the total land area of 57 million ha, only 26.6 million ha

Figure 1.
Map of Canada showing location of Prairie Provinces. Source: [8].

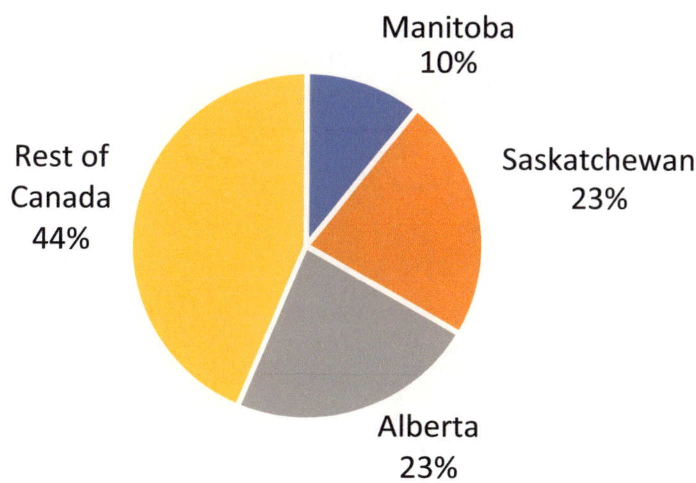

Figure 2.
Share of Canadian farm cash receipts by regions, 2017. Source: Data obtained from [9].

(or 46.6% of the total area) is under farms or agricultural operations. Most of the agricultural area is located in the southern portion of the region. Manitoba has most farms with cattle and field crops, whereas in Saskatchewan, field crops dominate. Alberta is a large producer of cattle and other livestock, besides grain and oilseeds. Climate change could bring forth many negative/positive impacts on the agricultural economy of the Prairie Region. One such factor, which is particularly relevant to the semiarid regions of Canada, is the occurrence of droughts. These events can have a highly negative impact on agriculture and through that on the rest of the economy.

Provinces/ region	Total area in 1000 ha			Percent of Canada		
	Total cropland and pastureland	Cropland including summer fallow	Tame and native pasture-land	Total cropland and pasture-land	Cropland including summer fallow	Tame and native pastureland
Manitoba	6450	4707	1743	11.1	12.2	9.0
Saskatchewan	23,470	16,963	6507	40.4	43.8	33.6
Alberta	19,097	10,484	8613	32.9	27.1	44.5
Prairie Region	49,017	32,154	16,863	84.4	83.1	87.2
Canada	58,027	38,342	19,342	100.0	100.0	100.0

Source: [10].

Table 1.
Regional characteristics of agricultural production in Canada, 2016.

3. Study methods

In order to organize the review of literature, a conceptual framework of direct and indirect changes hypothesized to climate change was developed (**Figure 3**). A change in the climate would first alter various climate attributes. These may include change in average temperature, change in amount of distribution of

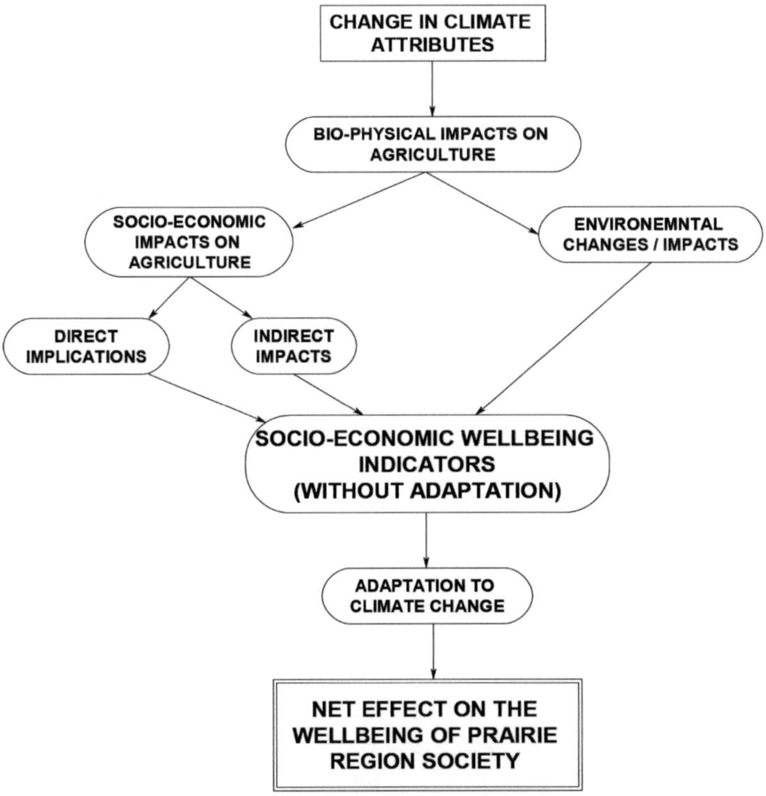

Figure 3.
Conceptual framework for estimating impacts of climate change on society.

precipitation (including the form of precipitation—snow vs. rain), and occurrence of extreme events in a region. These changes would directly affect the Prairie Region, particularly agricultural production. Some of these would include biophysical changes, such as change in yield, introduction of new crops, and suitability of a region for agricultural production, among others. These biophysical changes would translate into socioeconomic changes, first on farms and the agriculture industry, and then through economic linkages to other sectors of the economy, they would affect the entire regional economy. Since the region has a surplus for many crops and livestock products, its economic fortunes also depend upon agricultural commodity exports to other countries. As climate change affects other parts of the world, it may create food security issues in these countries and thus create new export opportunities for the region.

4. Results

4.1 Impacts on climate attributes

Climate change would alter several climate-related attributes that might affect the Prairie Region. These may include [11] (1) change in the temperature, (2) change in the level of precipitation and its form (more precipitation occurring as rain than snow), (3) inter-year variability in precipitation, and (4) change in the frequency of extreme events. In addition, an indirect economic effect on the region could be through sea level rise affecting agricultural production in other countries, affecting their food self-sufficiency.

Although global warming would affect all parts of the world, higher impacts are predicted for the northern hemisphere (which includes the Prairie Region). This region is expected to become warmer during the fall and winter seasons. In the agricultural belt, increases of 2–4°C will be experienced during the growing season [8]. During the September to November period, an increase of 2–3°C will be more common all over southern agricultural belt in Canada by 2041–2070. Increased temperatures would result in an increase in the duration of the growing season for crops and early seeding dates. Precipitation in the agricultural region is not projected to increase much but its variability is expected to increase. By the 2080s, projected precipitation increases may range from 0 to 10% in the far south, which through enhanced evapotranspiration will result in a moisture deficit during the growing season [12]. Associated with higher temperatures, some of the precipitation in the region would come as rain rather than snow. Depending on the saturation status of the soils, some of the precipitation would result in runoff and may create reduced moisture available for the crops. In addition, runoff from agricultural lands could create environmental problems such as eutrophication of water bodies.

Studies [13–16] have shown greater drought frequency and severity in the region mainly due to increasing temperatures. In the future, there would be a higher persistence of multiyear droughts in central and southern portions of Canadian Prairies. Multiyear severe droughts could have an effect on soil moisture and may eventually make some parts of the region not suitable for agriculture. More recently, flooding has also been a major occurrence. This creates problems for farmers not only through soil erosion but also through late seeding or totally no seeding in some areas. In Manitoba, for example, there is evidence to suggest that recent agricultural losses from flooding increased by over 300% since the historical period (1966–2015) [15].

4.2 Biophysical impact of climate change attributes

4.2.1 Impact on water resources

Water resources in the Prairie Region would be affected through a culmination of changes in key variables governing the hydrological cycle: temperature, evapotranspiration, precipitation, and snow and ice. Since water in most prairie rivers is provided, in part, by melting of glaciers located in the Rocky Mountains (supplemented with seasonal rainfall), higher temperature coupled with evapotranspiration will play an important effect on the water cycle. According to [17], studies based on hydrological models suggest that annual streamflow of the South Saskatchewan River may range from an 8% increase to a 22% decrease, with an 8.5% decrease being an average prediction. In addition, this study suggested that there is not a dramatic drying of the prairies to be anticipated under climate change and that in some cases streamflow will increase for certain climate scenarios and under moderate degrees of climate change.

Another factor that may affect water resources is the form of precipitation. It is expected that due to warmer winters, much of the current winter precipitation, which now falls as snow, could be in the form of rainfall. This has implication for soil moisture for crop and thus for crop growth. Winter warming will reduce snow accumulations in alpine areas and across the prairies. This will cause a decline in annual stream flow and a notable shift in stream flow timing to earlier in the year, resulting in lower late season water supplies. Continued glacier retreat will exacerbate water shortages already apparent in many areas of Alberta and Saskatchewan during drought years. Lower stream flow would have an impact on the output of hydroelectric plants, as well on the capacity of the irrigation reservoirs to provide water to producers.

Groundwater is the source of potable water for about 30% of Manitoba residents, 22% of Albertans, and 43% of Saskatchewan's population [18]. Future groundwater supplies will be affected in a similar manner as changes in surface water flows and frequency of extreme events. Increased rainfall in early spring and late fall will enhance recharge if soil water levels are high; otherwise, water will be retained in the soil, benefiting ecosystem and crop productivity. Drier soils due to higher rates of evapotranspiration result in decreased recharge, which would lead to a slow but steady decline in the water table in many regions.

Climate change may also have an impact on prairie wetlands. Many of these may be reduced in size or even totally dry up, which may have some impact on groundwater quality as well as recharge capacity.

Some studies, such as [19], have reported that the excess heat caused by climate change may influence the effectiveness of inputs for crops; for example, fertilizer productivity may be reduced. This may lead to reduced fertilizer use, which may be good news for water quality damages, such as nitrates in groundwater, eutrophication, among others, but would be bad news for crop yields.

4.2.2 Effect of climate change on soil moisture

Under climate change, soils are projected to be slightly moister and warmer in winter, but large increases in soil temperature and large decreases in soil water content are projected during the growing season. During the warmer summer conditions, the level of evapotranspiration will increase along with demand from plants. Both of these can result in a drought stress, particularly when soil moisture level is low [20]. Depending on the frequency and intensity of droughts, soil moisture is expected to decrease and the semiarid areas of the southwestern Saskatchewan and southeastern Alberta (an area known as Palliser Triangle) will increase.

4.2.3 Other environmental effects related to climate change

Extreme dry condition over a prolonged period would have other environmental effects. In fact, the implication of more frequent and widespread droughts is that they place a larger area at risk of desertification more often unless counteracting management strategies are enacted [21]. Development of the sand dunes in southwestern Saskatchewan has been suggested to be an effect of prolonged droughts in the region (see more details in [22]). This would result in a loss of agricultural land for meeting growing demands for humans and animals.

4.3 Direct impacts of climate change on agriculture

4.3.1 Biophysical impacts on crop production

Due to enhanced evapotranspiration, driven by higher temperatures, many regions will experience a moisture deficit despite higher (but variable) amounts of precipitation. Water stress during critical times for plants (e.g., flowering) is especially harmful and would affect plant growth and productivity. Yield responses are sensitive to climatic change and location and tend to reduce the beneficial effects stemming from elevated carbon dioxide (CO_2) levels [23]. Details of changes in yields of selected crops in Canada (applicable to the Prairie Region) are presented in **Table 2**.

Estimates of future yields under climate change have been variable depending on the climate model used, location, and global warming levels. Another uncertainty is created by the positive effect of higher level of CO_2 in the future. A study by [23] has reported an increase in the yields of canola and wheat with increase in global warming, while maize (corn) yield was simulated to increase or slightly decrease depending on the characteristics of the currently grown cultivar and differences among the crop models. The same study also indicated that future warming accompanied by increased CO_2 concentration would remain beneficial to crop yields at the global warming level of 2.0°C for Canada.

Agriculture in the Prairie Region could benefit from warmer and longer growing seasons and a warmer winter. Climate change may also bring opportunities, which could increase productivity and allow cultivation of new and potentially more profitable crops and tree species. A study [27] has indicated that under the projected changes in climate, area allocated to wheat will continue to decrease into the future by 2.7–4.6% in various soil zones, while the area left to summer fallow is projected to increase. The choice of wheat is preferred over pulses, feed, and forages, while the choice of specialty oilseeds (flaxseed, mustard seed, and canary seed) is projected to become preferred over wheat in the future. Change in crop mix has been suggested [28], but it still will be dominated by wheat, barley, and canola. Producers under the changed climate may also introduce some new crops, such as:

Pulse crops: Pulse area is likely to increase in the drier, more arid growing environments that are expected in 2050.

Soybeans: The transition to larger soybean area in the prairies is already underway with former marginal areas in southern Manitoba and Saskatchewan now growing the crop in a regular rotation.

Corn: The movement of corn is also underway to parts of southern Manitoba and southern Alberta, but the transition is expected to take a longer time than

Crop	Model*	Global warming level in °C	Yield increase percentage of 2006–2015 mean	
			With CO_2 fertigation	Without CO_2 fertigation
Canola	DayCent	1.5	1.5	−2.8
		3.0	2.4	−13.4
	DSSAT	1.5	6.1	−1.0
		3.0	7.7	−15.5
Wheat	DayCent	1.5	1.9	−2.3
		3.0	4.6	−11.3
	DSSAT	1.5	6.1	2.1
		3.0	22.8	1.8
Maize (Corn)	DayCent	1.5	−0.7	−1.5
		3.0	−5.6	−9.0
	DSSAT	1.5	3.0	1.2
		3.0	−4.8	−11.8

DSSAT is the Decision Support System for Agrotechnology Transfer. For details, see [24].
Source: Adapted from [25].
**DayCent is a daily time series biogeochemical model used in agroecosystems to simulate fluxes of carbon and nitrogen between the atmosphere, vegetation, and soil and is a daily version of the CENTURY biogeochemical model. For details, see [26].*

Table 2.
Projected future changes in crop yield under rain-fed conditions with and without carbon dioxide fertigation.

soybeans. Corn will also be limited by the dryness in parts of the southern prairies. Corn has large moisture requirements to produce economically attractive yields.

Sorghum and millet: Sorghum and millet are two possible crops to move into the drier areas of the prairies in 2050. These crops represent a possible feed grain for the driest areas, but sensitivity to frost will limit area even with increased growing season [25].

Negative impacts may result from changes in the timing of precipitation, increased risk of droughts and associated pests, and excessive moisture. Because of these positive and negative factors, crop yields will vary from region to region.

4.3.2 Livestock production

Temperature is considered the most important bioclimatic factor for livestock [29]. Warmer temperatures may create benefits and challenges to livestock operation in the Prairie Region [30]. The major benefit would be through higher winter temperatures, which would lower feed requirements, increase survival of the young, and reduce energy costs [31]. Hot environment impairs production (growth, meat and milk yield and quality, egg yield, weight, and quality) and reproductive performance, metabolic and health status, and immune response [32]. However, these changes would also produce challenges for livestock producers in the form of new pests and diseases for animals, and alien species for grasslands and pastures. The warming climate would bring challenges during summer months, when heat waves can kill animals, particularly chickens [33]. Heat stress can also affect the productivity of dairy animals, the meat quality for beef animals, and reproduction (particularly for dairy animals). In some areas, the process of desertification may reduce the carrying capacity of rangelands (as suggested by [34]) and the buffering ability of agro-pastoral and pastoral systems.

4.3.3 Socio-economic effect of extreme events on agriculture

Results from studies on the economic impacts of climate change on prairie agriculture are highly variable from region to region and from study to study. Some studies suggest that overall economic consequences will be negative and small, whereas others indicate positive and large impacts.

Socio-economic impacts of climate change-induced changes have been estimated to be positive. A Manitoba study [35] estimated that changes in crop revenues under current economic/technological conditions will range from a 7% loss in Alberta under one scenario to an 8% increase in Saskatchewan under a slightly different scenario. Manitoba, the least water-deficient province, has been projected to benefit from warming as producers shift to higher value crops, resulting in an increased gross margin of more than 50%. Many studies have predicted increased land values in the Prairie Region under climate change. Similarly, for Saskatchewan, a study [36] reported that climate change is beneficial for Canadian Prairie agriculture except for some southeast regions of Alberta. Comparing the results from direct impacts of climate and price changes on land value with the results from indirect impacts through area response estimation reveals that direct impacts would increase land values by 31%. Ayouqi and Vercammen [37] have suggested similar results, who applied three different climate change scenarios. They reported that except for the north part of Saskatchewan and the west part of Alberta in the medium climate change scenario, all other cases show increase in the farmland value. In fact, the farmlands of Canadian Prairies were estimated to gain a value between $1.14 and $4.1 billion annually (based on the estimation model and scenario).

Extreme events can have devastating impacts on crop yields and through these on the rest of the regional and national economy. The Prairie Region (in fact other parts of Canada as well) experienced a major back-to-back drought during 2001 and 2002. Estimated impacts of these droughts [34] indicate that crop yields were as little as half of average yields during normal or more suitable growing conditions. Repercussions of these droughts were severe and far-reaching, including: (1) Agricultural production levels, through crop production losses, were devastating for a wide variety of crops across the Prairie Region, particularly in 2001. Total value of production dropped an estimated $3 billion for the 2001 and 2002 drought years, with the largest loss in 2002 at more than $2.2 billion. (2) The Gross Domestic Product fell some $4.5 billion for 2001 and 2002, again with the larger loss in 2002 at more than $3.1 billion. (3) Employment losses exceeded 27,883 jobs, including nearly 17,803 jobs in 2002. (4) Net farm income was negative or zero for several provinces for the first time in 25 years. (5) Livestock production was especially difficult due to the widespread scarcity of feed and water. Many producers culled their herd after the first year of the drought. (6) Water supplies that were previously reliable were negatively affected, and several failed to meet the requirements. (7) Multisector effects were associated with the 2001–2002 drought, unlike many previous droughts that affected single to relatively few sectors. Impacts were felt in areas as wide-ranging as agricultural production and processing, water supplies, recreation, tourism, health, hydroelectric production, transportation, and forestry. (8) Long-lasting impacts included soil and other damage by wind erosion, deterioration of grasslands, and herd reductions. (9) Several government response and safety net programs partially offset negative socio-economic impacts of the 2001 and 2002 drought years [34]. In the future, drought frequency has been predicted to

Overall impacts	Major impact type and/or limitations
Opportunities for agriculture may result from continued expansion of the growing season, increased heat units, and milder shorter winters.	Agriculture could benefit from several aspects of warming climate, depending on the rate and amount of climate change and ability to adapt.
Adoption of new crops would change the nature of diversification of the region.	Results of assessments are wide ranging and depend upon the climate scenarios, impact model used, scale of application, assumptions made, and how adaptation is incorporated.
Regions would become more vulnerable to droughts and floods.	Region is susceptible to droughts and floods, and their frequency is expected to increase under climate change.

Source: Adapted from [11].

Table 3.
Sensitivity of prairie agriculture to climate change.

increase under climate change; as projected, yields of various crops would decrease, thereby increasing the vulnerability of producers, particularly in semiarid regions of the prairies.

In addition to droughts, flooding of agricultural lands would also increase under climate change. Already, such events have occurred in various parts of the Prairie Region, affecting crop production. In Manitoba, for example, 2016/2017, the excess moisture losses accounted for 71% of total crop losses estimated at $198.7 million [15]. Similar estimates for other two provinces are not available.

Although overall studies have been positive about impact of climate change on prairie agriculture, increased frequency of extreme events may dampen these beneficial impacts somewhat. A summary of these impacts is shown in **Table 3**. The region would have new opportunities open under the changed climate and some new crops may help the region in diversifying the regional agriculture mix. However, increased frequency of droughts and floods may make some regions more vulnerable in the future.

5. Indirect impacts of climate change on prairie agriculture

Economics of crop production in Canada will be a joint outcome of changes within Canada and those outside the Canadian boundaries (in both the exporting and importing countries). International markets will play an important role in determining the economic impact of climate change on the Prairie Region agriculture. As most of the crops are sold in the international market place, their prices would be significantly influenced by conditions not only within Canada but also in the rest of the world. Coupled with increased export levels, these changes could have a profound impact on the economic situation in the region. Climate change would affect other parts of the world differently as shown in **Table 4**. Different areas of the world would be affected differently depending on the degree of global warming. For example, a very mild warming (say 1°C) would bring moderate increase in cereal crop yields, but a major warming (5°C or higher) would have catastrophic effects.

A further review of the changes, based on [38], suggests:

i. For US agriculture, most crops would show gains in crop yields to certain thresholds of temperature increase, which may increase their potential for exports. This is different from the impacts on crop yields in Australia, where a reduction

Temperature rise (°C)	Impact on water resources	Impact on food
1	Small glaciers in the Andes disappear completely, threatening water supplies to people.	Moderate increases in cereal yields in temperate regions.
2	Potentially some 20–30% decrease in water availability in some vulnerable regions (e.g., Southern Africa and Mediterranean).	Sharp decrease in yields in tropical regions (5–10% in Africa).
3	In Southern Europe, serious drought occurs once in every 10 years; 1–4 billion more people suffer water shortages, while additional 1–5 billion gain water, which may increase flood risk.	Additional 150–500 million people at risk of hunger if carbon fertilization is weak. Agricultural yields in higher latitudes likely to peak.
4	Potentially some 30–50% decrease in water availability in Southern Africa and Mediterranean.	Agricultural yields decline by 15–35% in Africa and entire regions out of parts of Australia.
5	Possible disappearance of large glaciers in Himalayas, affecting one-quarter of China's population and hundreds of millions in India.	
More than 5	Catastrophic effects. Hard to capture by present-day models.	

Source: [38].

Table 4.
Highlights of possible climate change impacts on water and food in the world.

is predicted. This would somewhat reduce Canada's competition in some cereal export market, notably wheat. European countries are also expected to have a decrease in production, although results may vary from country to country. Exact details on the Argentinean situation were not found in the literature.

ii. On the developing countries scene, studies suggest a decrease in the potential production. Given that demand in many of these countries, caused primarily by population growth, would most likely increase, potential for Canada to export would likely exist.

iii. Canada is a major exporter of grains and oilseeds (and of pulses recently). The impact on Canadian agriculture, therefore, would also result from changes in the crop production and their demand in the rest of the world under future climate.

6. Adaptation options under climate change

Prairie producers are highly adaptable to changing conditions. However, such adaptations are sometimes complex and often costly. Under climate change, there may exist some beneficial changes (such as higher level of pulse production, increase in the area under cultivation, and its productivity), but if the rates of these change are faster than producers have experienced, they may pose more difficulties for adaptation. For example, production of new crops may become ergonomically feasible, but whether producers would be able to adapt sufficiently or quickly enough to these new realities is somewhat uncertain. Financial requirements for making such adaptations may be a major constraint.

A synthesis of research on adaptation options for agriculture has identified four main categories [39]: (i) technological developments, (ii) government programs and insurance, (iii) farm production practices, and (iv) farm financial management. In addition to these 'direct adaptations,' there are options, particularly information provision that may stimulate adaptation initiatives. Most adaptation options are modifications to on-going farm practices and public policy decision-making processes with respect to a suite of changing climatic (including variability and extremes) and nonclimatic conditions (political, economic, and social) [39]. Migration by producers has also been seen as an adaptive measure for extreme events. After the prolong droughts of 1930s, many prairie producers in southwest Saskatchewan left for better agricultural regions. Such behavior is consistent with the concepts of vulnerability, exposure to risk, and adaptive capacity, as developed in the climate change research community [40].

Alberta Agriculture [41] suggests that if conditions are drier than usual, producers might (i) expect concerns with lower than usual seed germination and/or plant growth, lack of feed, and shortage of water; (ii) conduct small area tests on new crop types that take advantage of drought tolerance; and (iii) put animals on a rotational grazing program to allow grazing land to rest and recover. The Saskatchewan [42] Strategy for climate change specifically commits to develop and implement an offset system that creates additional value for actions that result in carbon sequestration or reduced emissions, especially from soils and forests.

Other systems, such as mixed systems and industrial or landless livestock systems, could encounter several risk factors mainly due to the variability of grain and pasture availability and cost, and low adaptability of animal genotypes. Regarding livestock systems, optimizing productivity of crops and forage (mainly improving water and soil management) and improving the ability of animals to cope with environmental stress by management and selection could be the best adaptive measures. For an Alberta mixed farm, reported by [43], beef herd adaptation strategy affected farm profits more and costs less. Maintaining the herd size and with regular feeding plan, and purchasing extra fees provided the best adaptation. To guide the evolution of livestock production systems under the increase of temperature and extreme events, better information is needed regarding biophysical and social vulnerability, and this must be integrated with agriculture and livestock components.

Shelterbelts also provide several environmental benefits to the society under climate change. One of the most important ecosystem values from shelterbelts is their capacity to sequester carbon (C) [44]. Six major species of tree and brush used for this purpose were examined in this study. At maturity, these trees can sequester a large amount of carbon. For example, a hybrid poplar can sequester 3–5 t of C per year, whereas a Caragana shrub can store only 1.3–2.7 t of C per year (**Table 5**). However, the amount of C stored in the younger trees would be lower but eventually the amount would reach the levels shown in **Table 5**. In addition, shelterbelts also sequester carbon in the soil and in the understory.

If one uses the 2022 price (as fixed by the Government of Canada) of $50 per t of C, a hybrid poplar tree is worth $150–250 to the society. Unfortunately, this value is not internalized in the decision of the landowner since at present they do not receive any compensation for the stored or sequestered carbon. Under these conditions, decision to maintain shelterbelts on farms may be discouraged since they do not enter into their economic decision-making. Policies have to be developed to compensate the producer for the loss of this revenue [45].

Irrigation is one of the best adaptive measures under drought conditions [46]. In addition to stabilize production over the drought period, they increase returns from crop production and help mixed farms maintain livestock activities through forage production [47]. In addition, the level of nonagricultural sectors is also higher due to availability of water and its backward and forward linkages [48].

Type of shelterbelt	Name of the plant	Amount of carbon sequestered in mega grams per year per ha	
		Low value	High value
Tree	Hybrid poplar	3.29	5.18
Tree	Scots pine	1.44	3.26
Tree	White spruce	2.24	4.13
Tree	Green ash	2.02	3.92
Tree	Manitoba maple	2.80	5.26
Shrub	Caragana	1.31	2.67

Source: [44].

Table 5.
Level of carbon sequestered by shelterbelts by type.

There are many potential adaptation options available for marginal change of existing agricultural systems, often variations of existing climate risk management [49], but their implementation is likely to have substantial benefits under moderate climate change for some cropping systems, and there are limits to their effectiveness under more severe climate changes.

7. Knowledge gaps and areas for future research

Estimates of climate change impacts on crop production are wide ranging. Of course, different studies use different assumptions about the nature of key climate variables, along with assumptions of crop type, seeding dates, fertilization, and irrigation. Many other factors, such as insects, diseases, and weeds, would also change because of climate change. Our knowledge of these changes is very weak. Furthermore, much of the Canadian research has concentrated on cereals and to a limited extent on oilseeds (mainly canola). Research on other oilseeds, forages, fruits, and vegetables (including potatoes) has been less extensive. This is a serious gap in our knowledge for climate change impacts on crop production.

One of the changes associated with climate change is the level of atmospheric carbon dioxide. Under these conditions, a situation of carbon fertilization may occur, which may increase some crop yields (particularly for plants using the C3 carbon-fixation pathway, like wheat or canola). Higher levels of atmospheric CO_2 may also improve water-use efficiency. However, the picture is complex, since weeds may also be more vigorous under a carbon-enriched atmosphere. Warmer and longer growing seasons could be positive for crop growth and yield. However, very little research has been done in the context of the Prairie Region.

Livestock production would be affected by availability of forages and direct effect through feed efficiency. Grassland production is limited by moisture supply. Although a drier climate would suggest declining production and grazing capacity, actual changes in grassland production are likely to be modest, given a longer growing season, reduced competition from shrubs and trees, and increases in warm-season grasses that have higher water-use efficiency [50].

Northward shift of climate congenial to agricultural production has been predicted for future. However, for the Prairie Region, the exact nature of this change is not known. Furthermore, to what extent this change would translate into profitable agricultural production has yet to be researched.

Most past studies have taken into account effect of a change in temperature and total precipitation. No study was found that had included the effects of changes in the distribution of precipitation or in the form of precipitation (snow vs. rain). Other changes induced by climate change, such as pollination, heat stress days, and asymmetry in day and night temperatures, were also not included in these studies.

A study considering all of these factors together in an integrated framework is needed. This study must cover all regions of Canada and employ comparable set of assumptions with respect to climate change and related factors. There is a need for an integrated bio-physical-socio-economic assessment of climate change impacts for all regions of the prairies. Scope of the investigation needs to be national, regional, and international in nature since all these factors would shape the nature of impacts under the changed climate and the adaptation measures for prairie agriculture.

On the international aspect of climate change impacts, very few studies (since the 2005 study by [51]) were found that have reported implications of these changes in production and demand in various parts of the world and/or on trade flows. There hasn't been an assessment of impacts of these changes for the exports of agricultural products from Canada. There is a need to examine the potential for Canadian exports and imports of various crops under the changed demand and supply conditions around the world induced by climate change.

For purposes of implementing adaptations to climate change in agriculture, there is a need to understand the relationship between potential adaptation options and existing farm-level and government decision-making processes and risk management frameworks better. The relationships that determine technology development are somewhat unclear and more research is needed to understand these factors. Historically, federal and provincial governments have responded to drought with safety net programs to offset negative socioeconomic impacts [52] and, more recently, through development of drought management plans. More intense and longer droughts will be expensive challenges to safety net programs. Soil conservation is a prime example of a 'no regrets' strategy, since preventing soil loss is beneficial whether or not impacts of climate change occur exactly as projected. The Permanent Cover Program, administered by the Government of Canada, has reduced sensitivity to climate over a large area. The move in recent decades to more efficient irrigation techniques has dramatically increased on-farm irrigation efficiencies. However, the continued loss of water from irrigation reservoirs and open channel delivery systems due to evaporation, leakage, and other factors indicates the need for further improvement in the management of limited water resources. More research is needed on appropriated adaptation measures under the frequent extreme event occurrence.

8. Conclusion: resiliency of agriculture in the Prairie Region

As a high-latitude country, warming is more pronounced in the Prairie Region, which would result in longer frost-free seasons, higher degree-days, but more frequent extreme events (droughts and floods). Alberta could benefit the most from increased summer and winter precipitation, while Saskatchewan and Manitoba would experience little change or small increases. Warmer temperatures could also mean lower energy costs for farmers, as well as benefit livestock production in the form of lower feed requirements and increased survival rates of the young animals.

Prairie farmers are highly adaptive. Past changes, such as introduction of zero tillage, reduction in summer fallow area, and introduction of new cultural practices, are all indicative of this. If climate change were slow, producers would be able to adjust through proper adaptation measures. However, government policies have to be developed to assist producers in making these choices. The biggest threat to resiliency would be through occurrence of intense and longer drought periods.

Climate change could improve soil quality by enhancing carbon sequestration through "carbon agriculture," which includes no-till farming (where you grow things year-to-year without disturbing the soil), cover crops (which help spur microbial activity in the soil), and a grazing technique, such as rotational grazing. Changes to land-use through annual crop production, perennial crops, and grazing lands could all contribute to reducing greenhouse gas emissions.

Warmer temperatures also mean warmer summers, which could be problematic for livestock producers who have to deal with heat-wave deaths. Reduced milk production and reproduction are other impacts in the dairy industry, as well as reduced weight gain among beef cattle. Higher levels of atmospheric carbon dioxide (CO_2) and an increase in the use of pesticides and pathogens in livestock and crops can lead to increased weed growth. Under climate change, farmers may have a better choice of crops, which may lead to diversification, and thus assist in making agriculture resilient to climate change [53]. Cultural practices under the purview of climate-smart agriculture could also transform and orient agricultural systems to resilience under the new realities of climate change [54].

While there is a lot of uncertainty surrounding the future of Canada's agriculture industry, one thing is clear: we are likely to see more extreme weather events and higher average temperatures. Farmers must look at environmentally friendly farming practices to adapt to the effects of climate change and stay in business, thereby creating a resilient agriculture industry.

Author details

Surendra Kulshreshtha
University of Saskatchewan, Saskatoon, Canada

*Address all correspondence to: suren.kulshreshtha@usask.ca

References

[1] Intergovernmental Panel on Climate Change (IPCC). In: Stocker TF, Qin D, Plattner GK, Tignor M, Allen SK, Boschung J, et al., editors. Climate Change 2013: The Physical Science Basis. Contribution of Working Group I to the Fifth Assessment Report of the Intergovernmental Panel on Climate Change. Cambridge, United Kingdom and New York, NY, USA: Cambridge University Press; 2013. pp. 1535

[2] Prairie Climate Center. 2018. Climate Atlas of Canada. Winnipeg

[3] Warren F. Climate change impacts and adaptation: A Canadian perspective. In: Lemmen DS, Warren FJ, editors. Prepared by the Climate Change Impacts and Adaptation Directorate. Ottawa: Natural Resources Canada; 2004

[4] Easterling W, Aggarwal P, Batima P, Brander K, Erda L, Howden S, et al. Food, fiber and forest products. In: Parry ML, Canziani OF, Palutikof JP, van der Linden PJ, Hanson CE, editors. Climate Change 2007: Impacts, Adaptation and Vulnerability. Contribution of Working Group II to the Fourth Assessment Report of the Intergovernmental Panel on Climate Change. Cambridge, UK: Cambridge University Press; 2007. pp. 273-313

[5] Doring T, Viewger A, Pautasso M, Vaarst M, Finckh M, Wolfe M. Resilience as a universal criterion of health. Journal of Science and Food Agriculture. 2013;**95**(3):455-465

[6] Hammond B, Beradi G, Green R. Resilience in agriculture: Small and medium-sized farms in Northwest Washington state. Agroecology and Sustainable Food Systems. 2013;**37**(3):316-339

[7] Berardi G, Green R, Hammond B. Stability, sustainability, and catastrophe: Applying residence thinking to U.S. agriculture. Human Ecology Review. 2011;**18**(2):115-125

[8] Patel S, Paulsen C, Heffernan C, Saunders D, Sharma M, King M. Tuberculosis transmission in the indigenous peoples of the Canadian prairies. PLoS ONE. 2016;**12**(11):1-16

[9] Statistics Canada. Farm cash receipts. Annual, Canada, Province and Territory. Table: 32-10-0045-01 (formerly CANSIM 002-0001). Ottawa; 2018

[10] Statistics Canada. Land use. Table no. 32-10-0406-01 (Formerly CANSIM Table 001-0203). Ottawa; 2019

[11] Kulshreshtha S, Wheaton E, Amiraslany A, Hill H, Chipanshi A, Howard A. A literature survey of potential socio-economic impacts of climate change on Canadian crop production. A Report Prepared for Agri-Environment Services Branch, Agriculture and Agri-Food Canada, Regina. Saskatoon: Department of Bioresource Policy Business and Economics, University of Saskatchewan; 2010

[12] Lemmen D, Warren F, Lacroix J, Bush E. From Impacts to Adaptation: Canada in a Changing Climate 2007. Ottawa, ON: Government of Canada; 2008. p. 448

[13] Bonsal B, Aider R, Gachon P, Lapp S. An assessment of Canadian Prairie drought: Past, present, and future. Climate Dynamics. 2013;**41**(2):501-516. DOI: 10.1007/s00382-012-1422-0

[14] Chipanshi A, Findlater K, Hadwen T, O'Brien G. Analysis of consecutive droughts on the Canadian prairies. Climate Research. 2006;**30**(3):175-187

[15] Ashton B. A sustained multi-disciplinary approach to the development and application of extreme moisture tools and management techniques for producers (with an emphasis on excess moisture). Unpublished manuscript. Brandon: Brandon University; 2018

[16] Wheaton E, Bonsal B, Wittrock V. Possible future dry and wet extremes in Saskatchewan. Saskatoon, SK: The Water Security Agency, Saskatchewan Research Council; 2013

[17] Pomeroy J, Fang X, Williams B. Impacts of Climate Change on Saskatchewan's Water Resources. Centre for Hydrology. Saskatoon: University of Saskatchewan; 2009. p. 2009

[18] PPWB (Prairie Provinces Water Board). Groundwater Activities. 2017. Available from: https://www.ppwb.ca/information/5/index.html [Accessed: 7 May 2019]

[19] Huffman W, Jin Y, Zheng X. The economic impact of technology a climate change: New evidence from corn yields. Agricultural Economics. 2018;**49**:463-479

[20] Rennenberg H, Loreto F, Polle A, Brilli F, Fares S, Beniwal R, et al. Physiological responses of forest trees to heat and drought. Plant Biology. 2006;**8**:556-571

[21] Sauchyn D, Barrow E, Hopkinson R, Leavitt P. Aridity on the Canadian plains. Géographie Physique et Quaternaire. 2002;**56**(2-3):247-259. Accessed on-line at: https://www.erudit.org/en/journals/gpq/2002-v56-n2-3-gpq740/009109ar/

[22] Sauchyn D, Beaudoin A. Recent environmental change in the Southwestern Canadian plains. The Canadian Geographer. 1998;**42**(4):337-353

[23] Brklacich M, Stewart R. Wheat yield sensitivities in the Canadian Prairies to climate change. In: Rosenzweig C, editor. Climate Change and Agriculture: Analysis of Potential International Impacts. Madison, WI: ASA Special Publication 59. American Society of Agronomy; 1995. pp. 147-162

[24] Jones J, Hoogenboom G, Porter C, Boote K, Batchelor W, Hunt L, et al. DSSAT cropping system model. European Journal of Agronomy. 2003;**18**:235-265

[25] Qian B, Zhang X, Smith W, Grant B, Jing Q, Cannon A, et al. Climate change impacts on Canadian yields of spring wheat, canola and maize for global warming levels of 1.5, 2.0, 2.5 and 3.0°C. Environmental Research Letters. DOI: 10.1088/1748-9326/ab17fb. In press

[26] Del Grosso S, Parton W, Mosier A, Walsh M, Ojima D, Thorton P. DayCent national-scale simulations of nitrous oxide emissions from cropped soils in the United States. Journal of Environmental Quality. 2006;**35**:1451-1460

[27] Grise J, Kulshreshtha S. Farmers choice of crops in Canadian Prairies under climate change: An econometric analysis. Journal of Earth Science and Climatic Change. 2016;7(2):332-342

[28] Burnett B. Climate change impacts on crop selection and rotation in 2050. Winnipeg: University of Manitoba. Available from: https://umanitoba.ca/faculties/afs/ncle/pdf/Ag2050_Essay_Burnett.pdf [Accessed: 9 November 2018]

[29] Rotter R, van de Gejin S. Climate change effects on plants growth, crop yield and livestock. Climatic Change. 1999;**43**(4):651-681

[30] Kulshreshtha S. Climate change, Prairie agriculture, and Prairie economy: The new normal. Canadian Journal of Agricultural Economics. 2011;**59**:19-44

[31] Rosenzweiz C, Hillel D. Climate Change and the Global Harvest: Potential Impacts of the Greenhouse Effect on Agriculture. New York: Oxford University Press; 1998

[32] Nardone A, Ronchi B, Lacetera N, Ranieri M, Bernabucci U. Effects of climate changes on animal production and sustainability of livestock systems. Livestock Science. 2010;**130**:57-69

[33] National Drought Monitoring Center. Droughts in the United States. 1998. Available from: http://enso.unl.edu/ndmc/impacts/us/usag98.htm [Accessed: 23 September 2016]

[34] Wheaton E, Kulshreshtha S, Wittrock V, Koshida G. Dry times: Hard lessons from the Canadian drought of 2001-2002. The Canadian Geographer. 2008;**52**(2):241-262

[35] Arthur L, Abizadeh F. Potential effects of climate change on agriculture in the prairie region of Canada. Western Journal of Agricultural Economics. 1988;**13**(2):216-224

[36] Amiraslany A. The impact of climate change on Canadian agriculture: A Ricardian approach. Ph.D. Thesis. Saskatoon: University of Saskatchewan; 2010

[37] Ayouqi H, Vercammen J. Evaluating the impact of climate change on Canadian Prairie agriculture. PR-01-2014. Vancouver: University of British Columbia; 2004

[38] Stern N. The Economics of Climate Change—The Stern Review. New York: Cambridge University Press; 2007

[39] Smit B, Skinner M. Adaptation options in agriculture to climate change: A typology. Mitigation and Adaptation Strategies for Global Change. 2002;**2002**(7):85-114

[40] McLeman R, Smit B. Migration as an adaptation to climate change. Climatic Change. 2006;**76**(1-2):31-53

[41] Agriculture and Forestry. Agricultural Adaptation Strategies in Alberta. Edmonton; 2018

[42] Government of Saskatchewan. Prairie Resilience: A Made-in-Saskatchewan Climate Change Strategy. Regina; 2017

[43] Poudel S, Kulshreshtha S. Choice of beef herd adaptation strategy on Canadian Prairie mixed farms under extreme climate events. Sociological Study. 2016;**6**(3):147-163

[44] Amichev B, Bentham M, Kurz W, Laroque C, Kulshreshtha S, Piwowar J, et al. Carbon sequestration and growth of six common tree and shrub species in 50,000 km of shelterbelts in Saskatchewan, Canada. Canadian Journal of Soil Science. 2017;**97**:368-381

[45] Kulshreshtha S, Ahmed R, Belcher K, Rudd L. Economic-environmental impacts of shelterbelts in Saskatchewan, Canada. In: Casares J, Passerini G, Parillo G, editors. Environmental Impact 2018. Ashurst, U.K: Wessex Institute of Technology; 2018. pp. 277-286

[46] Sauchyn D, Kulshreshtha S. Agriculture adaptation through irrigation. In: Sauchyn D, Diaz P, Kulshreshtha SN, editors. The New Normal: The Canadian Prairies in a Changing Climate. Regina: Canadian Plains Research Centre, University of Regina; 2010

[47] Brown J, Irrigation development as an instrument for economic growth in Saskatchewan: An economic impact analysis. M.Sc. Thesis. Saskatoon: University of Saskatchewan; 2017

[48] Kulshreshtha S, Paterson B, Hart D, Nicol L. Irrigation's impacts on

economic growth in Alberta, Canada. Irrigation and Drainage Systems Engineering. 2006;**5**(1):156-161

[49] Howden M, Soussana J, Tubiello F, Chhetri N, Dunlop M, Meinke H. Adapting agriculture to climate change. PNAS. 2007;**104**(50):19691-19696

[50] Thorpe J. Vulnerability of Prairie Grassland to climate change. Saskatchewan Research Council. A Report Prepared for Prairie Adaptation Research Collaborative (PARC). Saskatoon, Saskatchewan; 2011

[51] Fischer G, Shah M, Tubiello F, van Velhuizen H. Socio-economic and climate change impacts on agriculture: An integrated assessment, 1990-2080. Journal of the Royal Society Biological Sciences. 2005;**360**:2067-2083

[52] Wittrock V, Koshida G. Canadian droughts of 2001 and 2002: Government Response and Safety Net Programs—Agriculture Sector. SRC Publication No. 11602-2E03. Saskatoon, SK: Saskatchewan Research Council; 2005

[53] Lin B. Resilience in agriculture though crop diversification: Adaptive management for environmental change. Bioscience. 2011;**61**(3):183-193

[54] Lipper L, Thomton P, Campbell B, Baedeker T, Braimoh A, Bwalya M, et al. Climate-smart agriculture for food security. Nature Climate Change. 2015;**4**:1068-1072

Chapter 8

Sugarcane Production under Changing Climate: Effects of Environmental Vulnerabilities on Sugarcane Diseases, Insects and Weeds

Sadam Hussain, Abdul Khaliq, Umer Mehmood, Tauqeer Qadir, Muhammad Saqib, Muhammad Amjed Iqbal and Saddam Hussain

Abstract

Sugarcane is an important crop for bioenergy and sugar, contributing to Gross Domestic Product (GDP) of Pakistan. Global warming and increasing greenhouse gas emission result in the increased intensity and frequency of extreme weather events. Temperature stress is a major environmental stress that limits the sugarcane growth, productivity and metabolism worldwide. Numerous biochemical reactions are involved in plant development, and these biochemical reactions are very sensitive to temperature stress. Now a day, temperature stress is a major concern for sugarcane production and approaches for high yield of sugarcane under temperature stress are important agriculture goals. Sugarcane plant adapts a number of acclimation and avoidance mechanism against different environmental stress. Plant survival under different stresses depends on ability to generate and transmit the signal and biochemical and physiological changes. In future, climate change is an important consequence for sugarcane production in the world because of its relative low adaptive capacity, poor forecasting system and high vulnerable to natural hazard. In this review we briefly describe climate change effects on sugarcane, sugar production in several countries especially in Pakistan, future challenges for sugar production under changing climatic scenario and propose strategies for mitigation negative impacts of climate change.

Keywords: climate change, diseases, drought, high temperature, weeds, sugarcane

1. Introduction

Natural process and anthropogenic activities result in global climate change and variability that affect the world during twenty-first century. According the fourth assessment report of intergovernmental panel on climate change (IPCC), estimate of temperature increase in the range 1.8–4°C in 2090–2099 relative to 1980–1999 and extreme events such as floods and drought are projected in future [1]. Due to combustion of fossil fuel, industrial processes and deforestation, atmospheric CO_2 concentration has increased by 30% since mid of eighteenth century [2] and projection indicates that CO_2 concentration would be double in high emission scenario by the end of this century. Increases in concentration of CO_2 and air temperature can be beneficial for some plant [3]. Abiotic stresses, change in rainfall pattern, frequency of extreme low and high temperature, flood and drought are projected in future [1].

Changing climatic conditions influences the population dynamics, life cycle duration and overall occurrence of majority of insects, pathogen and weeds of sugarcane. Weeds, pathogen and insects are among the agriculture pest that will be influence by climate change. Changes in temperature, rainfall and CO_2 levels, will affect pathogen, insects and weeds distribution and their competitiveness with wheat crop. C3 cultivars are performing well under high CO_2 concentration than C4 cultivar. Agriculture sector is sensitive to temporary weather changes and seasonal, annual and long term variation in climate. Agronomic practice, soil, seed, pest and diseases have significant influence on crop yield. Human induced climate change and environmental problems, provides a limiting factor. Sugarcane is a C4 crop, mainly grow in tropics and subtropics regions and important source of bioenergy and sugar in the world. Sugarcane is perennial crop cultivated on 20 million ha in subtropical and tropical region [4] with annual yield app. 1325 million tons stalks for sugar, energy, rum and chemicals [5]. Sugarcane is one of the world's major food-producing crops, providing about 75% of sugar produced in the world for human consumption [6]. Sugarcane is cash crop of Pakistan and makes contribute in 0.6% in total GDP. During 2015–2016, sugarcane crop cultivated on 1132 thousand hectares as compared 1141 thousand hectares previous year, with production of 65,475 thousand tons [7]. Decline in area is due to shifted sugarcane area to other crops. Total cultivated area and production of sugarcane in Pakistan is given in **Figure 1**.

Sugarcane mostly propagated by placing cutting and whole stalks in furrows. After each harvest, ratoons mostly grows from stubble and it is possible to harvest 20 successful ratoon crop from a single plantation [8] but environmental related factor such as pathogen infection, low winter temperature, weed competition, stalk borer injury and water deficit condition reduce the production one season to next [9]. Climate related and weather events such as temperature, precipitation and atmospheric CO_2 are the key factor for sugarcane production in the world [10].

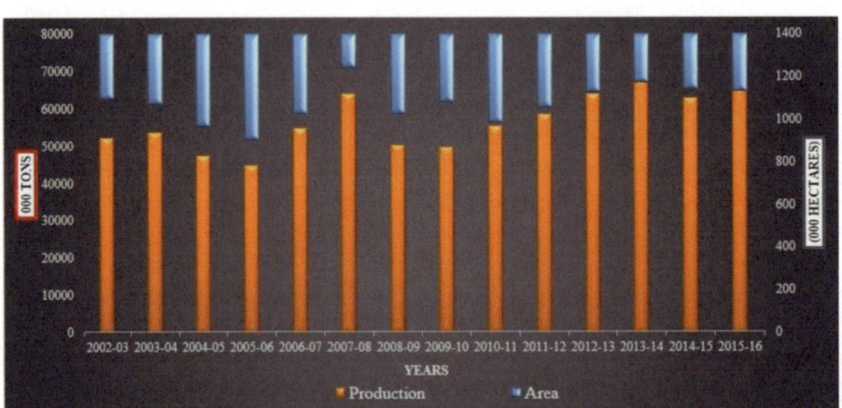

Figure 1.
Total area and production of sugarcane in Pakistan 2002–2016 (Pakistan Bureau of Statistics, 2016).

2. Temperature stress and sugarcane production

Sugarcane is a C4 species; increase in temperature in the range of 8–34°C increases the carbon dioxide assimilation and improve cane growth during winter but low temperature limit the photosynthesis and leaf growth rate ([11]; **Table 1**). Low temperature below 15°C limited the cultivation of sugarcane but temperature increase under changing climatic condition during low temperature period improve the sugarcane yield [19]. High temperature likely reduce the incidence and severity of frost and extending the growth during winter months, frost known to poor quality in sugarcane [20, 21]. However high temperature has negative effect on sprouting and emergence of sugarcane and ultimate low plant population [22]. Temperature above 32°C result in increased number of nodes, short internodes, higher stalk fiber and lower sucrose [23]. High night temperature usually more number of flowering and flowering in sugarcane ceases the growth of internodes and leaves ultimately reduce the sucrose and cane yield [20]. Increase in temperature under changing climatic conditions also alter the daily evaporation, may cause water stress and more frequent irrigation cycle will be done to meet the demand of evaporation and crop. Frequent irrigation result in over irrigation and create water logging and salinity problem which can reduce the sugarcane yield [24]. Temperature changes also affect the ripening of sugarcane. During winter, low temperature is very important for natural ripening. Under changing climate, elevated temperature reduces the ripening and quality of sugarcane [11].

Authors	Study traits	Effects
Morales et al. [12]	High temperature and photosynthesis	Photosystem II activity greatly reduce under high temperature
Marchand et al. [13]	Heat stress and photosynthesis	Heat stress reduce the amount of photosynthesis pigments
Rodriguez et al. [14]	Temperature stress and sucrose phosphate activity	Temperature stress reduce the activity of sucrose phosphate synthase
Warland et al. [15]	Temperature stress and yield	Small increase in temperature significant reduce the yield of crops
Johkan et al. [16]	Temperature stress and germination stage	Heat stress exert have negative impact during germination stage
Srivastava et al. [17]	High temperature and photosynthesis rate	High temperature reduce the net assimilation rate (NAR) and relative growth rate (RGR)
Omae et al. [18]	Temperature stress and leaf morphology	High temperature stress damage the leaf tip and margins, drying the leaves and observed necrosis

Table 1.
Temperature stress under changing climatic conditions and its effects on sugarcane crop.

3. Tolerance mechanism against high temperature stress

Plant response to temperature stress, vary with degree, duration and plant type. At high temperature, cell death or cellular damage may occur, which lead to catastrophic collapse of cellular organization [25]. Heat stress effect all the process includes germination, growth and yield [26] and the stability of various protein, cytoskeleton structure and efficiency of enzymatic reactions [24, 27]. Under high temperature stress, plant adopt various mechanism include long term phenological and morphological adaptations and short term avoidance mechanism such as transpirational cooling and

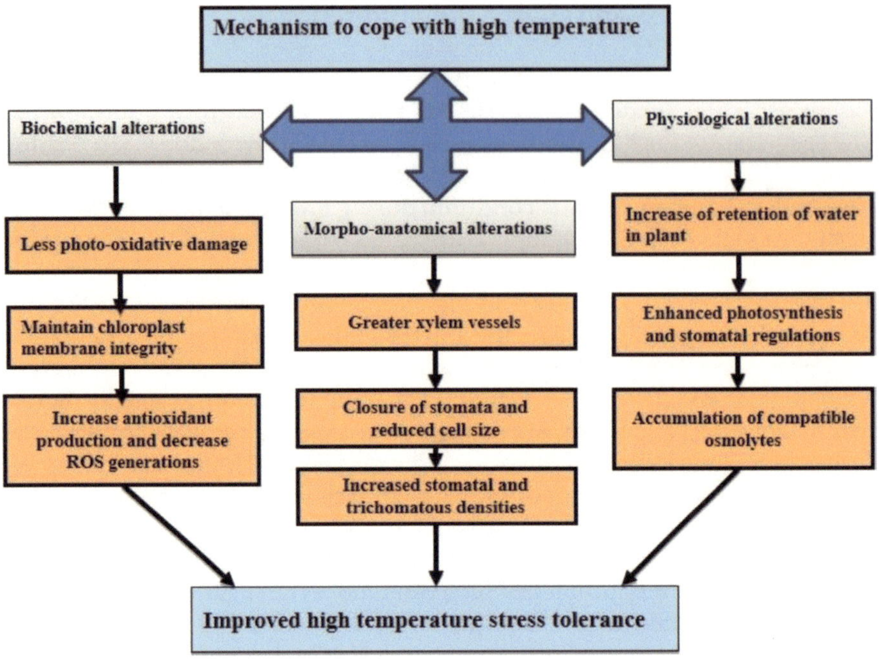

Figure 2.
Mechanism to cope with high temperature.

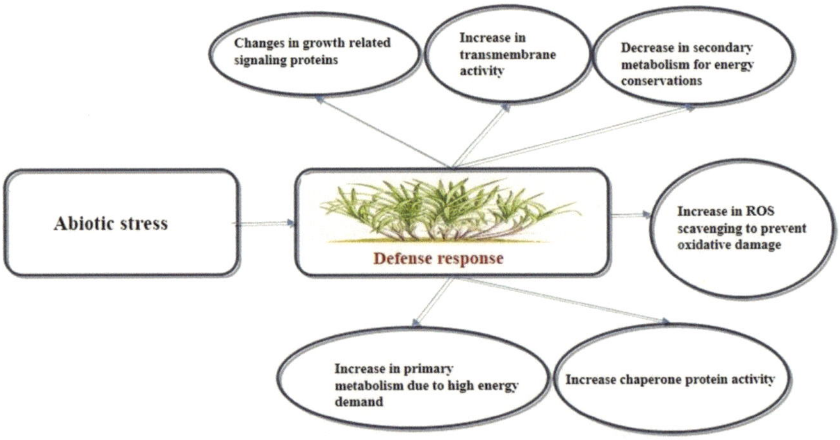

Figure 3.
Plant responses to abiotic stresses.

changing leaf orientations. Reduce water loss, closure of stomata, increased stomatal densities and alteration of membrane lipids compositions are the common feature which adopt by plant under stress [17], **Figure 2**. High temperature stress changes the degree of leaf rolling [28].

Low temperature affects the geographical distribution and planting seasons of crops, especially in tropical and subtropical regions [29]. Low temperature retard plant growth and development by reducing the metabolic process, leading to oxidative and osmotic stress [30]. Plant possesses the many strategies to response the temperature fluctuation such as cell remodeling and gene expression and metabolism reprogramming [20, 29]. Under low temperature stress, C-repeat binding factors, bind to dehydration responsive elements in gene promoters to active the COR

Figure 4.
National loss due to diseases of sugarcane.

(cold response genes), called as ICE-CBF-COR pathway [31, 32]. More study shows that miRNAs also play a critical role in this process [33, 34]. Ref. [19] observed the up-regulation of miR139 and down-regulation its target in both cold tolerant and sensitive verities of sugarcane. Some other cold related miRNAs such as miR156K and miR394 has been reported [4, 35] (**Figures 2–4**).

4. Increase in CO_2 concentration and sugarcane production

Increases in CO_2 concentration directly affect the photosynthesis and stomatal physiology, increase growth rate in many plants [33]. High concentration with increase temperature will alter the plant ability to grow and modify the distribution of weeds across globe and their competitiveness in different habitat [36, 37]. In C4 plats, internal mesophyll cell arrangement is different as compare to C3 plant; help in efficient transfer of CO_2 and increase photosynthesis and reduce the photorespiration [33]. In future under increasing CO_2 condition, C4 crop may become more vulnerable to increased competition from C3 weeds. Double concentration of CO_2 may decrease 30–40% in stomatal aperture and 25–40% transpiration loss in both C3 and C4 plant. Double concentration of CO_2 may decrease 30–40% in stomatal aperture and 25–40% transpiration loss in both C3 and C4 plant. In long term field study, LAI (Leaf Area Index) did not increase in any species under elevated CO_2 conditions [33]. Bowes [38] discus that under increasing cons. of CO_2, starch cons. Also increases in tissue reduce protein content. Elevated CO_2 cons. increase the growth and root-shoot ratio [39] and alter the photosynthesis activity in plants. Plants available nitrogen also reduces under elevated CO_2 [40]. Leaves carbon-nitrogen ratio increased under increasing CO_2.

5. Drought stress

Environmental stress reduces the crop productivity and plant growth and drought is the major abiotic stress, affecting crop productivity [41, 42]. Sugarcane crop is highly sensitive to water deficit [43] and water deficit reduce the crop productivity up to 60% [44, 45]. Under water deficit conditions, sugarcane providing a key impetus to develop bio-technological strategies [43]. Under drought conditions, plant adopt various tolerant strategies such as modulation of growth, changes in life cycle, evolution of stress perception for rapid expression of stress tolerance and balance allocation of resources for stress adaptation and growth [25, 46, 47]. Molecular Breeding and biotechnology techniques are helpful tools to enhance crop productivity under water deficit conditions [25].

6. Morphological and physiological response of sugarcane to drought stress

Physiological and morphological response of sugarcane to drought stress varies according to duration, intensity of stress, type of tissue affected and genotype of plant [26, 48]. In sugarcane, common water responses are stomatal closure, inhibition of stalk and leaf growth, leaf rolling, reduce leaf area [26] and cell elongation and division are interrupted [49]. Stem and leaf elongation are most seriously affected [50]. Under water deficit conditions, root development also influence [51]. Sugarcane crop have C4 photosynthetic pathway and under water stress, decrease in transpiration rate, stomatal conductance and photosynthesis rate occurs [52, 53].

Water stress, decline the photosynthesis activity by decrease in phosphoenolpyruvate carboxylase and Rubisco activity [43]. Sugar accumulation in leaves also change the photosynthesis rate [54] and high sugar content moderates carbon fixation [55]. Under water deficit conditions, increase level of trehalose sugar, reducing the damage to cell membrane [34]. In reduce CO_2 fixation conditions, increase in starch hydrolysis helped in sustain carbon supply, which facilitate growth recovery after stress [56].

6.1 Changing rainfall pattern and sugarcane production

Climate change can affect agriculture through rise in temperature, variation in precipitation, increase in CO_2 concentration and weather vulnerabilities like flood, drought etc. Held et al. [57] predicted the more droughts in future; small changes in rainfall in future [58]. Extreme changes in rainfall have impact on sucrose yield [59] and frequent drought has negative effect on sugarcane as crop requires more water. Water stress also alters the photosynthesis, respiration and stomatal conductance [35]. For mitigate the drought conditions, farmer likely to more irrigate and increased the salinity problem and risen the water table. Reduce precipitation during harvesting period is likely to increase harvesting efficiency [60]. Nitrogen is the most limiting factor for sugarcane production [61]. Wetter years are likely to cause flood which may leach the nitrogen and farmer are use high rates of fertilizer during wetter years. Water logging may reduce the oxygen availability for root system and inhibit the uptake of nutrients [62]. Increased precipitation also reduces the quality of cane by inadequate "dry off" period. Changing in precipitation also has prevalence of weeds, diseases and insect pest. Termite and nematodes is expected to increase under warm and dry conditions. Some weeds such as Cyperus spp. may decrease under frequent drought conditions [20]. Effect of drought on sugarcane production depends upon plant growth stage and duration of stress. Drought at early and mid-growth stages mainly reduce sucrose yield. During the late growth stage, Moderate drought, increase the sucrose content in stalks. In china drought is the most important stress for production [63].

Environmental stress makes crops more vulnerable to insects and pathogen attack and less competitive with weeds [64]. Frequency and intensity of rainfall also changed under changing climatic conditions. After application of herbicides, precipitations wash out the herbicides and reduce it efficiency. For weed seeds germination, moisture is require, so weeds have more competitive advantage over crops. Increased rainfall and changed intensity will reduce the uptake of soil applied herbicides [65, 66].

Under water deficit conditions, efficacy of sethoxydim was lower in goosegrass [67] and *Urochloa plantaginea* (signalgrass) not effectively controlled by ACCase inhibiting herbicides when applied during later stages [68]. Patterson [64] discus that under drought conditions, leaf pubescence and thickness increased and it will reduce the entry of herbicides into leaves. Water stress conditions affect the plant growth and efficacy of herbicides [69]. Most of studies done on impact of climate change on crop production but a little attention is given to identify the effects of climate change on weeds. Environmental stress changes the herbicides efficiency and contributes to loss in production. Under these changing climatic conditions it is necessary to understand the how environmental conditions affect the herbicides performance.

7. Increases temperature conditions and different weeds, insects and diseases of sugarcane

Climate change effect the agriculture sector in different direction, directly by changing in temperature and/or precipitation as well as through indirect by changing pest pressure and availability of pollination service [31]. Diseases are main threat to the food security and the responsible for 10% reduction in globally food production [70], **Figure 4**. Many abiotic and biotic stresses such as weed competition, soil nitrogen and water deficit can exert the stress on sugarcane and increase the herbivores attack. Among herbivores arthropods, sugarcane borer, stalk insect borer and Mexican rice borer are the most important for sugarcane production [71]. A change in temperature under climate change conditions will have effects on some weeds, insects and diseases of sugarcane [10]. Smut disease (caused by *Sporisorium scitamineum*) is likely to increase under high temperature conditions [60]. But Sanguino [72] discus the different diseases of sugarcane, such as *Ustilago scitaminea* (Smut), *Sugarcane mosaic virus* (SCMV) and *Xanthomonas albilineans* (leaf scaled disease) and told that all these diseases are systemic and only changed by direct human interference. However Chakraborty [30] told that leaf scaled may spread by severe storms.

Extreme weather events have caused more disease pathogen and overwintering pest and also increased the input cost for control them. Sugarcane leaf and orange rust are bid challenges for sugarcane production in Florida [32]. Reduced rainfall will also reduce the growth of crops and pasture and decrease canopy covers which favor the weed infestation [73]. *Eoreuma loftini* is a major sugarcane pest [74]; drought results in greater abundance of dry leaf tissue and number of *E. loftini* egg have been positively associated with number of dry leaves [75]. Under well water conditions sugarcane plant had 82.8–90.2% less *E. loftini* eggs than water deficit conditions [75]. In wet tropics, cyclones will disperse weeds seed through flood and wind. Moderate drought can increase the population of many herbivorous arthropods and cause the injury in crop [76] drought also enhanced host plant suitability for herbivores arthropods [77].

Summer weeds such as *Rottboellia cochinchinensis*, *Ipomoea plebeia* and *Digitaria sanguinalis* are expected to increase under high temperature conditions. Dormancy of some summer weeds will have been broken under high temperature and these weeds may appear in winter. Insect pests such as *Heteronychus licas* and *Margarodes* spp. Will not be spared by temperature changes. Matthieson [60] told that temperature changes creates favorable condition for introduce new pests and diseases in sugarcane. Due to climate change, shift in temperature will have effect on diseases, weeds and insect pests [78]. Matthieson [60] discus that attack of smut diseases is likely to increase under high temperature condition. High temperature induced the pollen sterility, poor anthesis and reduce the grain filling duration [79]. High temperature stimulates the stomata conductance and increases the permeability [80] and reduce the uptake of herbicides. Above discussion will help in identifying possible measure for weeds control under changing atmospheric conditions. So it is need to evaluate the climate change effects on weed flora and herbicides efficacy for adaptation and mitigation strategies.

8. Increasing concentration of CO_2 and weeds of sugarcane

Climate change is a continuous process, occur due to human and natural activities. Greenhouse gasses emission due to anthropogenic activities, accumulate the earth atmosphere and increase in concentration over time [81]. Concentration of CO_2 has risen to 387 μmol mol^{-1} till 2007 and expected to reach 600 μmol mol^{-1} till 2050 [27]. Modern agriculture techniques is main driving force which contribute in 30% of greenhouse gas emissions [82]. Weeds have negative effects on agriculture and public health. Weeds plants exert a variety of effects on sugarcane ecology under field conditions [83]. Uncontrolled annual summer weeds results 24% reduction in sugarcane stalks density, 19% in biomass and 15% reduction in commercial sugar production [84]. Under higher CO_2 concentration, C3 weeds generally increased their leaf area and biomass as compared with c4 weeds. Parthenium (*Parthenium hysterophorus*) is a C3 weed will be much more competitive under raised CO_2 environment [64]. Yield reduction due to weeds differ accordingly weeds density and species, when weeds emerge in high density, competitions will be highest. Crop and weeds competition affected by environmental conditions and have been shown to change with increasing CO_2. Temperature is a primary factor influencing the distribution of weeds at higher latitudes. Increase precipitation and temperature may provide suitable conditions for some species [37]. CO_2 enrichment reduce the effects of water stress and increased the leaf area in C4 grasses and increase the growth of C3 and C4 plants under stress. Increasing cons. of CO_2 from 300 to 600 ppm increased water use efficiency (WUE) in sunflower by 55%, in corn by 54%, in soybean by 48% and in redroot increased WUE by 76% [29] and greater stimulation of WUE in weeds than crops, provide competitive advantage to weeds. Rise in CO_2 concentration reduce the performance of many herbicides and plant growth, so it is needed to optimize herbicides application for better weed management in coming days. Climate also change the phenology and population of weeds. Most weeds species spread to new areas and researcher suggest that due to strong response of weeds to elevated CO_2, invasive species may become a threat in changing climate [85]. AT elevated CO_2, C3 plant have higher photosynthesis rate and response more favorable than C4 weeds. Alberto [86] discus the interaction between temperature and CO_2 and reported that elevated CO_2 favor the growth of barnyard grass at 37/29°C. It is essential to understand the factor that reduce the performance of any herbicides and insecticides, so for the successful management of chemicals it is necessary to understand its interaction with plants and environment. In plants, Herbicide absorption greatly depend on its interaction with environment. Soil

Authors	Study traits	Effects
Rodenburg et al. [66]	Changing rainfall pattern and herbicides uptake	Increase in rainfall, reduce the uptake of soil-applied herbicides
Zanatta et al. [69]	Water stress and herbicides efficiency	Water stress conditions reduce the herbicides efficacy
Matthieson [60]	Elevated temperature and diseases of sugarcane	Attack of Smut disease is increased under high temperature conditions
Showler and Reagan [84]	Weeds in sugarcane	Uncontrolled annual summer weeds resulted in 15% reduction in commercial sugar production and 19% reduction in biomass
Singh et al. [85]	Climate change and weeds	Invasive species of weeds may become a threat
Alberto [86]	CO_2 and different weeds	Elevated CO_2 favor the growth of barnyard grass at 37/29°C

Table 2.
Climate change and biotic stresses in sugarcane.

applied herbicides mainly influenced by soil temperature and moisture conditions. Environmental factor like temperature, humidity, moisture conditions and solar radiations also affects the efficacy of herbicides. Climate change also reduce the photosynthesis process which affects herbicides absorption and translocation (**Table 2**).

9. Conclusions

Environmental vulnerabilities are the major constraints for growth, development and productivity of sugarcane plant. The present rate of greenhouse gasses emission is believed responsible for gradual increase in temperature, changing rainfall pattern and environmental vulnerabilities and result in global warming. Plant response and adaptation to different stresses and effects of changing climatic conditions and weeds, insects and diseases, needs to be better understand for sugarcane crop. Sugarcane productivity under changing climatic conditions have been studied in recent years; however, a complete understanding of sugarcane production under changing climatic conditions remain elusive. Under different stress, sugarcane plant accumulate different mechanism and accumulate different metabolites such as antioxidant, osmoprotectants and heat shock protein and different metabolic process are activated.

10. Recommendations

- Under high temperature stress, exogenous application of protectant such as phytohormones and osmoprotectants, have beneficial effects on plants [87, 88].

- Soaking sugarcane nodal buds in exogenous pro (mM) and GB (20 mM) solution, restricted the H_2O_2 generation, improved the accumulation of soluble sugars and reduce the effects of heat stress on sugarcane crop [22].

- Managing the cultural practices, such as adequate sowing time and method, irrigation management and selecting the suitable cultivars, decrease the effects of different stresses.

- Under climate conditions, elevated temperature reduce the natural ripening, Chemical such as Ethrel, Fusillade super and round up may be helpful in ripening sugarcane.

- The effects of temperature and drought stress can be mitigated by growing the temperature and drought resistance varieties. A deep insight and research is need for new cultivars that adopt to temperature and drought stress conditions and greater water use efficiency.

- For mitigate the effects of climate change on sugarcane, scientist also use biotechnology for breeding the new cultivar to reduce the abiotic and biotic stresses.

- Health and environmental issue must also be given due consideration to address the new cultivar.

- Due to climate change, floods are projected in some areas, it is therefore important to adopt sugarcane production to such extreme conditions.

- Self-trashing varieties of sugarcane may be adopted, in order to complement harvesting without burning.

- Sugarcane residue may be used for weed suppression, increase the organic matter content in the soil and improve soil structure.

- Burning of sugarcane trash is a main cultural practice that effects of climate change. After harvesting trash is cleared from all ridges. Trash may use as mulch to control the water and wind erosion.

Author details

Sadam Hussain[1], Abdul Khaliq[1], Umer Mehmood[1], Tauqeer Qadir[1], Muhammad Saqib[1], Muhammad Amjed Iqbal[2] and Saddam Hussain[1*]

1 Department of Agronomy, University of Agriculture, Faisalabad, Punjab, Pakistan

2 Institute of Agricultural and Resource Economics, University of Agriculture, Faisalabad, Punjab, Pakistan

*Address all correspondence to: sadamhussainuaf@gmail.com

References

[1] Trenberth KE, Jones PD, Ambenje P, et al. Observations: Surface and atmospheric climate change: The physical science basis. In: Solomon S, Qin D, Manning M, et al., editors. Contribution of WG 1 to the Fourth Assessment Report of the Intergovernmental Panel on Climate Change. Cambridge, UK: Cambridge University Press; 2007. pp. 235-336

[2] Houghton RA, Lawrence KT, Hackler JL, Brown S. The spatial distribution of forest biomass in the Brazilian Amazon: A comparison of estimates. Global Change Biology. 2001;7(7):731-746

[3] Kimball BA. Carbon dioxide and agricultural yield: An assemblage and analysis of 430 prior observations. Agronomy Journal. 1983;**75**(5): 779-788

[4] FAO (Food and Agriculture Organization of the United Nations). Major Food and Agricultural Commodities and Producers. Available online: http://www.fao.org/ess/top/commodity.html?lang+en&item=156&year=2005 [accessed on 1 July 2015]

[5] Sharpe P. Sugar Cane: Past and Present. Carbondale, IL, USA: Southern Illinois University; 1998

[6] Souza AP, Gaspar M, Silva EA, Ulian EC, Waclawosky AJ, Nishiyama MYJR, et al. Elevated CO_2 increases photosynthesis, biomass and productivity, and modifies gene expression in sugarcane. Plant, Cell & Environment. 2008;**31**(8):1116-1127

[7] Government of Pakistan. Economic survey government of Pakistan. Finance Division, Economic Advisor's Wing, Islamabad. 2016. Available from: http://www.finance.gov.pk/survey/chapters_17/02-Agriculture.pdf

[8] Blackburn F. Sugar Cane. New York, NY, USA: Longman; 1984

[9] Edgerton CW. Stubble deterioration. Proceedings—International Society of Sugar Cane Technologists. 1939;**6**:334-341

[10] Chandiposha M. Potential impact of climate change in sugarcane and mitigation strategies in Zimbabwe. African Journal of Agricultural Research. 2013;**8**:2814-2818

[11] Gawander J. Impact of climate change on sugar-cane production in Fiji. World Meteorological Organization Bulletin. 2007;**56**(1):34-39

[12] Morales D, Rodríguez P, Dell'Amico J, Nicolas E, Torrecillas A, Sánchez-Blanco MJ. High-temperature preconditioning and thermal shock imposition affects water relations, gas exchange and root hydraulic conductivity in tomato. Biologia Plantarum. 2003;**47**(2):203

[13] Marchand FL, Mertens S, Kockelbergh F, Beyens L, Nijs I. Performance of high arctic tundra plants improved during but deteriorated after exposure to a simulated extreme temperature event. Global Change Biology. 2005;**11**(12):2078-2089

[14] Rodríguez M, Canales E, Borrás-Hidalgo O. Molecular aspects of abiotic stress in plants. Biotecnología Aplicada. 2005;**22**(1):1-10

[15] Warland JS, McDonald MR, McKeown AM. Annual yields of five crops in the family Brassicacae in southern Ontario in relation to weather and climate. Canadian Journal of Plant Science. 2006;**86**:1209-1215

[16] Johkan M, Oda M, Maruo T, Shinohara Y. Crop production and global warming. In: Global Warming Impacts-case Studies on the Economy,

Human Health, and on Urban and Natural Environments. InTech; 2011

[17] Srivastava S, Pathak AD, Gupta PS, Shrivastava AK, Srivastava AK. Hydrogen peroxide-scavenging enzymes impart tolerance to high temperature induced oxidative stress in sugarcane. Journal of Environmental Biology. 2012;**33**(3):657

[18] Omae H, Kumar A, Shono M. Adaptation to high temperature and water deficit in the common bean (Phaseolus vulgaris L.) during the reproductive period. Journal of Botany. 2012. Article ID 803413. 6 pages. DOI: 10.1155/2012/803413

[19] Ebrahim MKH, Vogg G, Osman MNEH, Kamor E. Photosynthetic performance and adaptation of *sugarcane* at suboptimal *temperatures*. Journal of Plant Physiology. 1998;**153**(6):587-692

[20] Clowes MJ, Breakwell WL. Zimbabwe Sugarcane Production Manual. Chiredzi, Zimbabwe: Zimbabwe Sugar Association; 1998

[21] Mathieson L. Climate Change and the Australian Sugar Industry: Impacts, Adaptation and R & D Opportunities. Australia: Sugar Research and Development Corporation; 2007

[22] Rasheed R, Wahid A, Farooq M, Hussain I, Basra SMA. Role of proline and glycine betaine pretreatments in improving heat tolerance of sprouting sugarcane (Saccharum sp.) bud. Plant Growth Regulation. 2011;**65**:35-45

[23] Bonnett GT, Hewitt ML, Glassop D. Effects of high temperature on the growth and composition of sugarcane internodes. Australian Journal of Agricultural Research. 2006;**57**(10):1087-1095

[24] Kahlown MA, Azam M. Individual and combined effect of water logging and salinity on crop yields in the Indus basin. Irrigation and Drainage. 2002;**51**:329-338

[25] Hu H, Xiong L. Genetic engineering and breeding of drought-resistant crops. Annual Review of Plant Biology. 2014;**65**:715-741

[26] Inman-Bamber N, Lakshmanan P, Park S. Sugarcane for waterlimited environments: Theoretical assessment of suitable traits. Field Crops Research. 2012;**134**:95-104. DOI: 10.1016/j.fcr.2012.05.004

[27] IPCC. Assessment Report. 2007. http://www.ipcc.ch/pdf/assessmentreport/ar4/syr/ar4_syr_spm.pdf

[28] Sarieva GE, Kenzhebaeva SS, Lichtenthaler HK. Adaptation potential of photosynthesis in wheat cultivars with a capability of leaf rolling under high temperature conditions. Russian Journal of Plant Physiology. 2010;**57**(1):28-36

[29] Carlson RW, Bazzaz FA. The effects of elevated CO_2 concentrations on growth, photosynthesis, transpiration and water use efficiency of plants. In: Singh JJ, Deepak A, editors. Environmental and Climatic Impact of Coal Utilization. New York: Academic Press; 1980. pp. 610-622

[30] Chakraborty S, Murray GM, Magarey PA, et al. Potential impact of climate change on plant diseases of economic significance to Australia. Australasian Plant Pathology. 1998;**27**:15-35

[31] Lobell DB, Burke MB, Tebaldi C, Mastrandrea MD, Falcon WP, Naylor RL. Prioritizing climate change adaptation needs for food security in 2030. Science. 2008;**319**(5863):607-610

[32] Zhao D, Glynn NC, Glaz B, Comstock JC, Sood S. Orange rust effects on leaf photosynthesis and

related characters of sugarcane. Plant Disease. 2011;**95**(6):640-647

[33] Drake BG, Gonzàlez-Meler MA, Long SP. More efficient plants: A consequence of rising atmospheric CO_2? Annual Review of Plant Physiology and Plant Molecular Biology. 1997;**48**:609-639

[34] Delorge I, Janiak M, Carpentier S, Van Dijck P. Fine tuning of trehalose biosynthesis and hydrolysis as novel tools for the generation of abiotic stress tolerant plants. Frontiers in Plant Science. 2014;**5**:147. DOI: 10.3389/fpls.2014.00147

[35] Gardener FP, Pearce R, Mitchell R. Physiology of Crop Plants. 1st ed. Iowa: Iowa State University Press; 1984. p. 328

[36] Bunce JA. Weeds in a changing climate. In: BCPC Symp. Proc. No. 77: The World's Worst Weeds. 2001. pp. 109-118

[37] Parry ML. The impact of climate change on European agriculture. In: Lewis T, editor. The Bawden Memorial Lectures 1973-1998, Silver Jubilee Edition. UK: British Crop Protection Council, Surrey; 1998. pp. 325-338

[38] Bowes G. Photosynthetic responses to changing atmospheric carbon dioxide. In: Baker NR, editor. Photosynthesis and the Environment. Advances in Photosynthesis. Vol. 5. Dordrecht: Kluwer; 1996. pp. 387-407

[39] Ziska LH, George K. Rising carbon dioxide and invasive, noxious plants: Potential threats and consequences. World Resource Review. 2004;**16**:427-447

[40] Zhang W, Parker KM, Luo Y, Wan S, Wallace LL, Hu S. Soil microbial responses to experimental warming and clipping in a tall grass prairie. Global Change Biology. 2005;**11**:266-277

[41] Lobell DB, Schlenker W, Costa-Roberts J. Climate trends and global crop production since 1980. Science. 2011;**333**:616-620. DOI: 10.1126/science.1204531

[42] Wang W, Vinocur B, Altman A. Plant responses to drought, salinity and extreme temperatures: Towards genetic engineering for stress tolerance. Planta. 2003;**218**:1-14. DOI: 10.1007/s00425-003-1105-5

[43] Lakshmanan P, Robinson N. Stress physiology: Abiotic stresses. In: Moore PH, Botha FC, editors. Sugarcane: Physiology, Biochemistry, and Functional Biology. Chichester: John Wiley & Sons, Inc.; 2014. pp. 411-434

[44] Basnayake J, Jackson PAN, Inman-Bamber G, Lakshmanan P. Sugarcane for water-limited environments. Genetic variation in cane yield and sugar content in response to water stress. Journal of Experimental Botany. 2012;**63**:6023-6033. DOI: 10.1093/jxb/ers251

[45] Gentile A, Dias LI, Mattos RS, Ferreira TH, Menossi M. MicroRNAs and drought responses in sugarcane. Frontiers in Plant Science. 2015;**6**:58. DOI: 10.3389/fpls.2015.00058

[46] Hirayama T, Shinozaki K. Research on plant abiotic stress responses in the post-genome era: Past, present and future. The Plant Journal. 2010;**61**:1041-1052. DOI: 10.1111/j.1365-313X.2010.04124.x

[47] You J, Chan Z. ROS regulation during abiotic stress responses in crop plants. Frontiers in Plant Science. 2015;**6**:1092. DOI: 10.3389/fpls.2015.01092

[48] Bartels D, Sunkar R. Drought and salt tolerance in plants. Critical Reviews in Plant Sciences. 2005;**24**:23-58. DOI: 10.1080/07352680590910410

[49] Machado R, Ribeiro R, Marchiori P, Machado D, Machado E, Landell M. Biometric and physiological responses to water deficit in sugarcane at different phenological stages. Pesquisa Agropecuaria Brasileira. 2009;**44**:1575-1582. DOI: 10.1590/S0100-204X2009001200003

[50] Inman-Bamber N, Bonnett G, Spillman M, Hewitt M, Jackson J. Increasing sucrose accumulation in sugarcane by manipulating leaf extension and photosynthesis with irrigation. Australian Journal of Agricultural Research. 2008;**59**:13-26. DOI: 10.1071/AR07167

[51] Smit M, Singels A. The response of sugarcane canopy development to water stress. Field Crops Research. 2006;**98**:91-97. DOI: 10.1016/j.fcr.2005.12.009

[52] Basnayake J, Jackson PAN, man-Bamber G, InLakshmanan P. Sugarcane for water-limited environments. Variation in stomatal conductance and its genetic correlation with crop productivity. Journal of Experimental Botany. 2015;**66**:3945-3958. DOI: 10.1093/jxb/erv194

[53] Medeiros DB, da Silva EC, Mansur Custodio Nogueira RJ, Teixeira MM, Buckeridge MS. Physiological limitations in two sugarcane varieties under water suppression and after recovering. Theoretical and Experimental Plant Physiology. 2013;**25**:213-222. DOI: 10.1590/S2197-00252013000300006

[54] McCormick AJ, Cramer MD, Watt DA. Regulation of photosynthesis by sugars in sugarcane leaves. Journal of Plant Physiology. 2008;**165**:1817-1829. DOI: 10.1016/j.jplph.2008.01.008

[55] Goldschmidt EE, Huber SC. Regulation of photosynthesis by endproduct accumulation in leaves of plants storing starch, sucrose, and hexose sugars. Plant Physiology. 1992;**99**:1443-1448. DOI: 10.1104/pp.99.4.1443

[56] Sales CRG, Ribeiro RV, Machado DFSP, Machado RS, Dovis VL, Lagôa AMMA. Gas exchange and carbohydrate balance in sugarcane plants under root stressful conditions. Bragantia. 2012;**71**:319-327. DOI: 10.1590/S0006-87052012000300001

[57] Held IM, Delworth TL, Lu J, Findell KL, Knutson TR. Simulation of Sahel drought in the 20th and 21st centuries. Proceedings of the National Academy of Sciences of the United States of America. 2005;**102**(50):17891-17896

[58] Huntingford C, Hugo LF, Gash JHC, Taylor CM, Challinor AJ. Aspects of climate change prediction relevant to crop productivity. Philosophical Transactions of the Royal Society of London B. 2005;**360**(1463):1999-2010

[59] Wheeler TR, Craufurd PQ, Ellis RH, Porter JR, Prasad PVV. Temperature variability and the annualyield of crops. Agriculture, Ecosystems & Environment. 2000;**82**:159-167

[60] Matthieson L. Climate change and the Australian Sugarcane Industry: Impacts, adaptation and R&D opportunities. In: SRDC Technical Report, Sugar Research and Development; Brisbane, Australia. 2007

[61] Wiedenfeld B, Enciso J. Sugarcane responses to irrigation and nitrogen in semiarid South Texas. Agronomy Journal. 2008;**100**(3):665-671

[62] Glover J, Johnson H, Lizzio J, Wesley V, Hattersley P, Knight C. Australia's Crops and Pastures in a Changing Climate—Can Biotechnology Help? Canberra: Australian Government Bureau of Rural Sciences; 2008

[63] Li Y-R, Wei Y-A. Sugar industry in China: R & D and policy initiatives

to meet sugar and biofuel demand of future. Sugar Tech. 2006;**8**(4):203-216

[64] Patterson DT. Weeds in a changing climate. Weed Science. 1995;**43**:685-701

[65] Bailey SW. Climate change and decreasing herbicide persistence. Pest Management Science. 2004;**60**:158-162. DOI: 10.1002/ps.785

[66] Rodenburg J, Meinke H, Johnson DE. Challenges for weed management in African rice systems in a changing climate. The Journal of Agricultural Science. 2011;**149**:427-435

[67] Pereira MRR, Souza GSF, Martins D, Melhorança Filho AL, Klar AE. Respostas de plantas de Eleusine indica sob diferentes condições hídricas a herbicidas inibidores da ACCase. Planta Daninha. 2011;**29**:397-404

[68] Pereira MRR. Efeito de herbicidas sobre plantas de Brachiaria plantaginea submetidas a estresse hídrico. Planta Daninha. 2010;**28**:1047-1058

[69] Zanatta JF et al. Teores de água no solo e eficácia do herbicida fomesafen no controle de Amaranthus hybridus. Planta Daninha. 2008;**26**:143-155

[70] Strange RN, Scott PR. Plant disease: A threat to global food security. Annual Review of Phytopathology. 2005;**43**:83-116

[71] Showler AT, Reagan TE. Ecology and tactics for control of three sugarcane stalk-boring species in the Western Hemisphere and Africa. In: Goncalves JF, Correia KD, editors. Sugarcane: Production and Uses. Hauppauge, NY, USA: Nova; 2012. pp. 1-15

[72] Sanguino A. Impacto potencial das mudanças clima'ticas sobre as principais doenças da cana-de-açu'car no Brasil. In: Ghini R, Hamada E, editors. Mudanças Clima'ticas: Impactos sobre Doenças de Plantas no Brasil. Brası'lia, Brazil: Embrapa; 2008. pp. 207-213

[73] Kriticos DJ, Sutherst RW, Brown JR, Adkins SW, Maywald GF. Climate change and the potential distribution of an invasive alien plant: *Acacia nilotica* ssp. *Indica* in Australia. Journal of Applied Ecology. 2003;**40**:111-124

[74] Morrill AW. Commercial entomology on the west coast of Mexico. Journal of Economic Entomology. 1925;**18**:707-716

[75] Showler AT, Castro BA. Influence of drought stress on Mexican rice borer (Lepidoptera: Crambidae) oviposition preference and development to adulthood in sugarcane. Crop Protection. 2010;**29**:722-727

[76] Showler AT. Drought and arthropod pests of crops. In: Neves DF, Sanz JD, editors. Droughts: New Research. Hauppauge, NY, USA: Nova; 2012. pp. 131-156

[77] Mattson WL, Haack RA. Role of drought in outbreaks of plant-eating insects. Bioscience. 1987;**37**:110-118

[78] Neumeister L. Crop protection: Anything can happen. In: PAN Asia and the Pacific; Penang, Malaysia. 2010

[79] Kadam NN, Xiao G, Melgar RJ, Bahuguna RN, Quinones C, Tamilselvan A, et al. Agronomic and physiological response to high temperature, drought, and elevated CO_2 interaction in cereals. Advances in Agronomy. 2014;**127**:111-156

[80] Sharma SD, Singh M. Environmental factors affecting absorption and bio-efficacy of glyphosate in Florida beggarweed (*Desmodium tortuosum*). Crop Protection. 2001;**20**:511-516

[81] Schneider SH. The climatic response to greenhouse gases. Advances in Ecological Research. 1992;**22**:1-32

[82] Kumar A, Kumar M. Climate change's impacts on weeds and herbicide efficacy: A review. International Journal of Current Microbiology and Applied Sciences. 2017;**6**(9):2846-2853

[83] Reagan TE, Way MO, Beuzelin JM, Akbar W. Assessment of varietal resistance to the sugarcane borer and Mexican Rice borer. In: Sugarcane Research: Annual Progress Report. Baton Rouge, LA, USA: Louisiana State University AgCenter, Louisiana State University; 2008

[84] Showler AT, Reagan TE. Effects of sugarcane borer, weed, and nematode control strategies in Louisiana sugarcane. Environmental Entomology. 1991;**20**:358-370

[85] Singh MC, Dubey SC, Yaduraju NT. Climate change and its possible impacts on weeds. International Journal of Science, Environment and Technology. 2016;**5**(3):1530-1539

[86] Alberto AM, Ziska LH, Cervancia CR, Manalo PA. The influence of increasing carbon dioxide and temperature on competitive interactions between a C4 crop, rice (*Oryza sativa*), and a C3 weed (*Echinochloa glabrescens*). Australian Journal of Plant Physiology. 1996;**23**:793-802

[87] Hasanuzzaman M, Nahar K, Fujita M. Plant response to salt stress and role of exogenous protectants to mitigate salt-induced damages. In: Ecophysiology and Responses of Plants under Salt Stress. New York, NY: Springer; 2013. pp. 25-87

[88] Hasanuzzaman M, Hossain MA, Fujita M. Nitric oxide modulates antioxidant defense and the methylglyoxal detoxification system and reduces salinity-induced damage of wheat seedlings. Plant Biotechnology Reports. 2011;**5**(4):353

Chapter 9

Methane Emission Assessment from Indian Livestock and Its Role in Climate Change Using Climate Metrics

Shilpi Kumari, Moonmoon Hiloidhari, Satya Narayan Naik and Raj Pal Dahiya

Abstract

Indian livestock farming is one of the significant anthropogenic sources of methane (CH_4) in the world. Here, CH_4 emission from Indian livestock and climate change impact in terms of two climate metrics, global surface temperature change potential (GTP) and absolute GTP (AGTP), to assess the surface temperature changes for 20 and 100 year time frame have been studied. CH_4 emission from Indian livestock was 15.3 Tg in 2012. GTP_{20} and GTP_{100} for livestock-related CH_4 emission in India in 2012 were 1030 and 62 Tg CO_2e, respectively. The study also illustrates that CH_4 emissions can cause a surface temperature increase of up to 0.7–0.036 mK over the 20 and 100 year time periods, respectively. Thus, the negative climate change impact is global in nature, not only restricted to India. GTP and AGTP can be used in climate change impact study and as a more policy relevant tool.

Keywords: CH_4 emission, climate change, global temperature change potential (GTP), absolute GTP (AGTP)

1. Introduction

With the growing awareness toward the detrimental impacts of climate change, identifying and controlling of potential sources of greenhouse gas (GHG) emission have become a universal priority. Livestock farming is one of the most prominent anthropogenic sources of GHGs [1–3]. The total global GHG emission from livestock is 7.1 gigatonnes CO_2e year^{-1}, which accounts for 14.5% of all anthropogenic emissions [4, 5]. India, China, Brazil, and the USA are major regional contributors of GHG emission from livestock [6]. The growing economy and increasing demand for livestock products such as meat and dairy products increase challenges on livestock production and thus risk for climate change [7]. Therefore, it is very important in the coming future to reduce GHG emissions from livestock and promote sustainable livestock farming [8].

For sustainable livestock farming, climate change impact assessment of GHG emission and effective climate mitigation policies development are needed. For climate impact assessment, different climate metrics are being used to assess the climatic impact of non-CO_2 GHGs in terms of CO_2 equivalent emission [9, 10]. These climate metrics are estimated in tonnes of CO_2e per year by multiplying each non-CO_2 GHG emission with their absolute value [11]. Different climate metrics are available with different time horizons such as 20, 50, and 100 years, and it can be used for different non-CO_2 GHGs [6]. The assessment may be applied instantaneously or may be integrated over a specified period of time [6]. In IPCC first assessment report, global warming potential (GWP) is proposed as a method for comparing the potential climate impact of different non-CO_2 GHGs with reference to CO_2 [12]. But later on, the use of GWP in climate impact assessment has not been encouraged by many scientists as GWP does not explain the magnitude of climate change, i.e., impact on temperature rise [12, 13]. Thus, [14] proposed the global surface temperature change potential (GTP) as an alternative metric to GWP to assess climate change impact of GHG emission on climate change to assess its potential impact on surface temperature rise.

The GTP is the ratio of the change in the global mean surface temperature due to pulse or sustained GHG emission relative to CO_2 at a given time period. The GTP is more useful for those GHGs which have lifetime less than CO_2 such as short-lived GHG: CH_4 [15–17]. In comparison with GWP, the GTP gives climate impact in terms of change in temperature, and so it is a more policy-relevant tool for climate change impact mitigation [13, 15].

The negative climate change impact due to CH_4 emission is global in nature, not only restricted to India. Thus, the present chapter is focused on livestock-mediated CH_4 emission estimation in India and also to assess its role in climate change impact in terms of global surface temperature change potential (GTP) and absolute global surface temperature change potential (AGTP) for potential rise in surface temperature to identify the role of Indian livestock in climate change impact. This study focuses to evaluate the impact of livestock-mediated CH_4 emission on surface temperature change. Thus, the study helps researchers and scientists to predict climate change impact evaluation in terms of potential rise in global surface temperature using climate metrics due to any anthropogenic emission sources in future.

2. Methodology

The methodology is divided into three sections as presented in flow chart (**Figure 1**).

2.1 Livestock database collection

The livestock population database is taken from the Department of Animal Husbandry and Statistics, India, for the year 2012 [18]. The livestock census covers all the states (28) and 7 union territories (UTs) as well as all the districts (649) of India [19]. Once, the database is collected, it is sorted and categorized into four categories: cattle, buffalo, goat, and sheep. The cattle group is further categorized into two categories: dairy and nondairy cattle. Other livestock categories including population of pigs, horses, mules, and ponies are comparatively small (less than 5% of total livestock population) and therefore not included in the research work here.

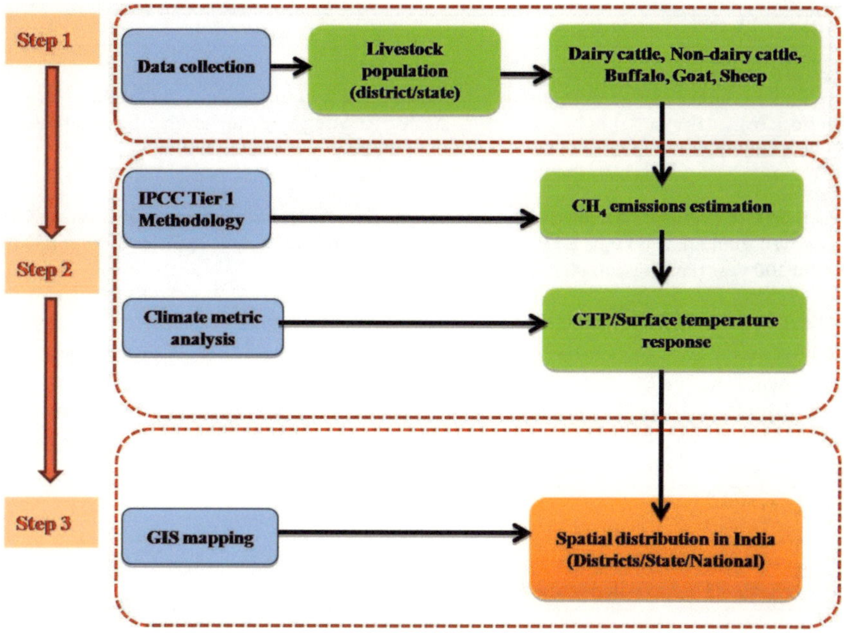

Figure 1.
Flow chart of methodology for estimation of CH_4 and climate metrics assessment. And results are represented in GIS mapping at district, state, and national level.

2.2 Estimation of CH_4 emission

Here, in IPCC guidelines, Tier 1 methodology is used for CH_4 emission estimation [20]. In IPCC Tier 1 methodology, country-wise livestock category-wise specific emission factors are available for enteric fermentation and manure management as shown in **Table 1**. The equation followed in CH_4 emission estimation is shown in **Table 2** as Eq. (1).

2.3 Other climatic metric assessments

The second objective of the present work of the book chapter is climate metric assessment of livestock-related CH_4 emission. Two climate metrics, viz., global surface temperature change potential (GTP) and absolute global surface temperature change potential (AGTP) and surface temperature response were applied for the CH_4 emission estimation from livestock at district, state, and national level.

Category		Enteric fermentation	Manure management
Cattle	Dairy cattle	58	5
	Non-dairy	27	2
Buffalo		55	4
Sheep		5	2
Goat		5	0.22

IPCC 2006 guidelines.

Table 1.
Specific CH_4 emission factor (kg CH_4 head^{-1} year^{-1}) of different livestock categories.*

Equations with their description
$E_d=\sum_{i=1}^{n}(p_i)\times EF_i$ (1) where, E_d is the CH_4 emission from enteric fermentation and manure management for the *i*th category of livestock (e.g., dairy cattle) in kg $year^{-1}$; p_i is the district wise population of *i*th category of livestock in million; and EF_i is the specific emission factor for *i*th category of livestock in kg CH_4 $head^{-1}$ $year^{-1}$
$GTP_{dt}=E_d\times GTP_t$ (2) GTP_{dt} is GTP of livestock-related CH_4 emission for *d*th district at time "**t**" (20 or 100 years), kg CO_2e; E_d is derived from Eq. 1; GTP_t is **GTP** at "**t**" time scale, which is equivalent to 67 for 20 year (GTP_{20}) and 4 for 100 year time horizon (GTP_{100}) [11]
$AGTP_{(CH_4)_t}=A_{CH_4}\times\sum_{j=1}^{2}\left[\left(\frac{\alpha\times c_j}{\alpha-d_j}\right)\times\left(e^{-t/\alpha}-e^{-t/d_j}\right)\right]$ (3) $AGTP_{(CH_4)_t}$ is the absolute global temperature potential of CH_4, K kg^{-1}, and t is 20 or 100 year time horizon; A_{CH_4} is radiative forcing of CH_4, 2.1×10^{-13} W $(kg\ m^2)^{-1}$; α is perturbation life or e-folding time of CH_4, 12 years; c_j is climate sensitive parameters and d_j response times [11]. c_1 and c_2 are 0.631 and 0.429, respectively; d_1 and d_2 are 8.4 and 409.5, respectively; $e^{-t/\alpha}$ is known as an impulse radiative flux (IRF), i.e., changes in instantaneous radiative flux due to pulse emission of GHGs
$\Delta T_t=E_d\times AGTP_{(CH_4)_t}$ (4) An annual CH_4 emission (kg) is multiplied by the **AGTP** values to arrive at the potential of temperature change (ΔT) in a given year (annual AGTP). In the equation, ΔT_t is temperature change response, K; E_d is CH_4 emission attributed by livestock, kg $year^{-1}$

Table 2.
Mathematical expression for CH_4 estimation and climate metric assessment used in methodology.

Climate metric GTP (CH_4) for two different time horizons, i.e., 20 and 100 years, is estimated as GTP_{20} and GTP_{100} as shown in Eq. (2) in **Table 2**. These two different assessments are highly significant for the GHGs, which have a shorter lifetime than CO_2 and more impact in a shorter time period than longer time horizon.

The AGTP estimates the temperature change (in Kelvin, K) at a time (t) associated with GHG emission as shown in Eq. (3) in **Table 2** [11, 12, 21]. The instantaneous surface temperature response (ΔT) is estimated by multiplication of annual CH_4 emission and AGTP [22]. Annual ΔT is used for evaluation of the direct temperature effects contributed by an annual rate of CH_4 emission over time from livestock as shown in Eq. (4) in **Table 2.**

2.4 GIS map generation

After the estimation of CH_4 emission and climate metric assessment from livestock CH_4 emission, GIS software, i.e., ArcGIS software, is applied to generation of spatial map for India up to state and district level. The GIS provides better understanding of results in the form of computerized spatial map. For GIS mapping, standard images have been collected from the National Remote Sensing Centre (NRSC), Government of India, for different districts and states of India. Once these standard images of the district level map and state level map of India have been collected, GIS mapping has been prepared. However, district level map could not be prepared for Jammu and Kashmir and represented at state level map, as their standard images up to district level are not available.

3. Results and discussion

The estimation of CH_4 emission from four different livestock categories, cattle, buffalo, goat, and sheep, in India are evaluated at districts, state, and national level using Eq. (1) mentioned in **Table 2**. In addition to CH_4 emission estimation, climate metrics, viz., global surface temperature change potential and absolute global surface temperature change potential and surface temperature response, are also estimated here (Eqs. (2)–(4), **Table 2**) to understand the climate change impact due to livestock-related CH_4 emission. The results are discussed below.

3.1 CH_4 emission

Using specific emission factors and IPCC Tier 1 methodology, the CH_4 emission in India was estimated to be 15.3 Tg CH_4 in 2012. CH_4 emission related to enteric fermentation is 92% of total CH_4 emission (14.20 Tg CH_4) and the rest 8% (1.16 Tg CH_4) of total CH_4 emission from manure management, respectively. Among the livestock groups, the highest CH_4 emission is contributed by the cattle group which is nearly 51% of total livestock CH_4 emission, and the lowest CH_4 emission is contributed by sheep (as shown in **Table 3**).

Among the 29 states, the top three most emitting states are Uttar Pradesh (2.89 Tg CH_4), followed by Rajasthan (1.52 Tg CH_4) and Madhya Pradesh (1.30 Tg CH_4), and the lowest is in Mizoram (0.018 Tg CH_4). The spatial representation of CH_4 emission at state level is represented through **Figure 2**. From the spatial diagram of livestock CH_4 emission, it is observed that the major emitting states are distributed across the western and the Indo-Gangetic plains of India. CH_4 emission contributions from all the eight northeastern states are only 3.88% of total national emission. The low CH_4 emission is due to less livestock population in comparison with the other states. Details of results of different category-wise livestock estimated CH_4 emission from each state also shown in **Table 4**.

Livestock categories	Enteric fermentation	Manure management	Total
Cattle	7.25	0.59	7.84
Buffalo	5.97	0.43	0.64
Sheep	0.68	0.03	0.71
Goat	0.3	0.13	0.43

Table 3.
National level CH_4 (Tg $year^{-1}$) emission from different categories of livestock.

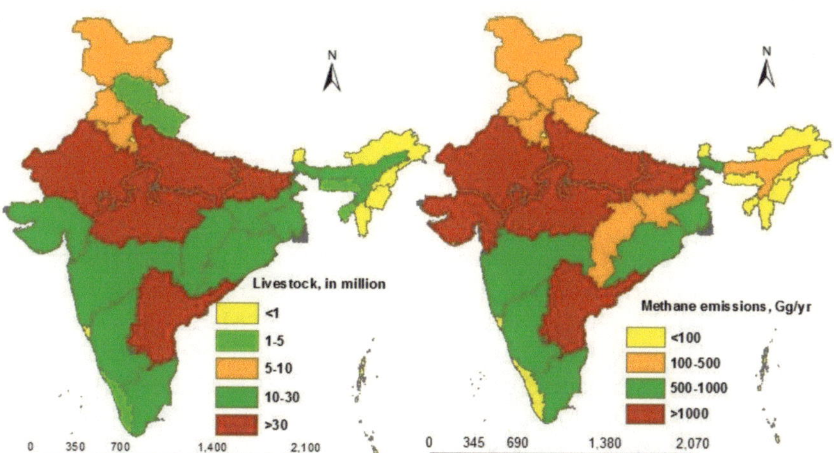

Figure 2.
Spatial distribution of CH_4 emission from livestock in India at state level.

State	Cattle	Buffalo	Sheep	Goat	Total
Andhra Pradesh	383	627	185	47	1242
Arunachal Pradesh	17	0	0	2	19
Assam	403	26	4	32	465
Bihar	508	446	2	63	1019
Chhattisgarh	373	82	1	17	473
Goa	2	0	0	0	2
Gujarat	417	613	12	26	1068
Haryana	78	359	3	2	442
Himachal Pradesh	93	42	6	6	147
Jammu and Kashmir	120	44	24	11	199
Jharkhand	328	70	4	34	436
Karnataka	410	205	67	25	707
Kerala	60	6	0	7	73
Madhya Pradesh	783	483	2	42	1310
Maharashtra	622	330	0	44	996
Manipur	10	4	0	0	14
Meghalaya	35	1	0	2	38
Mizoram	1	0	0	0	1
Nagaland	9	0	0	1	10
Orissa	442	43	0	34	519
Punjab	112	304	1	2	419
Rajasthan	586	766	64	113	1529
Sikkim	6	0	0	1	7
Tamil Nadu	392	46	34	43	515
Tripura	37	1	0	3	41
Uttar Pradesh	848	1807	9	81	2745
Uttarakhand	84	58	3	7	152
West Bengal	662	35	8	60	765
UTs	10	11	0	0	21

Table 4.
State-wise livestock category-wise CH_4 emission, Gg year^{-1} in the year 2012.

As there are significant variations in terms of livestock populations up to district level, CH_4 emission pattern also shows wide variations in India as shown in **Figure 3**. Banas Kantha, Gujarat (112 Gg CH_4); Paschim Medinipur, West Bengal (103 Gg CH_4); and Jaipur, Rajasthan (102 Gg CH_4) are top three districts in terms of livestock-related CH_4 emission. Furthermore, out of the total 15.3 Tg CH_4 emission in India, about 50% of the emission is contributed by 153 districts alone out of total 649 total districts. Within 153 districts, of the 4 livestock groups, maximum CH_4 emission (more than 50%) is contributed by buffalo in 84 districts followed by cattle (55 districts). Thus, this detailed GIS-based representation of the spatial distribution of CH_4 emission from livestock reveals that the highest emitting

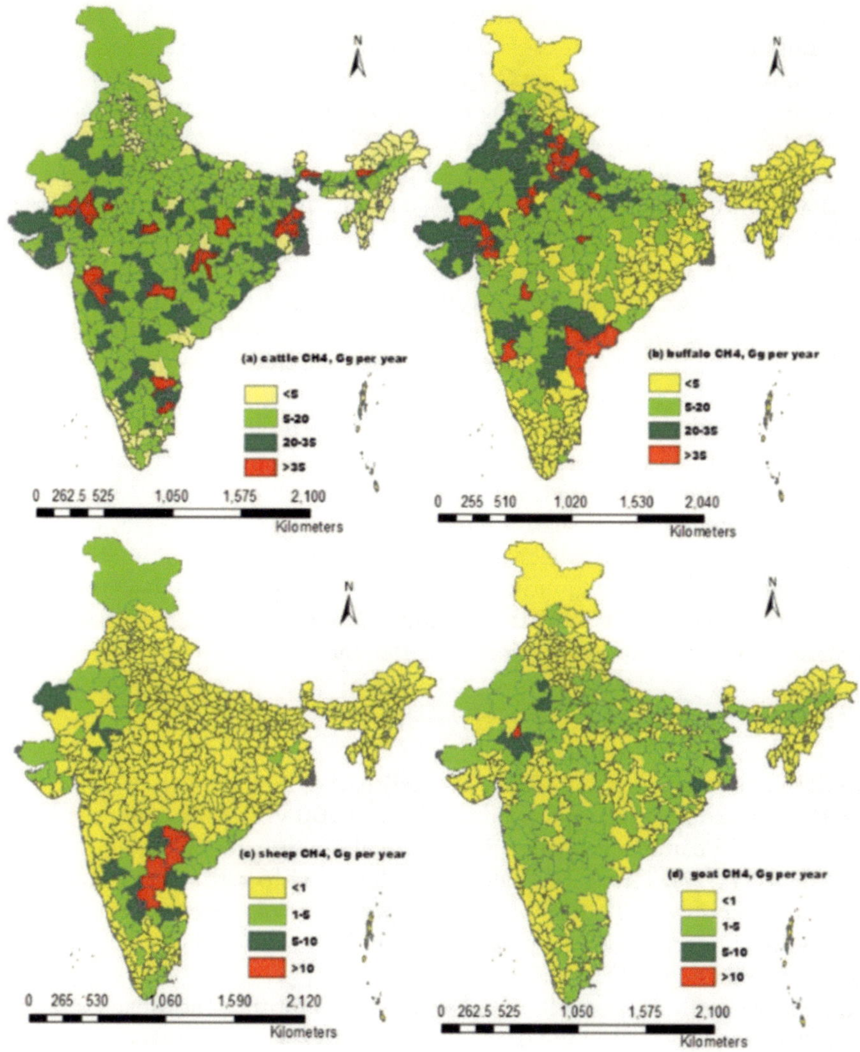

Figure 3.
CH_4 emission (Gg year^{-1}) from different categories of livestock at district levels in India, (a) emission from cattle, (b) emission from buffalo, (c) emission from sheep, and (d) emission from goat.

districts (emission >50% of total CH_4 emission) are located in the states of Uttar Pradesh, Gujarat, West Bengal, Rajasthan, Andhra Pradesh, and Maharashtra.

3.2 Climate metric assessment

The above estimation of livestock CH_4 emission is estimated further used to estimate its role in climate change using climate metrics in terms of GTP and AGTP. These are further elaborated to estimate surface temperature response (ΔT) from CH_4 emission due to Indian livestock. The results obtained from using Eqs. (2)–(4) (see **Table 2**) indicate significant contribution to GHG effect in global warming.

3.2.1 GTP of CH_4 emission

The estimated CH_4 emission data is used to calculate GTP at 20 and 100 year time horizon as GTP_{20} and GTP_{100}. GTP due to livestock CH_4 emission at 20 year time horizon is 1030 Tg CO_2e (GTP_{20}) while for 100 year time horizon 62 Tg CO_2e

(GTP_{100}). Among the livestock categories, cattle and buffalo are the major sources of CH_4 emission and hence for GTP. The GTP of cattle and buffalo together is worked out to more than 953.9 Tg CO_2e (GTP_{20}) and 56.9 Tg CO_2e (GTP_{100}), respectively, as given in **Figure 4**. The results also indicate that enteric fermentation is the major contributor (more than 90%) to GTP.

Similarly, at state level, GTP_{20} and GTP_{100} vary between 0.01–184 Tg CO_2e (GTP_{20}) and 0.007–18.0 Tg CO_2e (GTP_{100}), respectively, with the lowest in Mizoram and highest in Uttar Pradesh (**Table 5** and **Figure 5b** and **d**). At district level, GTP_{20} and GTP_{100} vary between 0.009–7.5 Tg CO_2e (GTP_{20}) and 3.75×10^{-6}–0.3 Tg CO_2e (GTP_{100}) (**Figure 5a** and **c**).

The GTP is a common unit of climate impact assessment per unit of GHG emissions. The results and findings of the climate metrics allow policymakers to develop GHG emission mitigation policies for different anthropogenic GHG emission sectors and for other non-CO_2 GHG gases [23]. The different time horizon for GTP measurement (e.g., 20 and 100 years) allows comparisons of the global warming impacts of a gas over a period of time [24, 25]. The larger the value of GTP, the higher will be the potential for temperature change by a given non-CO_2 GHG gas [15, 16, 26]. In the study, it is observed that climate change impact of CH_4 in GTP_{100} timeframe is smaller as compared to GTP_{20}, indicating that as the time horizon becomes longer, short-lived non-CO_2 GHG gases have less impact on GTP [10, 12]. This also suggests immediate requirements of mitigation measures for CH_4.

3.2.2 AGTP and surface temperature response (ΔT)

Similarly, climatic metric AGTP is also estimated, and it is worked out 4.56×10^{-14} and 2.28×10^{-15} K kg^{-1}, for 20 and 100 year time frames, respectively. The AGTP can be used to explore more about climate change impact assessment than GWP [27]. The AGTP value is further used to estimate surface temperature response (ΔT). The surface temperature response (ΔT) of CH_4 emission from the country for 20 year time frame is 0.70 mK (milli-Kelvin), and 100 year time frame is 0.036 mK.

At the state level, the highest global surface temperature response is observed resulting from CH_4 emission in Uttar Pradesh, with the lowest response resulting from CH_4 emission in Mizoram. CH_4 emission from livestock from different states can contribute to the surface temperature response (ΔT_{20}), ranging between 8.5×10^{-5} and 1.25×10^{-1} mK in 20 year time horizon. While in 100 year time horizon, ΔT_{100} varies from 4.23×10^{-5} to 6.50×10^{-3} mK for different states.

Potential rise in surface temperature due to Indian livestock sector that results from the annual CH_4 emission at district level is also evaluated here. At 20 year time

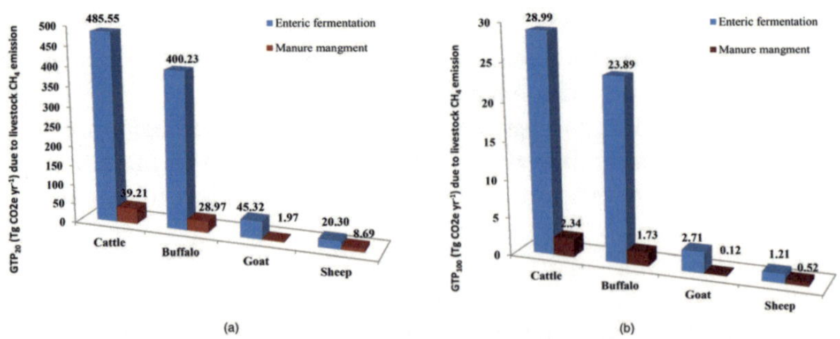

Figure 4.
Livestock category-wise GTP estimate for CH_4 emission at different time horizons (a) GTP_{20} and (b) GTP_{100}.

State	GTP_{20}	GTP_{100}
Andhra Pradesh	80.03	4.78
Arunachal Pradesh	1.29	0.08
Assam	31.09	1.86
Bihar	68.31	4.08
Chhattisgarh	31.65	1.89
Goa	0.17	0.01
Gujarat	71.30	4.26
Haryana	29.54	1.76
Himachal Pradesh	9.71	0.58
Jammu and Kashmir	12.86	0.77
Jharkhand	29.15	1.74
Karnataka	46.18	2.76
Kerala	4.87	0.29
Madhya Pradesh	87.75	5.24
Maharashtra	66.75	3.98
Manipur	0.98	0.06
Meghalaya	2.64	0.16
Mizoram	0.12	0.01
Nagaland	0.64	0.04
Odisha	34.75	2.07
Punjab	28.09	1.68
Rajasthan	101.29	6.05
Sikkim	0.44	0.03
Tamil Nadu	33.83	2.02
Tripura	2.72	0.16
Uttar Pradesh	183.79	10.97
Uttarakhand	10.12	0.60
West Bengal	51.12	3.05
UTs	1.54	0.09

Table 5.
State-wise GTP_{20} and GTP_{100} of CH_4 emission.

horizon, the ΔT_{20} varies from 1.53×10^{-7} to 0.005 mK due to Indian livestock sector. However, at 100 year time horizon, the ΔT_{100} varies from 7.66×10^{-9} to 0.0002 mK.

In addition to the above, the AGTP is also used to estimate the year-by-year response from a single year's CH_4 emission from livestock. The continuous analysis of AGTP is used to calculate the climate change impact on surface temperature using the annual AGTP calculation. The surface temperature change by the year (ΔT) is shown in **Figure 6**.

It is estimated that the surface temperature will keep rising till 2021 reaching the peak temperature rise (ΔT) 0.937 mK and would start decreasing thereafter. After few years of span beyond the year 2084, the surface temperature response would

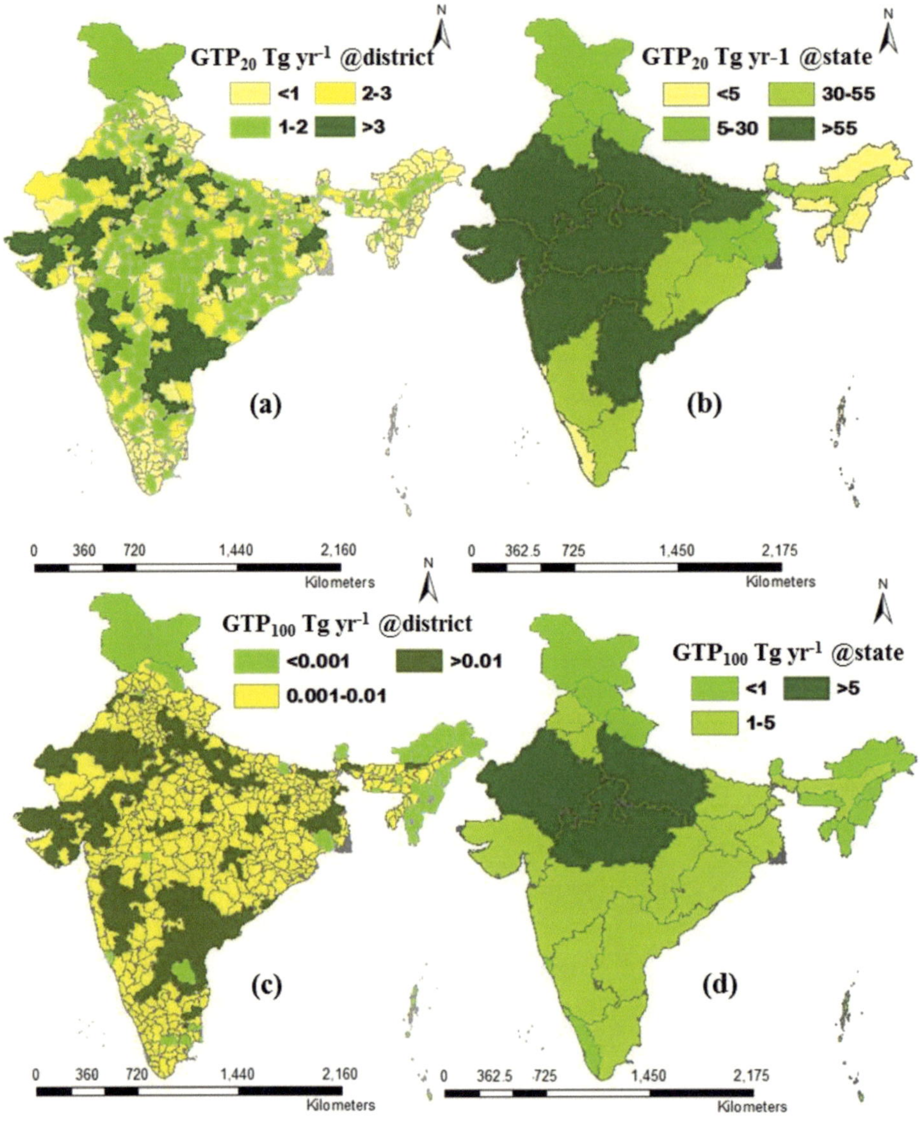

Figure 5.
GTP estimate of CH_4 emission in India: GTP_{20} of CH_4 in Tg CO_2e at (a) district and (b) state level; GTP_{100} of CH_4 in Tg CO_2e at (c) district and (d) state level.

asymptotically attain steady state. The continuous AGTP calculation is useful for policy makers when comparing multiple greenhouse gases. Due to high radiative forcing, CH_4 can cause large impacts on climate change on short time scales, but due to its short lifetime, that impact decreases more quickly than for longer-lived GHG gases. Although the potential rise in surface temperature due to different livestock size in states and districts is global in nature, their contribution from livestock is significantly variable with respect to different livestock sizes. Hence, estimating contribution from each state and each district will be useful for policy makers to develop decentralized mitigation policy. Thus, the surface temperature response gives significant information that CH_4 emission from livestock sector, even at small scale, can lead to significant climate change impact.

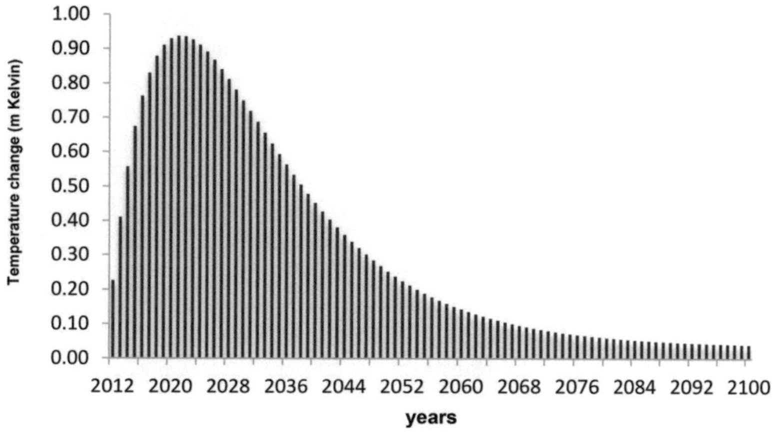

Figure 6.
Year-by-year surface temperature response (ΔT) due to constant rate of CH_4 emission, Tg year^{-1}.

3.2.3 Comparison between GTP and GWP

Here, CH_4 emission values are used to compare its GTP results with GWP values using GWP of CH_4, i.e., 34 [11]. The different values of GTP and GWP are given in **Table 6**. It is found that the results from GTP_{20} (1030 Tg CO_2e) to GTP_{100} (62 Tg CO_2e) drop off quickly compared to GWP_{20} (1292 Tg CO_2e) and GWP_{100} (430 Tg CO_2e). Both the climate metrics, GWP and GTP, are worked out in "CO_2 equivalents" but fundamentally different by construction, and therefore different numerical values can be expected [11]. If we look at the findings of GWP and GTP over the same period of time, GWP_{100} is higher than that of GTP_{100} due to the integrative nature of the GWP [11]. Also in the case of GTP_{20} and GTP_{100}, the GTP_{20} is 17 times higher than that of GTP_{100}, while GWP_{20} is only 3 times higher than that of GWP_{100}. The GTP calculation is based on assumptions about the climate sensitivity and heat uptake by the ocean and significantly varies with the change in these assumptions [11]. GTP is a metric which is used with reference to CO_2, and it is equal to the ratio of AGTP of reference gas and AGTP of CO_2. AGTP is the absolute GTP that gives temperature change per unit of GHG emission. As already discussed, GTP is an endpoint metric therefore for short GHG having half-life less than CO_2; its climate metric, taken for large time horizon, is less than that of climate metric calculated for short time horizon [11]. The differences in GTP and GWP could be due to the fact that the GTP accounts the atmospheric adjustment time scale of the

Category	Enteric fermentation				Manure management			
	GTP_{20}	GTP_{100}	GWP_{20}	GWP_{100}	GTP_{20}	GTP_{100}	GWP_{20}	GWP_{100}
Cattle	485.55	28.99	608.75	202.92	39.21	2.34	49.16	16.39
Buffalo	400.23	23.89	501.78	167.26	28.97	1.73	36.32	12.11
Goat	45.32	2.71	56.82	18.94	1.97	0.12	2.47	0.82
Sheep	20.30	1.21	25.45	8.48	8.69	0.52	10.90	3.63
Total	951.40	56.80	1192.80	397.60	78.84	4.71	98.85	32.95

Table 6.
Comparison between GTP_{20}, GTP_{100}, GWP_{20}, and GWP_{100} of estimated CH_4 emission from livestock.

component and the response time scale of the climate system, which is not considered in the GWP. Climatic impact assessment has been facing difficulties when comparing the effect of short- and long-lived GHGs. The GWP and GTP of long-lived gases are the same [10]. However, for short-lived GHGs, the GWP does not account the radiative forcing for a short period.

Therefore, the GTP has been proposed for the comparison of the impact of GHG emissions on temperature changes at a specific time in future rather than the radiative forcing over a period of time [23]. Hence, we can say that the GTP compares temperatures at the end of a given time period due to GHG emissions. In comparison to GWP, GTP extends the information from radiative forcing to rise in the surface temperature relative to that of CO_2 [11]. The GTP further extends the cause-effect chain by adding the temperature impact assessment in comparison with GWP and hence more relevant by comparing temperature changes [28]. The GTP is a function of time and used for analyzing the economic benefits from emission reduction. Therefore, it is useful to develop cost-effective policy for mitigation policies targeting temperature reduction.

Overall the results estimated here are compiled in **Table 7** in which the minimum, the maximum, and average are given.

3.3 Uncertainty analysis

The CH_4 emission estimation depends mainly on two factors, i.e., livestock population and CH_4-specific emission factors of different types of livestock categories. Both the factor could be a source of uncertainty. For the livestock population database, we rely on livestock census taken from the reports published by the Government of India [29], and emission factors are collected from the IPCC report [20]. During livestock census, the database collection based on only 5% of the total livestock population is used for sampling purposes during the census, which is then aggregated into 100% data. This creates uncertainty in the methodology. Also, in IPCC guidelines 2006, three types of estimation methodology are proposed, i.e., basic method IPCC Tier 1, intermediate method IPCC Tier 2, and complex method IPCC Tier 3. As the method becomes advance, uncertainty related to methodology decreases. As found by Patra [30], Tier 1 method overestimates the CH_4 emission by 15% compared to Tier 2 estimate. But, IPCC Tier 1 is readily available which covers for national or international level in combination with default emission

	CH_4 (Tg year^{-1})	GWP (Tg CO_2e)	GTP_{20} (Tg CO_2e)	GTP_{100} (Tg CO_2e)	ΔT_{20} (mK)	ΔT_{100} (mK)
Country level	15.3	523	1030	61.51	0.70	0.036
State level						
Minimum	0.12	4.06	0.01	0.00	0.00	0.00
Maximum	2.74	93.35	183.79	10.97	0.13	0.006
Average	0.43	14.93	29.22	1.74	0.02	0.001
District level						
Minimum	0.00	0.00	0.00	0.00	0.00	0.000
Maximum	0.11	3.82	7.53	0.45	0.002	0.003
Average	0.02	0.81	1.59	0.10	0.0005	0.0006

Table 7.
Results of CH_4 emission and other climate metrics at national, state, and district levels.

factors. Therefore, it is feasible for all countries. But, country-specific or even smaller region-specific emission factors would bring more precise information. However, such issues could not be considered in the present work and would require further investigation.

4. Conclusions

The findings of the study are CH_4 emission, high GTP and surface temperature response at district level, state level, and national level in India. The total CH_4 emission in India is 15.3 Tg in 2012, with the highest almost 92% of the emission occurring via enteric fermentation. The GTP due to CH_4 emission at 20 and 100 year time horizon in India is 1030 Tg GTP_{20} CO_2e and 62 Tg GTP_{100} CO_2e, respectively. The livestock emission in India has the potential to cause the surface temperature rise up to 0.69 mK and 0.036 mK over 20 and 100 year time period, respectively. At a state level, the emission can cause the surface temperature response (ΔT) to vary from 8.49×10^{-5} to 1.25×10^{-1} mK in 20 year time horizon and from 4.23×10^{-5} to 6.25×10^{-2} mK in 100 year time horizon. On the other hand, at district level, the ΔT varies from 1.53×10^{-7} to 0.005 mK in 20 years and from 7.66×10^{-9} to 0.0002 mK in 100 years' time frame. The GTP values of CH_4 for 20 and 100 years are 67 and 4, respectively. The AGTP values for the same time horizons are 4.6×10^{-14} and 2.3×10^{-15} K kg^{-1}. GTP is a metric, which is used in comparing multiple gases with reference to CO_2, whereas AGTP is the absolute GTP giving temperature change per unit of GHG emission. Temperature indices like GTP and AGTP both give the surface temperature change and response using pulse emission. GTP of any greenhouse gas is equal to the ratio of AGTP of the given gas and AGTP of CO_2. The AGTP measures the temperature change over the period of time after the GHG emission. It depends upon some factors such as climate sensitivity and ocean uptake of heat by the ocean. All of these factors response vary with the time horizon and may substantially modify climate metrics GTP and AGTP.

So, it follows a decreasing trend with an increase over the period of time from 20 to 100 years. GTP and AGTP follow the same pattern and also decrease with the year. These temperature indices GTP and AGTP both can be used to study the impact on surface temperature due to GHG emission with time. This finding helps to study the climate change impact on surface temperature from CH_4 emission, which can cause climate damage over a short period of time, even emitted in small quantity.

Acknowledgements

Shilpi Kumari is thankful to the University Grant Commission, Government of India for UGC-SRF, for providing research fellowship (JRF) (Sr. No. 2121120406 and Ref. No: 18-12/2011(ii) EU-V).

Author details

Shilpi Kumari[1*], Moonmoon Hiloidhari[2], Satya Narayan Naik[3] and Raj Pal Dahiya[1]

1 Centre for Energy Studies, Indian Institute of Technology Delhi, New Delhi, India

2 IDP in Climate Studies, Indian Institute of Technology Bombay, Mumbai, India

3 Centre for Rural Development and Technology, Indian Institute of Technology Delhi, New Delhi, India

*Address all correspondence to: shilpidas.iit@gmail.com

References

[1] Casey JW, Holden NM. The relationship between greenhouse gas emissions and the intensity of milk production in Ireland. Journal of Environmental Quality. 2005;**34**(2): 429-436. DOI: 10.2134/jeq2005.0429

[2] Garnett T. Livestock-related greenhouse gas emissions: Impacts and options for policy makers. Environmental Science & Policy. 2009; **12**(4):491-503. DOI: 10.1016/j.envsci.2009.01.006

[3] Kumari S, Dahiya RP, Kumari N, Sharawat I. Estimation of methane emission from livestock through enteric fermentation using system dynamic model in India. International Journal of Environmental Research and Development. 2014;**4**:347-352

[4] Gerber PJ, Steinfeld H, Henderson B, Mottet A, Opio C, Dijkman J, et al. Tackling Climate Change through Livestock: A Global Assessment of Emissions and Mitigation Opportunities. Food and Agriculture Organization of the United Nations (FAO). Rome; 2013

[5] Kumari S, Dahiya RP, Naik SN, Hiloidhari M, Thakur IS, Sharawat I, et al. Projection of methane emissions from livestock through enteric fermentation: A case study from India. Environmental Development. 2016;**20**: 31-44. DOI: 10.1016/j.envdev.2016.08.001

[6] IPCC. Climate Change: The Physical Science Basis. Contribution of Working Group I to the Fifth Assessment Report of the IPCC. Cambridge and New York: Cambridge University Press; 2013

[7] Steinfeld H, Gerber P. Livestock production and the global environment: Consume less or produce better? Proceedings of the National Academy of Sciences. 2010;**107**(43):18237-18238. DOI: 10.1073/pnas.1012541107

[8] Kipling RP, Bannink A, Bellocchi G, Dalgaard T, Fox NJ, Hutchings NJ, et al. Modeling European ruminant production systems: Facing the challenges of climate change. Agricultural Systems. 2016;**147**:24-37. DOI: 10.1016/j.agsy.2016.05.007

[9] Fuglestvedt JS, Berntsen TK, Godal O, Sausen R, Shine KP, Skodvin T. Metrics of climate change: Assessing radiative forcing and emission indices. Climatic Change. 2003;**58**(3):267-331

[10] Huntingford C, Lowe JA, Howarth N, Bowerman NH, Gohar LK, Otto A, et al. The implications of carbon dioxide and methane exchange for the heavy mitigation RCP2.6 scenario under two metrics. Environmental Science & Policy. 2015;**51**:77-87. DOI: 10.1016/j.envsci.2015.03.013

[11] IPCC (Intergovermental Panel on Climate Change). Climate change 2014. In: IPPC Fifth Assessment Report. Geneva, Switzerland: IPCC; 2014

[12] Shine KP, Fuglestvedt JS, Hailemariam K, Stuber N. Alternatives to the global warming potential for comparing climate impacts of emissions of greenhouse gases. Climatic Change. 2005;**68**(3):281-302

[13] Sarofim MC. The GTP of methane: Modeling analysis of temperature impacts of methane and carbon dioxide reductions. Environmental Modeling and Assessment. 2012;**17**(3):231-239. DOI: 10.1007/s10666-011-9287-x

[14] Shine KP, Fuglestvedt JS, Hailemariam K, Stuber N. Alternatives to the global warming potential for comparing climate impacts of emissions of greenhouse gases. Climatic Change. 2005;**68**(3):281-302

[15] Shine KP, Berntsen TK, Fuglestvedt JS, Skeie RB, Stuber N. Comparing the

climate effect of emissions of short-and long-lived climate agents. Philosophical Transactions of the Royal Society of London A: Mathematical, Physical and Engineering Sciences. 2007;**365**(1856): 1903-1914. DOI: 10.1098/rsta.2007.2050

[16] Peters GP, Aamaas B, Berntsen T, Fuglestvedt JS. The integrated global temperature change potential (iGTP) and relationships between emission metrics. Environmental Research Letters. 2011;**6**(4):044021. DOI: 10.1088/1748-9326/6/4/044021

[17] Persson UM, Johansson DJ, Cederberg C, Hedenus F, Bryngelsson D. Climate metrics and the carbon footprint of livestock products: Where's the beef? Environmental Research Letters. 2015;**10**(3):034005

[18] Government of India. 19th Livestock Census 2014. New Delhi: Ministry of Agriculture, Govt. of India, Department of Animal Husbandry, Dairying and Fisheries; 2014

[19] BAHS. Ministry of Agriculture and Farmers Welfare Department of Animal Husbandry, Dairying and Fisheries. India. 2012

[20] IPCC. In: Eggleston HS, Buendia L, Miwa K, Ngara T, Tanabe K, editors. IPCC Guidelines for National Greenhouse Gas Inventories, Prepared by the National Greenhouse Gas Inventories Programme. Japan: IGES; 2006

[21] Farquharson D, Jaramillo P, Schivley G, Klima K, Carlson D, Samaras C. Beyond global warming potential: A comparative application of climate impact metrics for the life cycle assessment of coal and natural gas based electricity. Journal of Industrial Ecology. 2017;**21**(4): 857-873. DOI: 10.1111/jiec.12475

[22] Giuntoli J, Agostini A, Caserini S, Lugato E, Baxter D, Marelli L. Climate change impacts of power generation from residual biomass. Biomass and Bioenergy. 2016;**89**:146-158. DOI: 10.1016/j.biombioe.2016.02.024

[23] Manning M, Reisinger A, Bodeker G. Global Warming Potentials and Alternate Metrics. New Zealand Climate Change Research Centre; 2009

[24] Boucher O, Friedlingstein P, Collins B, Shine KP. The indirect global warming potential and global temperature change potential due to methane oxidation. Environmental Research Letters. 2009;**21**(4):4, 044007. DOI: 10.1088/1748-9326/4/4/044007

[25] Joos F, Roth R, Fuglestvedt JS, Peters GP, Enting IG, Bloh WV, et al. Carbon dioxide and climate impulse response functions for the computation of greenhouse gas metrics: A multi-model analysis. Atmospheric Chemistry and Physics. 2013;**13**(5):2793-2825. DOI: 10.5194/acp-13-2793-2013

[26] Kumari S, Hiloidhari M, Kumari N, Naik SN, Dahiya RP. Climate change impact of livestock CH_4 emission in India: Global temperature change potential (GTP) and surface temperature response. Ecotoxicology and Environmental Safety. 2018;**147**:516-522. DOI: 10.1016/j.ecoenv.2017.09.003

[27] Fagodiya RK, Pathak H, Kumar A, Bhatia A, Jain N. Global temperature change potential of nitrogen use in agriculture: A 50-year assessment. Scientific Reports. 2017;**7**:44928. DOI: 10.1038/srep44928

[28] Shine KP. The global warming potential—The need for an interdisciplinary retrial. Climatic Change. 2009;**96**(4):467-472. DOI: 10.1007/s10584-009-9647-6

[29] BAHS 2014. Ministry of Agriculture and Farmers Welfare Department of Animal Husbandry, Dairying and Fisheries. India

[30] Patra AK. Prediction of enteric methane emission from buffaloes using statistical models. Agriculture, Ecosystems & Environment. 2014;**195**: 139-148. DOI: 10.1016/j.agee.2014.06.006

Chapter 10

Climate Change and Uncharted Social Challenge in Existing Urban Setup in Bangladesh

Reazul Ahsan

Abstract

The impact of extreme climate events on human settlements has been startling, demonstrated by events such as Hurricane Katrina in the USA, Cyclone Nargis in Myanmar and Cyclones Sidr and Aila in Bangladesh. People, particularly those living in vulnerable coastal zones, face forced displacement due to such extreme climate events and need to search for alternative livelihoods. In most cases they ended up in nearest cities for income and livelihoods. Bangladesh represents a region where a complex set of climate-driven outcomes is already evident, as land is inundated, and populations migrate in large numbers in search of livelihoods. Using the region as an example, this research examines climate change impacts ranging from the primary impacts on natural systems, through secondary population displacement and migration, to the eventual outcomes of rapid urbanisation and urban poor. Tertiary impacts are defined as social changes in the urban system and deprivation of social justice for those migrants. The scope of this paper is to understand the issues and challenges associated with climate-induced displacement and policy applications to ensure social justice for those migrant communities.

Keywords: climate migration, urbanisation, impact levels, social changes, change of urban system

1. Introduction

The concept of climate change is understood as a gradual change in global physical conditions, encompassing changes in air and sea temperatures, sea level rise, changes in precipitation levels and seasonal climate variations [1]. Such changes, often manifested dramatically as sudden violent storms, are considered long-term processes with added latent risk factors. Their impacts are likely to undermine the essentials of human lives and social systems. These essentials include safe access to food, fresh water, health, home, land and employment. In current world it has been evidenced that human action is the greatest triggering force for global climate change trend. Therefore, onus is on us to identify the likely changes and to deal with the consequences. Expected climate changes are no longer limited within our scientific understanding, such as change of temperature and precipitation levels but also has long term social impacts which could also be addressed as slow-onset impact of climate change. Rising sea levels are believed to further increase the multiple risks from extreme climate events, including cyclones and frequent and prolonged floods, tidal surges and increasing salinity intrusion [2].

Increasing sudden-onset natural hazards will result in substantial human displacement from different parts of the world by 2050, who will be known as climate migrants or climate refugee [3–5]. The IPCC Fourth Assessment Report [6] addresses natural hazards and resulting forced migration, suggesting such development could move migration to a new magnitude. In early 2008, another 15 million people in Sichuan, China, and 2.4 million in Myanmar were displaced because of natural disasters [7–8]. It is widely predicted that this will increase poverty and provoke a crisis in fresh water supply, shelter, food and energy, leading to socioeconomic disparity and playing a pivotal role in significant changes to social justice.

Urban slums are usually the first destination for displaced people. Approximately 1 billion people live in urban slums; another 50 million are likely to be displaced by climate factors and will add to urban slum populations in Asia, Africa and Latin America [9]. Here they have limited access to employment, water, shelter, sanitation and basic amenities. Often, such forced displacement trades physical safety for the basic human rights these people have previously enjoyed – namely, the right to development and the right to live in their own society and culture – until a point where survival is at stake [10].

Among different social and economic adaptation approaches, involuntary migration is notable, especially for vulnerable coastal communities in the developing world [11]. Parvin and Shaw [12] also noted that climate migrants are usually poor, locating themselves in congested inner-city areas for easier access to jobs or on the urban fringes because of lower rents. Such unplanned and largely illegal and certainly perilous land occupancy changes urban land-uses, extending uncontrolled urban peripheral growth and increasing urban poverty and poor quality of living standard. Socioeconomic inequalities are increasing in urban areas. Migrants who have lost almost everything due to climate change can become trapped in a downward spiral of urban poverty in their migration destination and become a social threat.

This study explores climate change as a potential trigger of new social challenge in existing urban setup and it is become a challenge for those vulnerable marginal communities, using Bangladesh as a case study. Bangladesh is a low-lying coastal zone, where internal migration has long served as an adaptive strategy, but where climate adaptation is seldom addressed in development policy.

2. Research design

South coast region of Bangladesh is home to 35 million people at a density of 738 persons/km^2. It is also projected that by 2050 total population would be 40–50 million in the same area [13]. These coastal communities, mainly dependent on subsistence agriculture and fishing on the fertile plains along rivers and the coast, and they are the prime victims of extreme climate events. To gain a deeper understanding of climate migration and human life changes in Bangladesh, this study conducted a field investigation in Khulna, the capital city of south coast, with a jurisdiction of 4394.46 km^2 and a home for 3 million growing at 3.8% per annum [14].

Climate induced migrants are coming from the poor communities along the coastal areas have located themselves in the urban fringe and slums, where they seek cheap accommodation and unstable access to low-paid employment.

With the support of local and national NGOs – namely, 'Pothikkrreith' – a mass-scale dialogue session with audio-visual presentation on climate change and displacement was conducted with such residents, around 700 people in Khulna. From that mass-scale dialogue and participant list, 200 respondents were purposely chosen as the targeted community. These targeted people had all migrated or been displaced – due to cyclones, flash floods, river erosion or drought – from different

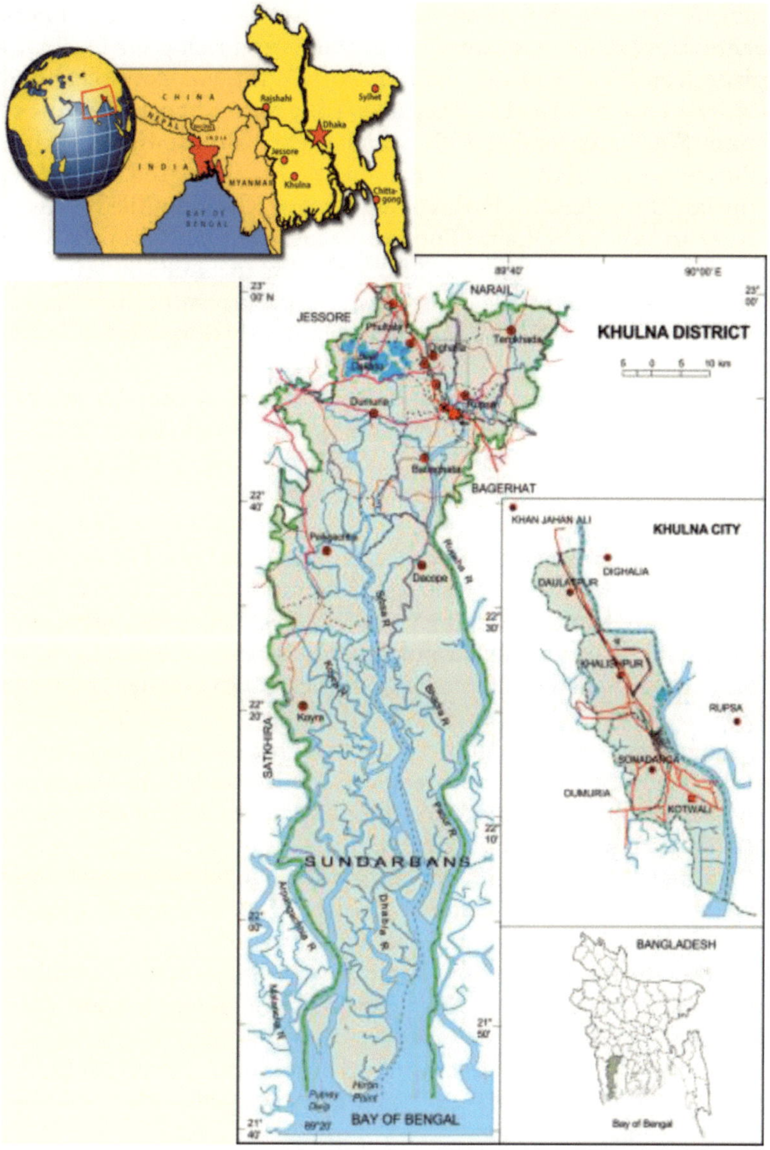

parts of the coastal districts, so that diversity could be reflected in the sample. Out of the 200 selected people, 100 respondents were then chosen through systematic random sampling, with every alternate person being chosen to provide answers to a questionnaire survey.

This research reflects the migrants' opinions about the challenges they face in an urban setting where they have very limited access to urban facilities and have encountered diminished social justice as an indirect impact of climate change, which has largely been overlooked under social crisis management strategies.

3. Climate change and Bangladesh

Bangladesh is located between 20°34′-26°-38’ N and 88°01′-92°41′ E. Bangladesh is bordered on the west, north and east by India, on the south-east by Myanmar and on the south by the Bay of Bengal [15]. The country is located at the unique juxtaposition of the composite, which makes Bangladesh highly exposed and vulnerable to extreme climate events. The impacts of changing temperatures, level of precipitation, more extreme weather events and rising sea level are already felt in Bangladesh and will continue to intensify [16, 17]. For example, hundreds of thousands of lives have been lost in Bangladesh due to recent catastrophic cyclones, notably Cyclone Sidr in November 2007, which caused almost 4000 deaths. Moreover, the frequency of cyclones in Bangladesh has increased more than five times over the last three decades [16]. Following Sidr, Cyclone Aila hit the coast on 25 May 2009 and about half a million people lost their land, homestead and livelihoods and had no option but to sail for a new life to an unknown place. Large areas of productive farmland were rendered useless as a result of inundation and subsequent salinity and in other cases were permanently flooded, forcing residents to migrate.

In their search for alternative livelihoods, displaced communities generally opt for nearby urban destinations where they have opportunities to earn and livelihoods are perceived to be available. Hence, they can be labelled as climate migrants and are really a sub-category of internally displaced persons.

It is estimated that each year between 300,000 and 400,000 people migrate to the country's capital to improve their economic prospects [18]. The poor internal migrants usually end up in the urban slums. The proportion of internal migration within urban slum areas varies, ranging from 53% in Dhaka (the capital) to 70% in Khulna and Rajshahi districts. The proportion of migrants is higher in the coastal belt districts because of extreme climate events including cyclones and frequent surges [19].

Figure 1 shows the number of natural disasters that have hit Bangladesh since 1993 and the total number of affected people. Concurrent natural disasters at high intensity have forced many victims to consider migration as their only viable adaptive response (data adapted from EM-DAT [20]).

Since 1993, almost every year the country gets affected with different natural disasters including flood, storms and cyclone. With high population density each disaster affects a huge number of population and that act as a natural push for displacement.

Figure 2 shows the migration trend and the underlying reasons for migration, based on results of a field survey of 100 heads of households who were identified as climate migrants.

Based on primary data, the largest numbers of migrants identified themselves as driven out by Cyclone Aila in May 2009, both by the cyclone itself and by subsequent salinity problems.

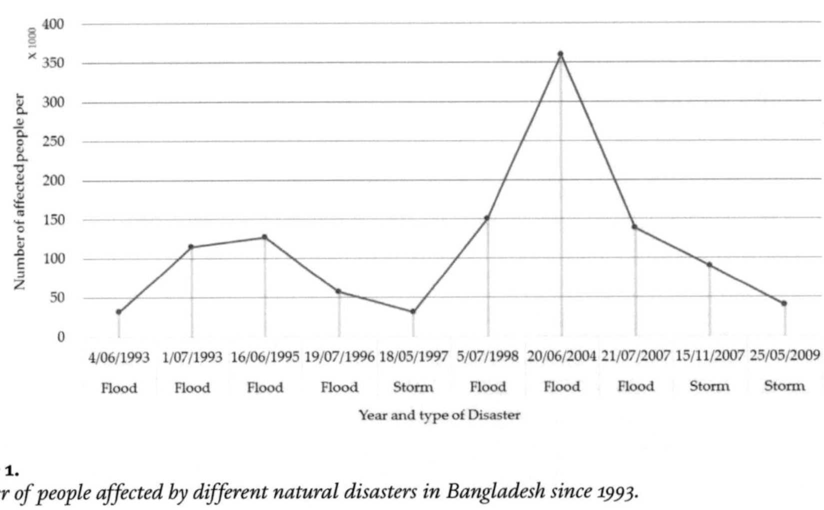

Figure 1.
Number of people affected by different natural disasters in Bangladesh since 1993.

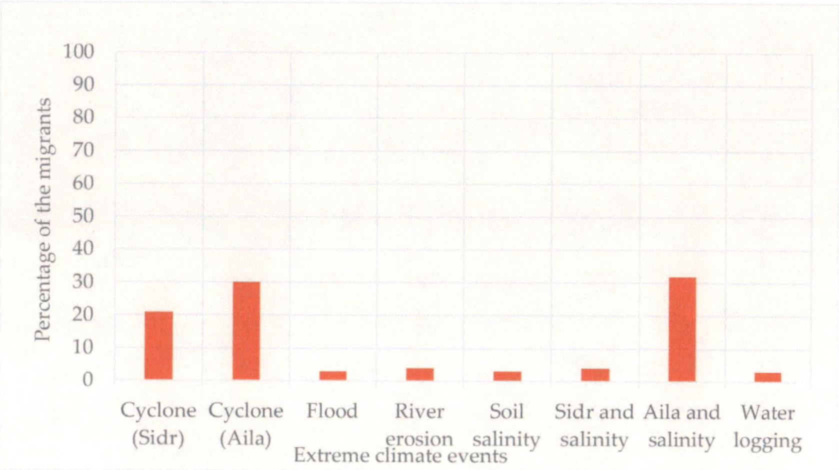

Figure 2.
Migration trend due to extreme climate events from 1980 to 2010 [21].

4. Climate migration and social changes

The suffering and hardship of the climate migrants is a never-ending challenge. They faced natural hardships and hazards at their point of origin, and these hardships do not end when they arrive at their urban destination. Here they experience multiple aspects of deprivation, which restrict their efforts to attain a decent standard of living. For example, more than 65% of residents in the urban slums of Khulna do not have any sanitation facilities, 45% of slum dwellers do not have any fixed place to dispose of their garbage, and 35% are outside the garbage collection system [22]. This increased influx of migrants, driven by extreme climate events, poses new burden for the existing urban services.

Concerning climate migration and social challenges, it was found that the migration destination offered them little or no improvement in their housing, as they could not afford the cost of formal housing. Few migrant households could gain access to established urban slums. However, a big number of the migrant communities have no alternatives but to established themselves as squatters in urban fringe areas, on marginal agricultural land, along rail corridors, next to the highway, or even in the natural drainage network, as well as in low-lying flood-prone areas and on river banks, using very informal construction materials. These various forms of informal settlements provided immediate shelter for some of those migrants but failed to ensure social security and justice for them in an urban setup. **Figure 3** shows the different land uses occupied by migrants in Khulna City, offering them some kind of shelter.

About half of the total respondents in Khulna located themselves in urban fringe areas. This urban fringe land has been preserved for further urban expansion or to support landless or homeless communities. **Figure 4** portrays the informal settlements/housing of those migrants in urban fringe.

This research demonstrated that climate actions not only forced migrants to find a new place of residence, but also forced them to accept any kind of low-paid occupation to survive in the urban areas. Migrant communities were forced to become involved with different and often new economic activities to support themselves. Since they often have no experience beyond farming or fishing, the city does not offer them many options. For example, most of the jobs available in urban areas are basic daily service jobs. Migrants lack job security and are particularly vulnerable as they might not get work every day. Migrants also often needed to work longer hours to meet the cost of living in the city.

Figure 3.
Location of climate migrants in urban areas [21].

Table 1, showing the migrants' occupation at their origin and their current occupation in the city since they migrated. Following table, clearly indicates they have little-option but to accept whatever work they can find in the city to survive, and in most cases, those are labor-intensive occupations, even for the women.

Table 1 indicates that almost half of the respondents working as a day labourer in the city were involved in different occupations at their place of origin. For example, they had been farmers, seasonal workers and small businessmen, but now forced to be a day labourer. Many of them have no experience about city life, earning and living style and standard in a complex urban system. That put them at the edge of vulnerability, when they are already displaced from their origin and vulnerable. Approximately 24 respondents were rickshaw pullers, and almost none of them had any previous experience in such labour-intensive jobs. Even the women who were housewives before migrating now forced to work as day labourers and domestic helpers to support their families to survive in the city. This change

Figure 4.
Typical informal settlements of migrant communities in urban fringe of Khulna [21].

Occupation at place of origin	Occupation at migration destination								
	Day labourer	Rickshaw puller	Petty business	Construction	Office	Van puller	Jobless	Housemaid	Total
Agriculture	11	4		1			2		18
Fisherman	8	2	2	1					13
Fishing business	8	7	4			1			20
Petty business	5	3	2	1	3				14
Land labourer	13	4		1		4			22
Housewife	1			1			1	1	4
Day labourer	3	3		2					8
Student		1							1
Total	49	24	8	7	3	5	3	1	100

Table 1.
Occupation of migrant household heads in urban areas compared to their previous occupation at their place of origin [21].

in economic status also affects social relations and self-esteem. The following case demonstrates how climate migration and occupational change can affect the social and family structures of migrant households.

> *Mariam is a 32-year-old woman who migrated with her husband and two boys from Kyora to Khulna due to Cyclone Aila. Mariam's family decided to migrate as they had lost everything, including their home and agricultural land at their place of origin. Her husband was working as a day labourer, her 8-year-old son worked in a shop and her 4-year-old son stayed at home with her. Her husband's and son's incomes were not enough to pay rent and buy food in the city, as they did not get work every day. Therefore, her husband forced her to try to earn extra income, but she failed, as she had never been to the city and was not skilled enough to fit into the urban job market. As she failed to support her family, her husband divorced her. Now Mariam is living on the urban fringe with her 4-year-old son and is serving as a housemaid. Forced migration has made her earn an income but has taken away her family.*

The lack of job security and opportunity or regular income for climate migrants in the informal sector has direct impacts on their household income.

The housing conditions of migrant are proportionate to their income and their access to urban facilities. They do not have access to established urban slums and minimum standard of life, like fresh water supply, sanitation and waste management. Therefore, they try to build their own place which is very informal in nature, mostly made of mud, hard plastic, cardboard, dry leaves and bamboo. None of these materials are formally used as construction materials. Houses or structures in the existing slums are constructed with similar materials and provide shelter for the urban poor and migrants but have very limited facilities such as access to clean water and toilets. **Figure 5** shows the typical condition of informal settlements of the migrants, which completely fail to meet the minimum standard of life of the residents.

The typical size of a slum house is 14 m^2 for a family of four. The average size of an individual squatter dwelling on urban public land and in fringe areas is 10 m^2, and occupants are always under threat of eviction either by the government or local musclemen, if they fail to pay the required tolls.

Figure 5.
Living standard and quality of life of the migrants [21].

In Khulna, climate migrants clustered on the urban fringe are highly exposed and remain vulnerable to floods and waterlogged soil. At the same time, complex urban governance systems have no options or policy strategies to recognise and support them with basic service facilities beyond their service jurisdiction. Therefore, migrants who are living in urban fringe and city coasts are deprived of access to these facilities and consequently are exposed to risks of an unhygienic and unhealthy life.

According to The Daily Star [23], Khulna WASA (Water Supply and Sewerage Authority) provided 90 million litres of water daily to meet the demand of 240 million litres per day – a daily shortage of about 150 million litres. The existing services are limited even to city residents, are uncommon in existing slum areas and are non-existent in the fringe areas. Tube wells are the main source of fresh water supply for migrants. Some migrants also use surface water, such as water from ponds, for everyday use, and obtain their drinking water from tube wells. **Figure 6** presents the various sources of water supply used for both drinking and domestic use by respondents.

More than 85 respondents obtained their drinking water from tube wells. For domestic use, most of the respondents used natural sources such as ponds, rivers and lakes. Those who were far from service facilities and could not afford a tube well had no choice but to drink from the natural sources such as ponds.

Climate migrants are considered an isolated community; they are not only deprived economically but also socially. They lose their dignity and cannot achieve a minimum standard of living at the migration destination. A ranking model has

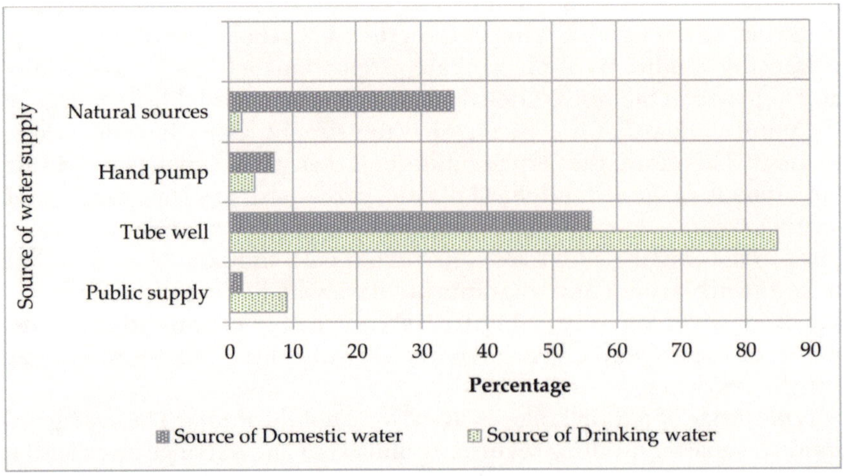

Figure 6.
Water supply facilities for respondents for drinking and domestic use [21].

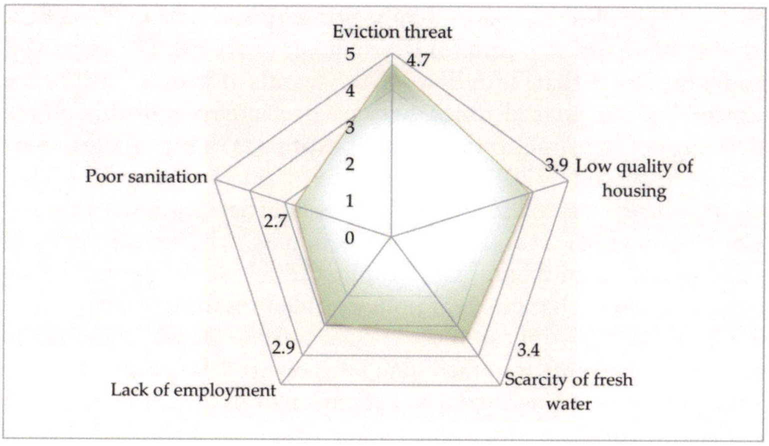

Figure 7.
Social challenges prioritised by the climate migrants in urban areas [21].

been used to understand the key challenges faced by the climate migrants in urban societies. In Khulna, a total of 46 respondents identified five different problems they faced as migrants and a 5-degree ranking model identified the challenges faced by the migrants in this new social context.

$$V = \left\{ \left(\frac{Ro}{Rt} \right) \times Di \right\}. \quad (1)$$

here, V = value under the 5-degree scale; Ro = number of respondents; Rt = total number of respondents; and Di = 5, a value of common scale.

Figure 7 presents the outcomes of the ranking model. According to the model, 'eviction threat' was one of the main problems faced by the migrant communities, hence, it is ranked 4.4 under the 5-degree scale.

Most of the migrants are always afraid of eviction from their home, since they were occupying land without any land right or tenure ship. Though they were paying toll or protection money to the local leaders and muscleman, but that is not a formal land right. They know government could evict them without any notice and they would be on the road again.

Respondents also ranked housing quality as and living standard another key problem. Water supply, employment and sanitation facilities were ranked third, fourth and fifth, respectively, based on the respondents' opinions under the 5-degree scale, as shown in **Figure 5**. In view of such clear-cut evidence of precarious living conditions, it is important to develop and design innovative and effective rights-based domestic solutions to climate displacement in Bangladesh.

5. Limitation to understand climate migration in urban governance

The Bangladesh Government introduced the first National Adaptation Programme of Action (NAPA) in 2005 to address climate change and a possible adaptation approach. However, NAPA treated migration as an undesirable outcome of climate change. Under the Programme of Action, Project No. 11 states: "Affected community would not migrate to cities for job and livelihood" [24] (p. 36), and "Social consequences of mass scale migration to cities would to some extent be halted" [24] (p. 36). Furthermore, internal migration due to climate change is addressed as a negative impact of livelihood in NAPA 2005 [25].

In 2008, the Bangladesh Climate Change Strategy and Action Plan (BCCSAP) was prepared to focus on long- and mid-term goals of NAPA. The revised BCCSAP 2009 documents predict that "hundreds of thousands of people" will be forced to migrate, especially from coastal zones, because of decreasing livelihood opportunities and lowering agricultural productivity. It estimates that globally 6–8 million people could be displaced by 2050 just because of climate change [25]. The slums in big cities have been referred to as a highly possible destination for those who migrate. Considering the fast and unplanned ongoing urbanisation in Bangladesh, this poses an impending problem [26]. Despite the acknowledgement of significant migration due to climate change, clear policy guidelines for migrants have not been addressed, neither in NAPA nor in BCCSAP. Also, there are no guidelines for the national and local government to face such long-term social consequences of climate change. It could be argued that the limitation of understanding about cumulative impacts of climate change on social and institutional systems and the lack of capacity to respond to cumulative impacts climate change (migration, change of social structure) are the main barriers to incorporating climate change adaptation within the mainstream development process.

According to the Climate Investment Fund [27], adaptation policies that can address all the different impact levels should have three major components: (i) a coordination between local and national levels in order to understand impacts and vulnerabilities, (ii) a participatory approach where a broad range of stakeholders are involved rather than a single line ministry from the national level, and finally, (iii) an ongoing implementation of the adaptation process. Climate migrants have limited options to be noticed by the policy makers. At the same time the policy makers have limited understanding to differentiate the climate migrants from the traditional economic migrants. Stakeholder participation where victims such as climate migrants can also be incorporated, as well as NGOs who can be the voice of the vulnerable communities, is important as it can help to explain how the adaptation process should be implemented case-by-case or at policy level.

The policy gap and lack of understanding of climate migration and its urban impacts is the key limitation of the government to address this emerging social challenge.

6. Conclusions

The Australia and Switzerland based International Organisation called Displacement Solutions [28] indicates "climate displacement in Bangladesh will only worsen as climate change increases the frequency of extreme natural hazards that are already leading to displacement across Bangladesh". This research provides clear evidences in support of this statement. Climate events in Bangladesh induce forced migration from coastal areas towards urban areas. This research confirms that too little is being done in Bangladesh to address the negative socio-cultural and economic changes of such migration in urban areas. Furthermore, lack of active urban governance very limited initiatives has yet been taken by to ensure quality of life and social justice for those vulnerable communities. Those migrants suffer deprivation of basic rights—not because of some nasty state action, but following from adverse natural conditions, brought about by human actions [3]. TO address this proliferation social challenges and changes it is important to have a details climate risks reduction plan to cover such slow-onset impact of climate change and the rights of climate migrants. Climate migration is not an impossible problem to address, but it needs to be seen as more than increasing numbers of urban poor.

DOI: http://dx.doi.org/10.5772/intechopen.83409

Acknowledgements

I would like to express my deepest appreciation to all those who provided me the possibility to complete this report. A special gratitude I give to Mr. Mukto Quazi Chief Executive of Pothikkrreith, who organised the audio-visual presentation in urban slums in Khulna to identify my target respondents for this research. Without is support it would be hard to complete the field survey. Thank you.

Author details

Reazul Ahsan
Ujigawa Open Laboratory, Research Center for Fluvial and Coastal Disaster, Disaster Prevention Research Institute (DPRI), Kyoto University, Japan

*Address all correspondence to: r.ahsan@unsw.edu.au

References

[1] Bernstein L, Bosch P, Canziani O, Chen Z, Christ R, Riahi K, et al. Climate Change 2007: Synthesis Report. Geneva: IPCC; 2007. p. 104

[2] Ahmed AF. Community based adaptation to climate change in Bangladesh: The global initiative at local level. In: Proceedings of the 5th Annual Community Based Adaptation to Climate Change; 14-15 December 2011; Melbourne. 2011. pp. 1-16

[3] Egeland J. Internal Displacement: Human Tide the Real Migration Crisis [Internet]. 2007. Available from: https://www.christianaid.org.uk/sites/default/files/2017-08/human-tide-the-real-migration-crisis-may-2007.pdf [Accessed: 2018-11-06]

[4] IOM Migration Research Series No. 31: Migration and Climate Change [Internet]. 2008. Available from: https://www.iom.int/news/iom-migration-research-series-no-31-migration-and-climate-change [Accessed: 2018-11-06]

[5] Mayers N. Environmental refugee: An emergent security issues. In: Proceedings of the 13th Economic Form; 23-27 May; 2005; Prague. 2005

[6] David J, Anton I, Chunzhen L, Annette M, Samuel R, Terry LR, et al. Assessment of observed changes and responses in natural and managed system. In: Climate Change 2007: Impacts, Adaptation and Vulnerability. 1st ed. New York: Cambridge University Press; 2007. pp. 80-132. ch 1

[7] IFRC Sichuan Earthquake. Emergency appeal [Internet]. 2010. Available from: http://www.ifrc.org/docs/appeals/08/MDRCN00329.pdf [Accessed: 2011-05-97]

[8] Lom C. IOM Feature Stories. Asia and Oceania. [Internet]. 2008. Available from: http://www.iom.int/jahia/Jahia/media/feature-stories/f [Accessed: 2011-09-20]

[9] UN-habitat cities and climate change adaptation. UN-habitat donors meetings [Internet]. 2008. Available from: http://mirror.unhabitat.org/downloads/docs/5883_19704_Cities%20and%20Climate%20Change%20Adaptation.pdf [Accessed: 2012-07-23]

[10] Dhar RL. Job search processes for tribal people from Jharkhand and West Bengal. South Asian Research. 2014;**34**:191-208. DOI: 10.1177/0262728014544932

[11] Gray CL, Mueller V. Natural disasters and population mobility in Bangladesh. Proceedings of the National Academy of Sciences of the United States of America (PAN). 2012;**109**(16):6000-6005. DOI: 10.1073/pnas.1115944109

[12] Parvin GA, Shaw R. Climate disaster resilience of Dhaka City corporation: An empirical assessment at zone level. Risk, Hazards and Crisis in Public Policy. 2011;**2**(2):2-30. DOI: 10.2202/1944-4079.1069

[13] Hossain R. Climate change environmental forced migrants and rural-urban migration: Are big cities prepared in Bangladesh. In: Proceedings of the Urban Poverty and Climate Change-Infrastructures of Development Conference; 26-28 January 2009; BRAC University; Dhaka. 2009. pp. 12-18

[14] Bangladesh Bureau of Statistics District Population census [Internet]. 2014. Available from: http://203.112.218.65:8008/WebTestApplication/userfiles/Image/District%20Statistics/Khulna.pdf [Accessed: 2018-11-08]

[15] Agrawala S, Ota T, Ahmed AU, Smith J, Aalst MV. Climate change in

Bangladesh: Focus on coastal flooding and the Sundarbans. In: Proceedings of the Environment Directorate Development Co-Operation. Paris: Organisation for Economic Co-operation and Development (OECD); 2003. pp. 1-70

[16] Haque A, Jahan S. Regional impact of cyclone Sidr in Bangladesh: A multi-sector analysis. International Journal of Disaster Risk Science. 2016;7:312-327. DOI: 10.1007/s13753-016-0100-y

[17] Ahsan R, Kellett J, Karuppannan S. A theoretical framework for analyzing climate impact. In: Proceedings of the World Planning Congress; 4-8 July 2011; Perth, Australia. 2011. pp. 2-17

[18] Rana MP. Urbanization and sustainability: Challenges and strategies for sustainable urban development in Bangladesh. Environment Development and Sustainability. 2010;**3**:237-256. DOI: 10.1007/s10668-010-9258-4

[19] Marshall R, Rahman S. Internal Migration in Bangladesh: Character, Drivers and Policy Issues. Dhaka: United Nations Development Programme (UNDP); 2013 [Internet]. Available from: http://www.bd.undp.org/content/dam/bangladesh/docs/Publications/Pub-2013/Internal%20Migration%20in%20Bangladesh%20UNDP%20Final.pdf [Accessed: 2018-07-28]

[20] EM-DAT Emergency Events Database. US Office of Foreign Disaster Assistance (OFDA) and the Centre for Research into the Epidemiology [Internet]. 2010. Available from: https://www.emdat.be/classification#Climatological [Accessed: 2011-05-03]

[21] Ahsan R. Climate migration and urban changes: A study of adaptation in Bangladesh [thesis]. University of South Australia; 2013

[22] Nazrul I. Slums of Urban Bangladesh: Mapping and Census. 1st ed. Dhaka: USAID; 2006. pp. 11-54

[23] AFP. Fresh water supply demand in Khulna City. In: The Daily Star. 2011

[24] MoEF, editor. NAPA. National Adaptation Programme of Action. Dhaka: Ministry of Environment and Forest, Bangladesh Government; 2005. p. 36

[25] Martin M, Kang YH, Billah M, Siddiqui T, Black R, Kniveton D. Policy Analysis: Climate Change and Migration Bangladesh. Dhaka and Brighton: Sussex Centre for Migration Research; 2013

[26] Forest MoEa, editor. BCCSAP. Bangladesh Climate Change Strategy and Action Plan. Dhaka: Ministry of Environment and Forest; 2009

[27] CIF. Programming and Financing Modalities for the SCF Targeted Program, the Pilot Program for Climate Resilience (PPCR). Washington, DC: Climate Investment Fund. [Internet]; 2009 Available from: https://www.climateinvestmentfunds.org/sites/cif_enc/files/meeting-documents/ppcr_programming_and_financing_modalities.pdf [Accessed: 2013-04-12]

[28] DS. Climate Displacement in Bangladesh. Geneva and Sydney: Displacement Solution; 2012

Chapter 11

Environmental Impact Evaluation of Rubber Cultivation and Industry in Malaysia

Sumiani Yusoff, Zameri Mohamed and Aireen Zuriani Ahmad

Abstract

Over the last 10 years, contribution of Malaysian rubber industry to Malaysia export earnings has increased significantly from RM 15.5 billion in 2003 to RM 33.7 billion in 2013. The main objectives of this study are to provide a comprehensive inventory and detailed quantification of the environmental impact and greenhouse gases emission (GHGs) for the major part of Malaysian rubber industry comprising the cultivation of rubber tree from cradle to grave as well as Standard Malaysian Rubber (SMR) production from cradle to gate. This study was conducted through questionnaire surveys in order to create a very comprehensive life cycle inventories tables representing the actual activities in the Malaysian rubber industry. The results from the questionnaire survey indicated that the GHGs emission from the average annual activities in the cultivation of rubber trees from cradle to grave in Malaysia is 315.54 $GgCO_2eq$ and it represents 0.11% from the 2011 Malaysia GHGs emission. The average annual GHGs emission from the production of SMR in Malaysia in this study is 229.41 $GgCO_2eq$ and it represents 72.7% from the average annual GHGs emission from the cultivation of rubber trees from cradle to grave in Malaysia.

Keywords: life cycle assessment (LCA), greenhouse gases emission, Malaysian rubber industry, Standard Malaysian Rubber (SMR)

1. Introduction

The history of rubber cultivation in Malaya started in the late 1877 when nine seedlings from a batch of about 2700 germinated seeds at Kew Botanic Gardens near London were dispatched and planted in Kuala Kangsar, Perak [1]. Since the first rubber plantation in Malaya was established in 1896, the rubber industry has grown tremendously into the present Malaysia. There were 218,900 hectares of rubber planted area in Malaya in 1910 [1] as compared to 1.066 million hectares of rubber planted area in Malaysia in 2014 [2].

Over the last 10 years, contribution of Malaysian rubber industry to Malaysia export earnings has increased significantly from RM 15.5 billion in 2003 to RM 33.7 billion in 2013 [3]. Malaysia has become the world fifth largest producer of natural rubber with the production of 0.67 million tons in 2014 [2]. Due to its importance, Malaysian rubber industry is included in Malaysia National Key

Economic Area (NKEA) [4]. Malaysia National Key Economic Area (NKEA) is an important driver of economic activities that has a potential to directly contribute to Malaysian Economic Growth measurable by Gross National Income (GNI) indicator and will assist Malaysia in achieving a high income status by 2020 [4].

2. Literature review

Malaysian rubber industry has always been regarded as an environmentally sustainable industry. Rubber trees play an important role as a carbon dioxide sequester from the atmosphere at a rate comparable to if not better than the natural forest [5]. After the process of falling down, rubber trees are converted into renewable rubber wood for furniture based industry. The term renewable or environmentally friendly associated with the rubber wood arises from the fact that the rubber wood represents a relatively sustainable alternative as compared to the tropical woods extracted from natural forest [6].

2.1 Environmental management in Malaysian rubber industry

As one of the Malaysian industries that contribute significantly to the economic development of the country, the Malaysian rubber industry also generated a significant amount of waste [7]. These wastes are subjected to various regulations under the Malaysian Environmental Quality Act 1974. The open burning of rubber plantation wastes in the form of rubber tree stumps after land clearing are governed under the Environmental Quality (clean air) Regulations 1978 Part III (burning of wastes). The practice of open burning is only allowed for specific cases after obtaining special permission from Department of Environment Malaysia (DOE) [8].

The Malaysian government also gazettes the Environmental Quality (prescribed premises) of Raw Natural Rubber Regulations (1978) in making sure that all the raw effluents from the raw rubber processing activities in Malaysia are treated and meet the legal discharge standard before they are allowed to be discharged into the watercourse. The rubber products manufacturing factories in Malaysia are subjected to Environmental Quality (sewage and industrial wastes) Regulations (1979) and Environmental Quality (scheduled waste) Regulations (1989) [9].

2.2 Climate change and Malaysia greenhouse gases emission

Climate is an integral part of environment and climate change in more ways than one is a measure of abuse and mismanagement of this environment through time [10]. According to [11], human influence on the climate change is clear and the more we disrupt our climate, the more we risk severe, pervasive and irreversible impacts on human and natural system.

Malaysia has developed two policies which are The National Policy on Climate Change and the National Green Technology Policy to collectively guide the nation towards addressing climate change holistically, ensuring climate-resilient development, developing a low carbon economy and promoting green technologies [12]. Moreover, low carbon economy is one of the key initiatives proposed by the Malaysian government in the fight against the issue of global warming and climate change [13].

On 13 July 1994, Malaysia has ratified the United Nation Framework Convention on Climate Change (UNFCCC) and Kyoto Protocol on 4 September 2002 [14]. As part of the obligations under Article 4 of the UNFCCC, the Government of Malaysia submitted its Initial National Communication in July 2000 and the Second

National Communication was submitted in January 2011 [15, 10]. Malaysia greenhouse gases (GHGs) emission for the year 2011 was 290.230 million tons CO_2eq and the removal was 262.946 million tons CO_2eq with a net sink of 27.284 million tons CO_2eq (**Table 1**).

Malaysia emissions per gross domestic product (GDP) for the year 2000 were 0.62 t CO_2eq/thousand RM [12]. Malaysia's commitment to address the GHGs emission in the context of sustainable development was announced by the Prime Minister during the 15th Conference of the Parties (COP 15) to the UNFCCC on 17th December 2009 [15, 16]. At the COP 15, the Prime Minister had announced Malaysia's voluntary reduction which was up to 40% in terms of carbon emission intensity of GDP by the year 2020 compared to year 2005 conditional on receiving the transfer of technology and finance support from developed countries [15].

2.3 Life cycle analysis (LCA) study for Malaysian rubber industry

Life cycle analysis (LCA) methodology is relatively a new approach in Malaysia. Majority of the LCA studies in Malaysia at present are conducted to highlight the environmental sustainability of the oil palm industry. The LCA studies on the oil palm industry in Malaysia covered all the sectors within the industry starting from the planting material production up to the biodiesel and other oil palm based products.

All the LCA studies from the oil palm industry in Malaysia have one common objective which is to dispel the misinterpretation of the oil palm industry as a very unsustainable industry by international non-governmental organization.

Life cycle analysis (LCA) methodology is the most relevant environmental management tool to measure the environmental impact and quantify the greenhouse gas emission from the Malaysian rubber industry. The LCA study conducted for the Malaysian rubber industry will definitely be a very useful source to identify the environmental hotspots in the Malaysian rubber industry and help in solving solutions to diminish these hotspots for the betterment of the Malaysian rubber industry.

Based on the findings of the LCA study from the Malaysian rubber industry, certain recommendations, policy or standard operating procedures may be introduced by Malaysian Rubber Board (MRB). The findings from the LCA study for Malaysian rubber industry will also be very beneficial for decision makers across the whole chain of the Malaysian rubber industry.

Sector	Emissions (million tons CO_2eq)	Sink (million tons CO_2eq)
Energy	218.914	
Industrial processes	18.166	
Agriculture	15.775	
Land use, land-use change and forestry (LULUCF)	2.490	−262.946
Waste	34.885	
Total	290.230	−262.946
Net total (after subtracting sink)	−27.284	

Source: [16].

Table 1.
Malaysia GHGs inventory for 2011.

According to [17], there was an earlier LCA study for the production of natural rubber latex concentrate and skim block rubber in North Sumatera, Indonesia involving two latex concentrate factories. The objectives of the study by [17] is not only confined to produce life cycle inventories and environmental impact data from the life cycle impact assessment stage, but the objective was further expanded to include the assessment on the level of eco-efficiency for the production of natural rubber latex concentrate and skim block rubber by utilizing the values obtain from the life cycle impact assessment analysis based on Eco-Indicator 99 methodology [17]. However, this LCA study for the Malaysian rubber industry is the first study of its kind carried out in Malaysia.

2.4 Lack of data on the rubber cultivation in Malaysia from cradle to grave perspective

Conducting LCA study for the natural rubber cup lump production and SMR block rubber production is the right step towards providing support to the Malaysian SMR block rubber industry. This may contribute more details and transparent information regarding the environmental impacts and the GHGs emission in the production of Standard Malaysian Rubber (SMR) block rubber from cradle to gate approach. The information from this study on LCA for the production of SMR block rubber will be very valuable for the international tires manufacturers especially in Europe to incorporate it as the verified background data in their LCA study for the tire production from cradle to grave approach.

The detailed information on the GHGs emissions from the LCA study for the production of SMR block rubber will also be very useful in assisting the Malaysian based rubber products to get certified by the newly launched Standard and Industrial Research Institute of Malaysia (SIRIM) Environmental Declaration Carbon Footprint Type III. The SIRIM Environmental Declaration Carbon Footprint Type III is part of the MyHIJAU Mark and is eligible for Malaysian Government Green Procurement Program.

2.5 Lack of detailed information on GHGs emission and the possibility of setting up voluntary carbon trading for Malaysian rubber industry

In short, it is timely that the GHGs emission related to the Malaysian rubber industry is properly studied and documented extensively for the benefits of the Malaysian rubber industry and Malaysia as whole. The results from the quantification of GHGs emission work for the Malaysian rubber industry using LCA approach will notably help in filling the information gap as described above. The results from this LCA study on the GHGs emission for the Malaysian rubber industry can also be used to project the environmental sustainability of the rubber planting activities in Malaysia as compared to other two major crops in Malaysia which are planting of oil palm and paddy cultivation.

2.6 Climate change and sustainable development

Climate change is summarized by [18] as the extraordinary warming of the earth from increased concentrations of greenhouse gases (GHGs). The current anthropogenic emission of GHGs is the highest in history and is driven largely by human activities through infrastructure development, industries, agriculture and motor vehicles [10, 11]. The atmospheric concentrations of carbon dioxide, methane and nitrous oxide at present are unprecedented at least for the last 800,000 years [11]. According to Van der et al. in [19], it is estimated that 12–15% of the global anthropogenic carbon dioxide emissions is originated from the deforestation and forest degradation.

Climate change is more than just a warming trend as the increasing temperature through continued emission of GHGs will cause further warming and long lasting changes in all components of the climate system [10, 11]. The consequences of the climate change are likely to be harmful to humans and natural environment in the form of changes in major wind patterns, amount and intensity of precipitation and increased frequency of severe storms and weather extremes [18, 10].

Agriculture industry would be the most affected sectors of climate changes as compared to other economic sectors since it has a strong linkage and dependence on the climate and the environmental factors as suggested in [20]. Rise of temperature, changes in sowing and harvesting dates, water availability and rainfall patterns are among climatic factors that can influence the agricultural productivity [21]. Baharuddin stated in [20] that an increase in rainfall is prejudicial for rubber plantations which suffer losses in the form of loss of tapping days and crop washouts.

3. Methodology

The main goal of the study is to provide comprehensive inventories, detailed quantification of the environmental impact and GHGs emission for the cultivation of rubber tree from cradle to grave in Malaysia. Therefore, this study is required to quantify the GHGs emission and recommended strategies for improvement based on the individual Life cycle inventory (LCI) for the cultivation of rubber trees from cradle to grave. The environmental impacts and hotspots identification for the study was carried out using SimaPro software version 7.3.3 developed by Pre Consultants B.V. Eco-indicator 99 was selected as the impact assessment methodology.

For this study, the survey only represents the rubber smallholders under the supervision of rubber related agencies in Malaysia. Individual rubber smallholders are excluded from this survey as there are great difficulties in getting verified information from this group of rubber smallholders on their agronomic practices as these smallholders normally did not have any proper written record on their agronomic practices and few of them are even illiterate. Amongst the main three government agencies in Malaysia which are responsible in supervising and managing the small plot of rubber planted area owned by the rubber smallholders, only The Federal Land and Development Authority (FELDA) and Federal Land Consolidation and Rehabilitation Authority (FELCRA) agreed to take part in this study while Rubber Industry Smallholders' Development Authority (RISDA) did not allow this study to be conducted in the rubber planting areas owned by the rubber smallholders under their supervision. Based on the discussion with FELDA and FELCRA management and supported by [22, 23] data, there are 21 FELDA schemes and 274 FELCRA projects that are currently in the mature rubber stage in Peninsular Malaysia.

3.1 Life cycle assessment (LCA)

Life cycle assessment (LCA) is an environmental management tool that enables quantification of environmental burdens and their potential impacts over the whole life cycle of a product, process or activity [24]. Primarily, LCA has been introduced in product manufacturing for the purpose of tracing direct impacts and impacts associated with a product throughout the entire life cycle from cradle to grave for the purpose of getting a holistic overview of the environmental burden associated with the products [25].

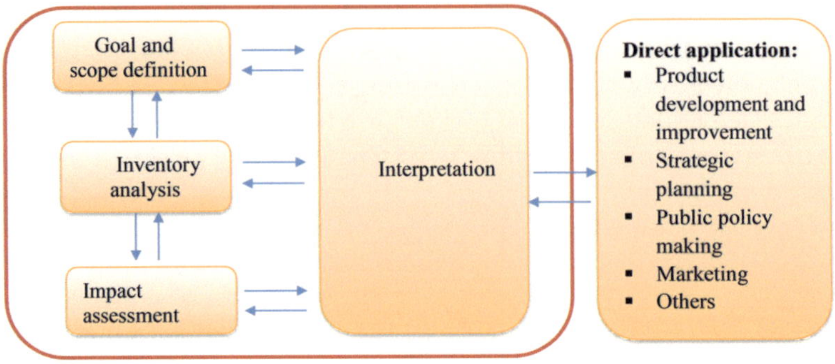

Figure 1.
Stages of LCA.

There are four phases in LCA studies namely goal and scope definition, inventory analysis, impact assessment and interpretation. The relationship between the phases is illustrated in **Figure 1**.

3.2 Goal and scope definition

This is the first phase of any LCA study and according to [26], the goal must clearly mention the intended application, the reasons for carrying out the study and the intended audience. The scope of any LCA study should be sufficiently well defined to ensure that the breadth, depth and the details in which the study is conducted are both compatible and sufficient to address the stated goals [26]. The functional unit, system boundary, allocation procedures, assumptions and limitation are parts of the scope.

3.3 Inventory analysis

The Life cycle inventory (LCI) phase is the second phase of any LCA study. Inventory analysis involves data collection and calculation procedures within the system boundary for inclusion in the inventory as relevant inputs and outputs of a product system [26, 27]. According to [28], LCI can be defined as an objective, data-based process of quantifying energy and raw materials requirements, air emissions, waterborne effluents, solid waste, and other environmental releases incurred throughout the life cycle of a product, process, or activity.

All calculation procedures in the inventory analysis for any LCA study must be transparently documented and the assumptions used must be clearly stated and explain [27]. Generally, there are two types of inventory data, i.e., the foreground data that have to be collected independently according to the purpose of carrying out LCA analysis and the background data which are usually collected from literatures and software [29]. Data validity check must be conducted during the process of data collection for inventory analysis to make sure that the data quality requirements have been fulfilled [27]. For the data collected from public sources, the sources must be referenced [27].

3.4 Impact assessment

The Life cycle impact assessment (LCIA) phase is the third phase of LCA and its purpose is to evaluate the significance of potential environmental impacts based on the LCI results [26]. The LCIA phase is important in providing the information for the life cycle interpretation phase [26].

4. Results and discussion

4.1 Life cycle impact assessment (LCIA) on GHGs emission in the production of mature rubber tree from immature rubber stage

The total GHGs emission value in maintaining the healthy growth of one immature rubber tree for a year for this study is 1.08 $kgCO_2eq$ as shown in **Figure 2** and **Table 3**. The highest contributor which represented 51.6% from the total GHGs emission value in maintaining the healthy growth of one immature rubber tree for a year is the emission of nitrous oxide from the usage of ammonium sulfate at 5.60E−01 $kgCO_2eq$ (**Figure 2**).

While the second highest contributor to the total GHGs emission value in maintaining the healthy growth of one immature rubber tree for a year is ammonium production with the percentage of 22.4%. Meanwhile, glyphosate production was recorded as the third highest contribution at 17.7% (**Figure 2**). The remaining three processes are considered as insignificant contributors towards the total value of GHGs emission to maintain the healthy growth of one immature rubber tree for a year (**Figure 2**).

Figure 2 obviously showed that the reduction in the usage of ammonium sulfate and glyphosate will definitely reduce the total GHGs emission value in maintaining the healthy growth of one immature rubber tree for a year. This can be achieved through the reduction in the immaturity rubber stage period and through incorporating the manual weeding method in weed management.

The GHGs emission value in maintaining the healthy growth of one immature rubber tree for 6 years duration during the immature rubber stage is 6.51 $kgCO_2eq$ and this represent 14.6% from the total value of GHGs emission for the cultivation of one rubber tree from cradle to gate of 44.68 $kgCO_2eq$.

The GHGs emission in maintaining the healthy growth of immature rubber trees in Malaysia per year which based on 0.379 million hectares of immature rubber area in Malaysia at the average stand of 410 rubber trees per hectare and with 51.8% of this area is fertilized at the recommended dosage is summarized in **Table 2**.

Based on **Table 2**, as compared to the Malaysian 2011 GHGs emission of 290,230 $GgCO_2eq$ in [30], the GHGs emission value from the perspective to maintain the healthy growth of immature rubber trees in Malaysia for 6 years, immature rubber stage and 1 year average for immature rubber stage is considered as insignificant.

The GHGs emission value of 524.69 $GgCO_2eq$ with duration of 6 years for immature rubber stage in maintaining the healthy growth of immature rubber trees in Malaysia is very low and represent only 3.3% from the 2011 Malaysian agricultural sector GHGs emission of 15775.3 $GgCO_2eq$ (**Table 2**). The GHGs emission value of 87.45 $GgCO_2eq$ based on the average 1 year for immature rubber stage is considered

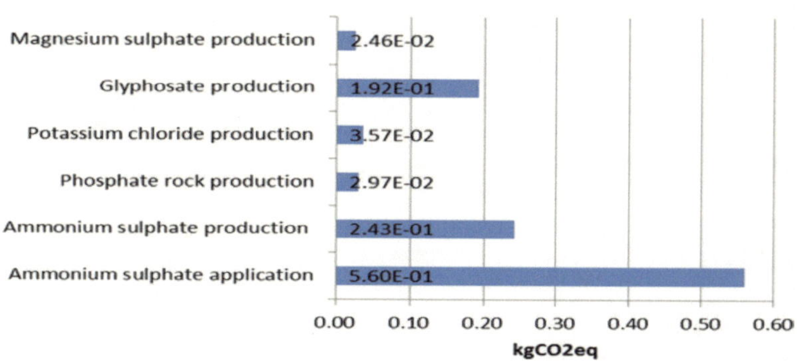

Figure 2.
GHGs emissions in maintaining the healthy growth of one immature rubber tree for a year.

as insignificant as compared to the GHGs emissions value from Malaysian agricultural sector in 2011 (**Table 2**).

Table 3 shows the list of GHGs emission and its corresponding values in contributing to the total GHGs emission value in maintaining the healthy growth of one immature rubber tree for a year.

4.2 Life cycle impact assessment (LCIA) on GHGs emission for natural rubber cup lump production from cradle to gate

The total GHGs emission value for the production of 1 kg natural rubber cup lump (56% DRC) is 4.89E−02 kgCO_2eq and its represent 0.11% from the total GHGs emission value for the cultivation of one rubber tree from cradle to grave (**Figure 3**).

	Immature rubber stage (6 years)	Average 1 year for immature rubber stage
GHGs emission (GgCO_2eq)	524.69	87.45
Percentage from Malaysia 2011 GHGs emission (%)*	0.18	0.03
Percentage from agriculture sector in Malaysia 2011 GHGs emission (%)*	3.33	0.55

**Source: [16].*

Table 2.
GHGs emission to maintain the healthy growth of immature rubber trees in Malaysia.

GHGs	Weight in kgCO_2eq
Nitrous oxide	5.63E-01
Carbon dioxide	4.91E-01
Methane	2.87E-02
Methane, tetrafluoro-, CFC-14	6.34E-04
Sulfur hexafluoride	5.38E-04
Ethane, hexafluoro-, HFC-116	1.34E-04
Methane, tetrachloro-, CFC-10	7.70E-05
Ethane, 1,1,1,2-tetrafluoro-, HFC-134a	4.05E-05
Methane, chlorodifluoro-, HCFC-22	3.82E-05
Ethane, 1,2-dichloro-1,1,2,2-tetrafluoro-, CFC-114	1.92E-05
Methane, bromochlorodifluoro-, Halon 1211	1.06E-05
Methane, bromotrifluoro-, Halon 1301	9.76E-06
Methane, dichlorodifluoro-, CFC-12	8.30E-07
Methane, trifluoro-, HFC-23	6.64E-07
Ethane, 1,1,2-trichloro-1,2,2-trifluoro-, CFC-113	1.31E-07
Ethane, 1,1-difluoro-, HFC-152a	1.74E-08
Chloroform	1.48E-08
Methane, trichlorofluoro-, CFC-11	1.09E-09
Methane, dichloro-, HCC-30	2.21E-10
Methane, monochloro-, R-40	2.00E-10
Ethane, 1,1,1-trichloro-, HCFC-140	7.90E-11
Methane, dichlorofluoro-, HCFC-21	2.13E-11
Methane, bromo-, Halon 1001	2.86E-17
Total GHGs Emission	1.08

Table 3.
GHGs emission profile to maintain the healthy growth of one immature rubber tree for a year.

Figure 3 has clearly described that the trend from the GHGs emission for the production of 1 kg natural rubber cup lump (56% DRC) is basically identical to the GHGs emission for the cultivation of one rubber tree from cradle to grave.

The application and production of ammonium sulfate are the two main processes responsible for 77.9% from the total GHGs emission value for the production of 1 kg natural rubber cup lump (56% DRC) (**Figure 3**). Potassium chloride production and glyphosate production recorded the contribution of 8.1 and 7.5% respectively while the remaining 13 processes are considered as minor contributors towards the total GHGs emission value for the production of 1 kg natural rubber cup lump (56% DRC) (**Figure 3**).

It is found that in Malaysia, the GHGs emission from the production of 1,193,946 tons of natural rubber cup lump (56% DRC) is 58.4 3 GgCO_2eq and this only represent 0.02% from the Malaysian 2011 GHGs emission of 290,230 GgCO_2eq [30]. Based on this value, the contribution of the GHGs emission from the production of 1,193,946 tons of natural rubber cup lump (56% DRC) in Malaysia is considered as insignificant as compared to the Malaysian 2011 GHGs emission.

4.3 Life cycle impact assessment (LCIA) on GHGs emission for the production of SMR block rubber from cradle to gate

Figure 4 indicates that the total GHGs emission value for the production of 1 kg Standard Malaysian Rubber (SMR) block rubber from this study is 0.407 kgCO_2eq.

Electricity generation, methane emission from the effluent treatment system, production of natural rubber cup lump from cradle to gate and transportation of raw material from the source to the Standard Malaysian Rubber (SMR) block rubber factories are the four key process contributors representing 95.9% from the total GHGs emission value in the production of 1 kg SMR block rubber from cradle to gate (**Figure 4**).

From **Figure 4**, it is noticeably reported that the reduction in the electricity consumption during the production of SMR block rubber, elimination in the methane mission from the effluent treatment system, reduction in the total GHGs

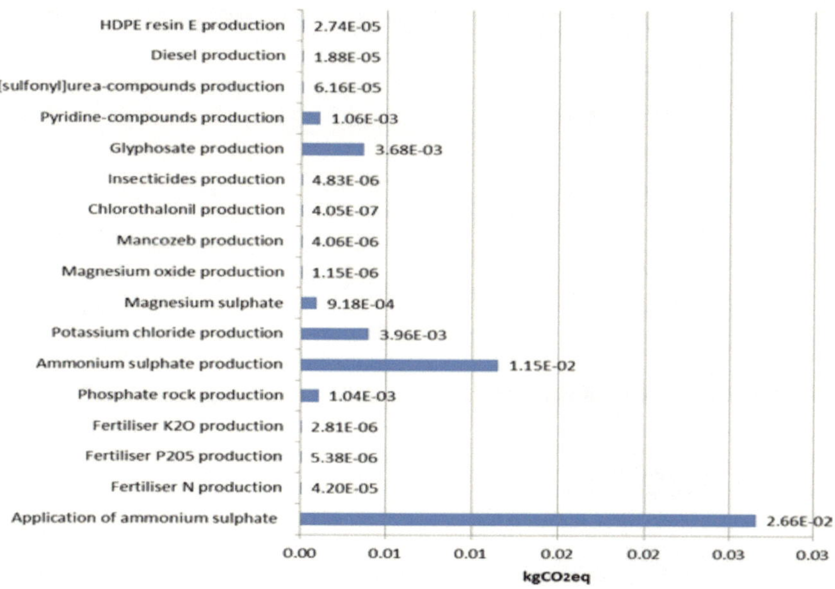

Figure 3.
GHGs emission values for the production of 1 kg natural rubber cup lump (56% DRC).

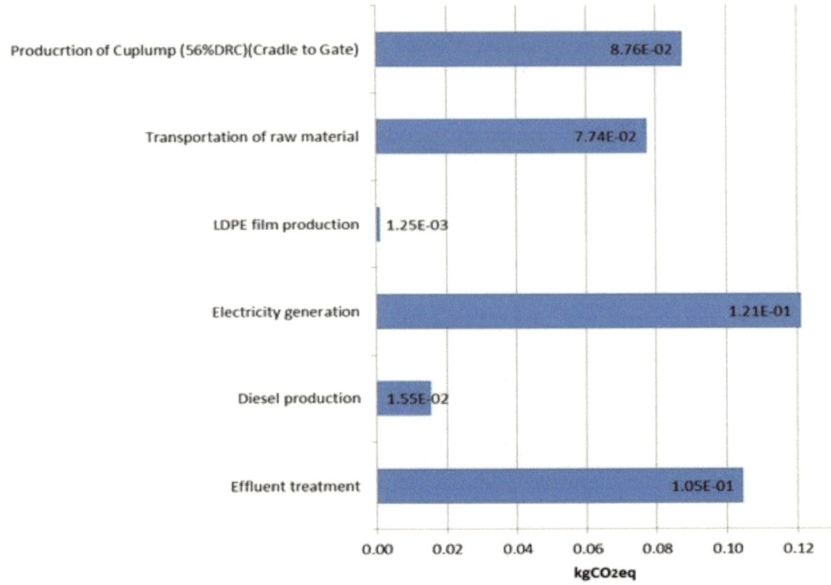

Figure 4.
GHGs emission values for the production of 1 kg Standard Malaysian Rubber (SMR) block rubber from cradle to gate.

GHGs	Weight in $kgCO_2eq$
Carbon dioxide	2.40E−01
Methane	1.18E−01
Nitrous oxide	4.90E−02
Methane, tetrafluoro-, CFC-14	6.16E−05
Sulfur hexafluoride	4.18E−05
Methane, chlorodifluoro-, HCFC-22	2.25E−05
Ethane, hexafluoro-, HFC-116	1.30E−05
Methane, bromotrifluoro-, Halon 1301	9.13E−06
Methane, bromochlorodifluoro-, Halon 1211	6.92E−06
Ethane, 1,1,1,2-tetrafluoro-, HFC-134a	4.39E−06
Methane, tetrachloro-, CFC-10	3.75E−06
Ethane, 1,2-dichloro-1,1,2,2-tetrafluoro-, CFC-114	2.24E−06
Methane, dichlorodifluoro-, CFC-12	7.84E−07
Methane, trichlorofluoro-, CFC-11	3.39E−07
Methane, chlorotrifluoro-, CFC-13	1.39E−07
Methane, trifluoro-, HFC-23	6.21E−08
Ethane, 1,1,2-trichloro-1,2,2-trifluoro-, CFC-113	1.23E−08
Methane, dichloro-, HCC-30	1.09E−08
Ethane, 1,1-difluoro-, HFC-152a	1.37E−09
Chloroform	1.31E−09
Methane, monochloro-, R-40	1.68E−11
Ethane, 1,1,1-trichloro-, HCFC-140	6.50E−12

GHGs	Weight in $kgCO_2eq$
Methane, dichlorofluoro-, HCFC-21	1.99E−12
Methane, bromo-, Halon 1001	2.84E−18
Total GHGs emission	0.407

Table 4.
GHGs emission profile for the production of 1 kg SMR block rubber from cradle to gate.

emission from the production of natural rubber cup lump (56% DRC) from cradle to gate and the reduction of fossil fuels based usage in the transporting of raw material from the source to SMR block rubber factory will definitely scale down the total GHGs emission from the production of 1 kg SMR block rubber from cradle to gate.

The GHGs emission from the production of SMR block rubber from cradle to gate had the potential to be reduced through the elimination of methane release from the effluent treatment system. The methane release from the treatment of SMR block rubber factory effluent can be eradicated through changing the current effluent treatment system of facultative/anaerobic ponding system to a fully aerobic system. At present, the methane emission from the effluent treatment plant in the block rubber factories are not subjected to any environmental regulations.

The GHGs emission from the production of 562,967 tons of natural rubber cup lump based SMR block rubber in Malaysia is 229.41 $GgCO_2eq$ and this only represent 0.08% from the Malaysian total GHGs emission of 290,230 $GgCO_2eq$ in 2011 [30].

The list of GHGs emission and its corresponding values in contributing to the total GHGs emission value for the production of 1 kg SMR block rubber from cradle to gate is shown in **Table 4**. Carbon dioxide, methane and nitrous oxide are the three major GHGs that contribute 99.96% from the total GHGs emission value in the production of 1 kg SMR block rubber from cradle to gate (**Table 4**).

The GHGs emission from the production of natural rubber cup lump (56% DRC) in Malaysia for the average period of 1 year for cradle to gate is 58.43 $GgCO_2eq$ and it represents 18.5% from the total GHGs emission for the cultivation of rubber trees from cradle to grave based on average 1 year perspective.

5. Conclusions

In summary, with the implement of Life cycle analysis (LCA) methodology for the Malaysian rubber industry in this study, it can be concluded that the reduction in the utilization of ammonium sulfate fertilizer to its optimum level has the potential to reduce the GHGs emission for the cultivation of rubber trees in Malaysia from cradle to grave perspective. Meanwhile, the reduction in the immaturity rubber stage period and incorporating of manual weeding method in the weed management have the potential to reduce the GHGs emission for the production of mature rubber trees for gate to gate boundary in Malaysia.

The GHGs emission from the production of SMR block rubber in Malaysia for the average period of 1 year for cradle to gate boundary have the potential to be reduce through increasing the supply of local natural rubber and making sure the natural rubbers are free or have a very minimum amount of contaminants. The GHGs emission from the production SMR block rubber in Malaysia for the average period of 1 year for cradle to gate boundary also has the potential to be reduce through replacing the current effluent treatment system to a fully aerobic system.

This study is hoped to be a part of the continuous effort in meeting sustainability goal in the Malaysian rubber industry and stringent environmental market regulations worldwide.

Acknowledgements

This author wishes to extend sincere gratitude to University of Malaya, Malaysian Rubber Board, for their support and other stakeholders as well as government agencies for their cooperation. The University of Malaya Research Grant (UMRG) (RPG23A-16SUS) also was acknowledged.

Author details

Sumiani Yusoff[1]*, Zameri Mohamed[2] and Aireen Zuriani Ahmad[3]

1 Institute of Ocean and Earth Sciences, University of Malaya, Kuala Lumpur, Malaysia

2 Malaysian Rubber Board, Sungai Buloh, Selangor, Malaysia

3 University of Malaya Eco-Campus Secretariat Office, Kuala Lumpur, Malaysia

*Address all correspondence to: sumiani@um.edu.my

References

[1] Chan H, Yeang HY, Lim CL, Rahman FA, Ishak S, editors. Milestones in Rubber Research. Kuala Lumpur: Malaysian Rubber Board; 2000

[2] Malaysian Rubber Board. Rubber Pocket Book. Kuala Lumpur: Malaysian Rubber Board; 2016

[3] Malaysian Rubber Board. One Nation Rubber Strategy. Kuala Lumpur: Malaysian Rubber Board; 2014

[4] Sumormo S. National key economic area (NKEA): Shaping the Malaysian rubber industry for 2020. Rubber Technology Developments. 2012;**12**(2):3-8

[5] Malaysian Rubber Board. The Malaysian Natural Rubber Industry [Internet]. 2017. Available from: http://www.lgm.gov.my/GreenMaterial/TheMalaysianNRIndustry.pdf [Accessed: 7 July 2017]

[6] Usubharatana P, Phungrussami H, Sukhodomchot N. Is parawood envrionmental firendly material? What should be known before answer? Research Journal of Applied Sciences. 2013;**8**(4):252-261

[7] Verasamy D, Isa Z, Karim MZA. Overview of waste management and recycling activities of Malaysian rubber industry. In: Proceedings of 17th Symposium of Malaysian Chemical Engineers (SOMChe2003); 29-30 December 2003; Penang, Malaysia: School of Chemical Engineering, Universiti Sains Malaysia; 2003. pp. 698-702

[8] Yew FK. Zero burning in rubber cultivation. Planters Bulletin. 1998;**1**:29-34

[9] Isa Z. Management of effluent and control of malodour from natural rubber industries. In: Short Course on Environmental Management for Rubber Industry. Selangor: Akademi Hevea Malaysia; 2004

[10] Sani S. The politics of climate change: International responses and Malaysia. Bangi, Selangor: Institut Alam Sekitar dan Pembangunan (LESTARI), UKM; 2009

[11] IPCC. In: Core Writing Team, Pachauri RK, Meyer LA, editors. Climate Change 2014: Synthesis Report. Contibution of Working Groups I, II and III to the Fifth Assessment Report of the Intergovermental Panel on Climate Change. Geneva, Switzerland: IPPC; 2014. pp. 151

[12] Ministry of natural resources and environment Malaysia. Second national communication to the UNFCCC; Putrajaya, Malaysia; 2011

[13] Amin AQ, Filho WL. An overview of prospects and challenges in the field of climate change in Malaysia. International Journal of Global Warming. 2011;**3**(4):396. doi:10.1504/ijgw.2011.044402

[14] UNDP. Third National Communications (TNC) to the UNFCCC and Biennial Update Reporting (BUR) for Malaysia [Internet]. 2015. Available from: http://www.my.undp.org/content/malaysia/en/home/operations/projects/environment_and_energy/third-national-communications--tnc--to-the-unfccc-and-biennial-u.html [Accessed: 29 July 2015]

[15] Ministry of natural resources and environment Malaysia. Malaysia Biennial update report to the UNFCCC. Putrajaya; 2015

[16] Theseira GW. Roadmap to achieving the 40% carbon intensity emissions reduction. In: REDD+ Stakeholder

Consultation Workshop; Kuala Lumpur: Sunway Putra Hotel; 2013

[17] Maulina S. Enhancement of eco-efficiency through life cycle assessment in natural rubber latex concentrate processing [Doctor of Philosophy]. Universiti of Malaya; 2014

[18] Harris PG. Climate change and American foreign policy: An introduction. In: Harris PG, editor. Climate Change and American Foreign Policy. New York: Palgrave Macmillan US; 2000. pp. 3-25

[19] de Blécourt M, Brumme R, Xu J, Corre MD, Veldkamp E. Soil carbon stocks decrease following conversion of secondary forests to rubber (*Hevea brasiliensis*) plantations. PLOS ONE. 2013;**8**(7):e69357

[20] Al-Amin A, Filho W, Maria De la Trinxeria J, Hamid Jaafar A, Abdul Ghani Z. Assessing the impacts of climate change in the Malaysian agriculture sector and its influences in investment decision. Middle-East Journal of Scientific Research. 2011;**7**(2):225-234

[21] Ahmed A, Masud MM, Al-Amin AQ, Yahaya SRB, Rahman M, Akhtar R. Exploring factors influencing farmers' willingness to pay (WTP) for a planned adaptation programme to address climatic issues in agricultural sectors. Environmental Science and Pollution Research. 2015;**22**(12):9494-9504. DOI: 10.1007/s11356-015-4110-x

[22] FELCRA. Maklumat Asas Projek Felcra Berhad Sehingga Disember 2011. 2015. Available from: http://www.felcra.com.my/ms/pejabat-wilayah [Accessed: 6 January 2015]

[23] Ghani MNA. Maklumat Keluasan Sawit dan Getah Felda. 2015

[24] Azapagic A. Life cycle assessment and its application to process selection, design and optimisation. Chemical Engineering Journal. 1999;**73**(1):1-21. DOI: 10.1016/S1385-8947(99)00042-X

[25] Yusoff S. Feasibility of life cycle management for improving environmental management-case study on Malaysian palm oil industry [Doctor of Philosophy]. University Malaya; 2006

[26] ISO 14040. Environmental management-life cycle assessment-principles and framework: ISO. 2006

[27] ISO 14044. Environmental management-life cycle assessment-requirements and guidelines: ISO. 2006

[28] Svoboda S. Note on Life Cycle Analysis [Internet]. 1995. Available from: http://www.umich.edu/~nppcpub/resources/compendia/CORPpdfs/CORPlca.pdf

[29] Narita N. LCA data in inventory analysis. In: Workshop on the Development of LCI for Crude Palm Oil; Shah Alam: SIRIM; 2005

[30] Ministry of Natural Resources and Environment Malaysia. Environmental Facts [Internet]. 2015. Available from: http://www.nre.gov.my/en-my/Environment/Pages/Environment-Fact.aspx [Accessed: 4 August 2015]

Chapter 12

Salinity Stress in Arid and Semi-Arid Climates: Effects and Management in Field Crops

Sajid Hussain, Muhammad Shaukat, Muhammad Ashraf, Chunquan Zhu, Qianyu Jin and Junhua Zhang

Abstract

Salinity stress is one of the most vital abiotic stresses which results in significant damages of agricultural production, particularly in arid and semi-arid areas of the world. Salinity causes by high accumulation of soluble salt, especially NaCl in soil and water. Salinity hampers the growth and survival of many field crops such as rice, wheat, maize, cotton, sugarcane, and sorghum. It affects the plant growth by three ways such as osmotic stress linked with an increase of phytotoxic ions, ionic stress e in the cytosol, and oxidative stress facilitated by reactive oxygen species (ROS). These stresses caused by salinity hinder the water uptake, causes ion imbalance, ROS production, and hormonal imbalance, and results in the decline of photosynthesis activities reduce the plant growth and final yield. However, the sensitivity of field crops depends on the nature of cultivar and growth stages. There are many strategies to cope with salinity stress which are the development of salinity tolerant crop cultivators by using genetic and molecular techniques such as QTLs and CRISPR CAS9 technique, nutrients management strategies, use of hormones regulators (AVG, 1-MCP, D-31). This chapter will give a brief idea to the scientist to understand the effects of salinity on field crops and their management strategies.

Keywords: salinity, field crops, physiology, yield, sodium chloride, reactive oxygen species

1. Introduction

The plant growth, development, and yield are negatively affecting by abiotic stresses such as drought, salinity, chilling, and high temperature. About 50% of plant productivity is under the influence of these abiotic stresses [1]. Among these abiotic stresses, salinity is considered as one of the most harmful agents for the plant life cycle. Salinity is an excess amount of salt in the soil, water, and plant. Salinity is frequently an underrated problem in the agriculture sector. It is estimated that salt affected area (sodic and saline) about 6% irrigated and 20% of world's total cultivable land is under the influence of salinity [2]. The irrigated areas of many countries are affected due to salinity in the world (**Table 1**) [3]. Salinity problem is caused by the natural and anthropogenic activities and increasing with time. It is also estimated that 50% of the cultivable land will effect due to salinity by 2050 [2]. On the contrary side, with the current speed of population increase in the world,

Country	Salt-affected area of irrigated in the world	
	Mha	%
China	6.7	15
India	7.0	17
Soviet Union	3.7	18
United States	4.2	23
Pakistan	4.2	26
Iran	1.7	30
Thailand	0.4	10
Egypt	0.9	33
Australia	0.2	9
Argentina	0.6	34
South Africa	0.1	9
Subtotal	29.6	20
World	45.4	20

where, mha = million hectare, % = percentage area. Source: Ghassemi et al. [3].

Table 1.
Global estimate of secondary salinity in irrigated lands of the world.

Soil types	ECe (dS/m)	ESP	SAR	pHs
Normal soil	<4	<15	<15	4.5–7.5
Saline soil	>4	<15	<15	<8.5
Sodic soil	<4	>15	>15	>8.5
Saline-sodic soil	>4	>15	>15	>8.5

Whereas, ECe = electrical conductivity, ESP = exchangeable sodium percentage, SAR = sodium adsorption ratio, and pHs = negative log of H^+ ion [12].

Table 2.
The USDA classification system of salt affected soils.

also need to produce more food up-to 70% till 2050 to feed the increasing mouths of the world [2].

Many major field crops such as wheat (*Triticum aestivum* L.), rice (*Oryza sative* L.), maize (*Zea mays* L.), sorghum (*Sorghum bicolor* (L.) Moench), cotton (*Gossypium hirsutum*), and sugarcane (*Saccharum officinarum*), etc. show negative response towards salinity. However, plant performance and grain yield may not decrease until a 'threshold' salinity level is reached. Threshold levels of salinity are generally defined as the maximum amount of salt that a plant can tolerate in its root zone without impacting growth (**Table 2**). Plant physiology is very susceptible to high salinity in its rhizosphere and affects germination rate, growth stages, and ultimately plant yield [1]. Similarly, many other growth hampering effects on plants due salinity are low net CO_2 assimilation to plant tissues, leaf area, leaf cell enlargement, dry matter production, and relative growth, poor development of spikelets (rice and wheat), boll (cotton), etc. [4, 5]. There are many reasons for hampering of plant production under salinity. Generally, salinity affects plant growth in three ways, such as osmotic stress, ionic stress or ion imbalance, and oxidative stress [6]. Osmotic stress disturbs the salt water balance, which results in a high concentration of salts and loses of water in plant cell sap and tissues. This imbalance causes ion toxicity within plant tissues, and plant shows leaf burn or wilting symptoms due to Na^+ and Cl^- accumulation. Ionic stress also causes nutrients disequilibrium and results in reduce final germination percentage (FG %), decrease vegetative and reproductive growth, decline yield, and yield components of the plant under salinity. Similarly, ionic stress in the plant due to salinity causes reduction of photosynthesis activity, alteration of enzymatic activities, oxidative stress, disrupted the biochemical membrane structure and function, destroy the ultrastructural cellular components, and hormonal imbalance are the primary reason for the reduction of overall plant's growth and development [7, 8].

For better plant performance under salinity stress, natural adaptation responses at physiological, molecular, and cellular levels to tolerate salinity stress is of great concern. These adaptations are an osmotic adjustment, closure of stomata, Na^+ exclusion from older leaves, maintenance of K^+/N^+ equilibrium, and cytosolic K^+, transpiration efficiency, and increased antioxidant defense system are very important for ideal plant growth under salinity. Besides these, various other management strategies have been embraced on a scientific basis to improve plant growth efficiency under salinity. These strategies are genetic modification, identification, sequencing of gene, microarray analysis, and plant transformation, and agronomic strategies to reduce salinity stress by soils reclamation via water and nutrients management, seed priming, and usage of hormone regulator to create homeostasis in hormonal production under salinity. These management strategies are being useful for stress management, including salinity. In this book chapter, we will review the latest information about the 'Salinity Stress in Arid and Semi-Arid Climates: Effects and Management in Field Crops, which could be a good advantage to the scientific community and farmers for the understanding of salinity issue in the field crops and their solution.

2. Salinity stress and its causes

The ecological anxieties (biotic and abiotic stresses) have turned into essential threats to plant growth, development, and survival. Among these ecological anxieties, abiotic stresses, for example, drought, chilling or high temperature, and salinity inactively influencing the growth, biomass generation, and yield of many field crops. These threats are ending up more deteriorated by regular or human-made activities, which result in the excessive soluble salts accumulation in the underground water and soil. As concern salinity stress, about 20% of the world's land, and about 33% of the world's irrigated zone is under the impact of salinity [9]. Besides, salinity influenced areas are expanding at a rate of 10% yearly. The expanding of salinity issues are because of low precipitation, high surface evaporation, weathering of native rocks, irrigation with saline water, and poor agronomic practices. Salt influenced soils have various sorts that negative effect on agricultural production, for instance, irrigation-induced salinity and 'transient' dry-land salinity have been arranged in detail with different perspectives considered by [10], and illuminate that salinity in the soil is one of the vast abiotic stress that hamper the agricultural production in the world. The estimation has been done that >50% of the agricultural land would be affected by agricultural till 2050 [11].

Salinity is the issue of almost all the continents and under a wide range of climates. However, the salinity issue is more in arid and semi-arid climate contrasted with the humid climate where yearly precipitation is not as much as evapotranspiration in the world. It is need of great importance to comprehend the mode and sources of salinity with classification, and its role in the plant life cycle. The characteristic critical source of salinity is the primary minerals in exposed layers of the earth crust by weathering process with the assistance of atmospheric

CO_2. The weathering of these primary mineral rocks in the earth crust is the primary source of all the dissolvable salts present in the soils and ocean. However, there are several other anthropogenic sources of salinity in the soil or water. Under arid to and semi-arid climates, the products from the weathering procedure of mineral and rocks accumulate in the soil and result in the advancement of salt-influenced soils (saline or sodic soil). Though, under a humid atmosphere, salt could not collect in-situ and filter down through the soil and transport to the close-by streams and waterways and caused the salinity in the water bodies [12]. The US Salinity Lab staffs (1954) group the salt-influenced soils (**Table 2**). These salt-affected soils types have unique nature of soluble salts. For instance, saline soil has $Cl^-$and SO_4^{2-} and CO_3^{2-} present and sodic or alkali soil has HCO_3^- of Na^+, and in exceptional cases with high CO_3^{2-} concentration with the capacity of alkaline hydrolysis. So also, saline sodic soil has predominant soluble salts of Na^+ with Cl^- and SO_4^{2-} with an average intensity of $NaHCO_3$ and Na_2CO_3 in a trace concentration. An ordinary soil has maximum nutrients for development and improvement of the plant. On the inverse, Salinity is one of the significant environmental element influencing plant growth and production. As indicated by FAO report, a saline soil is characterized as having a high concentration of soluble salts for the most of Sodium (Na^+), Calcium (Ca^{2+}), magnesium (Mg^{2+}) chloride (Cl^-) and sulfate (SO_4^{2-}). Magnesium sulphate ($MgSO_4$) and sodium chloride (NaCl, table salt), are among the most well-known soluble salts which are sufficiently high to influence plant growth and development.

2.1 Salinity effects on plant growth

Salinity influence crops in these ways: osmotic effect, specific ion effect, ion imbalance, and oxidative stress [6]. Salinity decline water uptake limit of plant, and causes a decrease in plant development. It might be explicit salt effects. If a high concentration of salt enters the plant, this high concentration of salt will increase at last ascent to a toxic level in older leaves causing early senescence and diminished the photosynthetic leaf area of a plant to a dimension that cannot support plant development [14]. Salinity seems to influence plant growth mechanism in two different ways, water relations, and ionic relations. Firstly, plants face water stress, which in cause decline leaf expansion. Secondly, long-term salt stress in soil and plant, plants involvement (Na^+ and Cl^-) ionic stress, which can prompt early senescence of older leaves [15] (**Figure 1**).

Plants experience the ill effects of the presentation of salinity until maturity [16]. Generally, the markers of salinity impacts in plants are impeded growth and small plants with fewer and smaller leaves. Munns [16] depicted salinity consequences for various plant development stages under a different period of the plant growth mechanism and development. After a couple of minute's introduction of salinity stress, dehydration and shrinkage of the cell begin, and following a couple of hours after the fact recovers their original volume. Regardless of this recovering of the original volume, cell elongation and cell division are diminished, prompting slower rates of root and leaf development. On the following days, a diminishing in cell division and lengthening change into slower leaf inception and size. Plants that are harshly salt influenced regularly build up obvious salt damage. As exposure of salinity extends to half a month, secondary shoot growth is influenced, and following a couple of months, clear changes observed in development and injury between salt-stressed plants and control. To comprehend these time-sensitive changes in light of salinity in plant development stages, the 'two-phase growth response to salinity idea created by [16]. The first phase of growth decline occurs within minutes after exposure to salinity. The decline of growth is because of the osmosis stress, osmotic changes

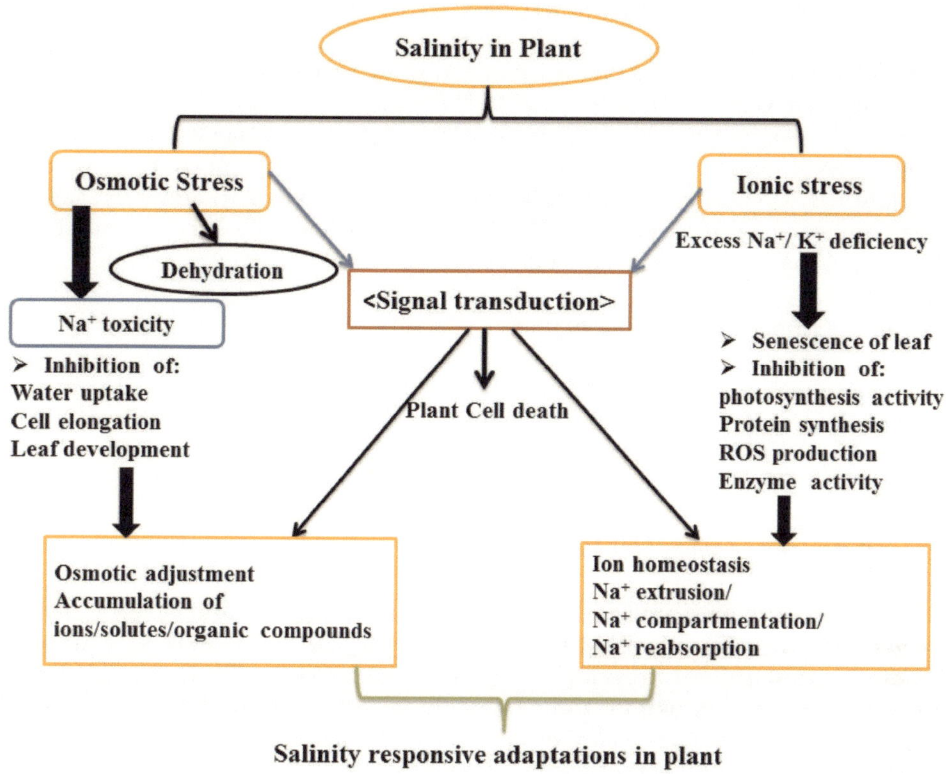

Figure 1.
Salinity response adaptations in plant. Extracted from Kumar et al. [17] and Hussain et al. [18].

outside the root surface, causing changes in osmotic impacts. In the wake of taking some days, weeks or even months the other slower impact (explicit salt impact), bringing about the aggregation of salt in leaves, basically in older leaves and salt toxicity in the plant. This salt toxicity in the plant can cause the death of leaves and decrease the total photosynthetic leaf area. Thus, there is a decrease in the availability of photosynthate to the plant and influence the overall carbon (CO_2) balance essential for sustainable plant growth and development [16] (**Figure 1**).

2.1.1 Salinity and ion toxicity in plant

The important harmful effect of salinity is the sodium and chloride ions accumulation in plant tissues and soil [19]. The higher concentration of soluble salts in the soil profile may cause physiological drought to plant, that is, reduction in uptake of water due to salt accumulation in the root zone [20]. The entrance of sodium and chloride ions into the plant cell from soil causes ion imbalance in plant and soil and excessive uptake of these ions by plant causing many problems related to the physiology of plant's tissues such as root, leaf, grain, fruit, or fiber [21]. Similarly, the reduction of plant osmotic potential, excessive uptake of Na^+ and Cl^- in the cell, and disruption of cell metabolic functions is due to ion toxicity [21]. Excessive sodium ion in plant tissues harms the cell membrane and plant organelles, and as a result, cell death of plant [22]. These physiological changes in the plant include the membranes disruption, reactive oxygen species (ROS) production, reduction of photosynthesis rate (Pn), and scavenging of antioxidants [21]. Consequently, the accumulation of soluble salts in the rhizosphere is one of the main reasons for low crop productivity.

2.1.2 Salinity and nutrient imbalance in plant

Salinity has direct effects on nutrients imbalance between soil and plant. The most important harmful effect of salinity is the sodium and chloride ions accumulation in plant tissues and soil [19]. High sodium ion (Na^+) concentration has an antagonistic effect on potassium (K^+) ions [23]. Moreover, N uptake reduction by the plant has also been observed under high salt conditions [24]. Similarly, salinity has an antagonistic effect on P, K^+, Zn, Fe, Ca^{2+}, and Mn while it has a synergistic effect on N and Mg in field crops such as rice [23, 25].

2.1.3 Salinity and oxidative stress in plant

The production of reactive oxygen species (ROS), like oxygen radical (O^{2-}), superoxide (OH^-), and H_2O_2 under salinity is high [30]. These oxidative species can interrupt the routine functions of various cellular plant modules. For example, DNA, proteins, and lipids, are interfering metabolism of the plant [26].

2.1.4 Salinity and hormonal response in the plant

The phytohormones are naturally produced in a chemical form called plant growth regulators. The phytohormones are active signal compounds which show response against salinity stress and reduce the plant growth [27]. Under salinity stress, the ethylene, cytokinin, and gibberellic acid concentration decreased, and abscisic acid contents increased. This alteration of hormones effects plant growth, such as germination, tiller formation, and reproductive growth. For example, poor development of rice and wheat spikelets, boll of cotton, etc.

3. Effects of salinity on field crops

Fulfill the food demand and livelihood of the increasing population by 2050, a remarkable increase about 50% more yield in the form of grain, fiber, sugar, etc. is required from major field crops such as wheat, rice, and maize sorghum, cotton, and sugarcane [28]. However, the purpose of competing for the demands of human beings on the globe while combating abiotic stresses, including salinity. The different crop has different responses against abiotic stresses such as salinity. Salinity suppresses the crop plants growth, development, and productivity. The sensitivity of the crops varies from low to high concentration of soluble salts or EC (**Table 3**). At low salt concentrations, yields are slightly affected or not affected at all in some crops [29]. Whereas the most plants, glycophytes, including the most crop plants, decrease yield towards zero or even plant death as soluble salt concentrations increase by 100–200 mM NaCl due to low resistance and tolerance capacity of plants [30] (**Table 3**).

3.1 Salinity and rice (*Oryza sativa* L.)

Rice is monocot and belongs to a C3 plant with salinity responsive behavior as compared to other field crops [37]. Rice is vital to the lives of billions of people around the globe. Rice is grown in many parts of the world, especially in Asia, Latin America, and Africa, and taken as a chief food item for more than 50% population of the world [38]. Rice is among the first five major carbohydrate crops for the population of the world, particularly for Asian countries. Only Asia contributes 90% of total rice cultivation in the world. From this 90%, China contributes 30%, India

Crop type	Tolerance based on	Threshold EC levels (dS m^{-1})	25% yield loss (dS m^{-1})	50% yield loss (dS m^{-1})	Zero yield (dS m^{-1})	Ranking	References
Wheat	Grain yield	6–8	6.3	10	16–24	MT	[31]
Rice	Grain yield	3	3.2	3.5–4	8–16	S	[32]
Maize	Ear FW	1.8	2.5–6.8	8.6	15.3	MS	[33]
Sorghum	Grain yield	6.8	7	10	30	MT	[34]
Cotton	Seed cotton	7.7	8.3 7	17.0	16–24	T	[35]
Sugarcane	Shoot DW	1.7	3.9	13.3	16–24	MS	[36]

Where EC = electrical conductivity, FW = fresh weight, DW = dry weight, S = sensitive, MS = moderately sensitive, MT = moderately tolerant, and T = tolerant.

Table 3.
Salt tolerance classification of major field crop.

(21%), and Pakistan (18%) respectively, while remaining 30% is contribution belongs to Japan, Thailand, Indonesia, and Burma [39]. Rice is a high yielding crop. However, the current average yield is 8–10 t/ha for *indica* rice, 10–15% yield is lower than its potential [40]. This rice production gap is due to many reasons, such as environmental stresses (biotic or abiotic), management strategies, and nutrients deficiencies.

Among the abiotic stresses, especially salinity is among the essential causes of this low yield. The morphological characteristics of rice are severely affected by salinity [41]. Rice plant responds differently against salinity compared to other field crops. The intent of salinity in rice plant life cycle varies from growth stages, and cultivar to cultivar, that is, the early seedling growth stage is more sensitive than the tillering stage in rice plant [14]. The threshold level of salt stress for rice is 3 dS m^{-1} [42]. However, a significant reduction in seedling growth and fresh weight were observed with increased salt stress from 1.9 to 6.1 dSm^{-1} and 5 to 7.5 dSm$^{-1,}$ respectively [43]. Many studies also exposed that salinity stress decrease rice stand density and production of seedling biomass, which shows the high sensitivity against salinity [4, 44].

The first organ of the rice plant that keeps in contact with soluble salt is a root [45]. The root is responsible for the entrance of hydrogen peroxide (H_2O_2) and solutes by through different pathways such as symplastic, apoplastic, and transcellular, respectively. So, transport of water and solutes through the apoplastic pathway is vital in rice [46]. Mostly Na^+ transport in rice shoots via the apoplastic passage where Na^+ transports by apoplast through Casparian tubes [47]. As a result of this Na^+ accumulation, a significant reduction in numbers of root per plant, root length, and shoot length occurred under increased salinity [48]. Based on these proofs, the reduced root and shoot lengths are considered two indicators of rice plant response to salinity.

Moreover, cell division and cell elongation in rice plant are severely affected by salinity, which results in a reduction of the root, leaf growth, and yield [16]. Rice plant shows response very soon after the exposure of salinity stress and affects plant growth. For example, rice leaf mortality boosted with increased salinity in almost all rice cultivars at early seedling stage [14]. Some rice cultivars showed leaf mortality up to 0–300% after 1 week of salinity exposure [16]. Salinity effect cause panicle sterility and poor development of inferior and superior spikelets, which result in the reduction of rice grain yield [4]. Many rice cultivars showed panicle sterility at pollination and fertilization stages due to some genetic mechanisms and nutrient deficiencies resulting from salinity stress [49], which leads to a decrease in grain setting rate, pollen viability, and decline of the stigmatic surface.

3.1.1 Salinity and rice physiology

Plant physiological traits are susceptible to the high soluble salts in its rhizosphere. Salinity has bunch of adverse effects on physiology of rice plants, such as hinder the net photosynthesis (Pn), stomatal conductance (Gs), transpiration rate (Tr), photosynthetically active radiation (PAR), degradation of pigment and relative water content (RWC) as well as affect the water use efficiency (WUE) [50]. As far as photosynthesis activity is a concern, rice plants under salinity have decreased photosynthetic efficiency through the complex of photosystem II (PSII). Furthermore, chlorophyll contents in rice leave tissues are damaged by the excessive accumulation of Na^+ and Cl^-, which hamper the primary electron transport in PSII [51]. The chlorophyll contents (chl a, b, and carotenoids) in rice leaves were significantly declined under salinity [52]. High salinity also reduces the quantum yield of the complex PSII, and to decrease K^+/Na^+ ratio. All these factors cause adverse pleiotropic effects on rice physiology and development at the molecular and biochemical levels [53], and cause abnormal rice growth, development, and ultimately plant death [19].

3.1.2 Salinity and ion imbalance in rice plant

Ion imbalance is the ultimate effect of salinity. Under salinity, the severe competition of Na^+ and Cl^- with K^+, Ca^{2+}, and NO_3^- occurs. Generally, high NaCl concentration in the soil and plant decrease the reduce N, P, K, Ca, Mg, and Mn in rice root and shoot, and increases Na^+ and Cl^-, and increases Na^+/K^+ and Na^+/Ca^{2+}, Ca^{2+}/Mg^{2+}, and Cl^-/NO_3^- ratio leads to specific ion (Na^+ and Cl^-) toxicity in plant's organelles [54, 55]. Similarly, boron (B), silicon (Si), and zinc (Zn) availability decreased to the rice plant, and increased cadmium (Cd) toxicity subjected to salinity [56, 57].

3.2 Salinity and wheat (*Triticum aestivum* L.)

Wheat is a worldwide staple food belongs to the *Poaceae* family. Wheat ranks as the first position in grain production globally. About 36% population of the world consume Wheat as a staple food and provides carbohydrates (55%) and 20% of the food calories (20%), and protein contents (13%), which is higher than other cereals crops worldwide [58]. However, wheat production is severely affected by salinity. Wheat is susceptibility to salinity starts at 6 dS m^{-1}. Under salinity, water potential in soil lower down and Na^+ concentration within plant tissues increases, and as a result wheat plant faces osmotic and ionic stresses. Salinity stress having passive impacts on agronomic, physiology, and chemical characteristics of the wheat plant. As salinity level crosses the threshold level (6 dS m^{-1}) of the wheat plant, germination rate, net photosynthesis rate, transpiration rate decrease, and yield, and increases the Na^+ and Cl^- in the wheat plant which disturbs the normal metabolism of the plant [58]. Similarly, water use efficiency (WUE), production of reactive oxygen species (ROS) and scavenging of antioxidants are attributes of the wheat plant affected by salinity.

3.2.1 Salinity and agronomic attributes of wheat

Salinity stress hinders the germination rate (GR) and speed of germination, which is the vital process of the plant cycle, and an important indicator of growth and yield components of the plant, but depend on nature of cultivar. For example, at 125–200 mM NaCl and 12.5–16 dS m^{-1} salinity levels, germination time increased and decreased the GR and germination index [59–61]. During the germination process under salinity, seed faces the osmotic stress, which imbalance the enzymatic activities necessary for nucleic acid and protein metabolism, hormonal imbalance, and ultimate the disturb the seed reserves [62]. Along with these germination characteristics, salinity also affects the other agronomic parameters such as root length, shoot length, root and shoot dry weight, plant height, leaf area, tillering dynamics, and spikes numbers per plant at the early seedling stage. At the early growth stage of the wheat plant, plant shows high sensitivity at 120, 125, 150 mM NaCl, and 16 dS m^{-1}, even seedlings death occurs [11, 63]. Furthermore, wheat seedlings also reduce its growth; even exposure to salinity stress is for a few days (7–10 days) at 100 mM NaCl salt level. Similarly, yield components such as the number of spikes per plant, spikes length, and the number of spikelets per spike, above ground biomass, 1000-grain yield, harvest index, and grain yield per plant decreased with increased salinity stress [64]. However, when the wheat plant cross the threshold level of salinity (6 dS m^{-1}), wheat grain yield reduces at the rate of 7.1% with increasing salinity of per dS m^{-1} and significant yield reduction occurs at 15 dS m^{-1} [65].

3.2.2 Salinity and wheat physiological traits

Photosynthesis activities such as net photosynthesis rate (Pn), stomatal conductance (Gs), transpiration rate (Tr), intercellular CO_2 concentration, and water use efficiency (WUE) affected by salinity. The Pn badly influenced by the high accumulation of Na^+ and Cl^- in the chloroplast tissues [66]. These parameters (Pn, Tr, Gs, intercellular CO_2 concentration) are reduced under 150 mM NaCl salinity level. Similarly, a decrease in photosynthesis pigments was observed at 320 mM NaCl concertation, and after 10 days exposure of NaCl, the chlorophyll contents (chl a, b, and carotenoids) decreased [67]. WUE and RWC also affect by osmotic stress caused by salinity. Water potential lower down with increased salinity levels and as a result relative RWC in the wheat plant decreased by 3.5% in the salt tolerant cultivar and 6.7% in salt-sensitive cultivar after 6 days exposure of NaCl (100 mM NaCl) [68]. Along with this leaf water potential and WUE also decrease in 150 mM NaCl and 16 dS m^{-1} salinity levels. For example, water content percentage in root reduced and increased in shoots and spike of wheat cultivar Banysoif 1. Similarly, at 320 mM NaCl, the RWChas decreased in leaves of wheat cultivar *T. monococcum* seedlings [69].

3.2.3 Reactive oxygen species production and scavenging of antioxidants

Reactive oxygen species (ROS) increase under salinity in the plant. However, when plant faces high salinity, the production of ROS reduces the scavenging system and stops the oxidative stress. This change occurs in plants due to the reduction of CO_2 availability in leaves and inhibits fixation of carbon, and excitation energy enhance which expose the chloroplast, all these happened due to stomatal closure. ROS such as H_2O_2, superoxide ($O2^{\bullet -}$), hydroxyl radical ($OH^{\bullet}$), and singlet oxygen ($^1O_2$) are produced under increasing salinity stress in the plant [62, 65]. Osmotic stress caused by salinity is the leading cause of ROS production and results in the cellular damage by oxidation of lipids, proteins, and nucleic acid. The oxidative stress is caused by an imbalance in ROS production and scavenging of antioxidants in plant tissue. As a result of ROS production, phytotoxic reactions in plants occur such as lipid peroxidation, protein degradation, as well as DNA mutation [70]. For example, exposure of salinity levels 5.4 and 10.6 dS m^{-1} for about 2 months caused a significant increase in lipid peroxidation and hydrogen peroxide (H_2O_2) in seedlings of the wheat-sensitive cultivar [71]. Similarly, H_2O_2 (60%) and MDA (73%) increased at 300 mM NaCl salinity level, and decreased ascorbic acid (AsA) content (52%) in wheat seedlings [67]. For a short period salinity exposure such as after 5 days, MDA contents increased by 35%, and after 10 days, MDA contents increased by 68% at 100 mM NaCl salinity level in wheat leaves. Along with these, the concentration of salt levels in term of EC levels such as 2, 4, 8, and 16 dS m^{-1} EC effects the lipid peroxidation, MDA increased significantly and varied from cultivar to cultivar.

Plants also have an anti-oxidative system to compete against adverse salinity conditions. Therefore, under unfavorable conditions (salinity) plant produce anti-oxidant enzymes in an excessive amount such as superoxide dismutase (SOD), POD, CAT, GR, and APX, etc. which reduce the damage caused by salinity. A study showed that, under increased salinity stress, the SOD, CAT, POD, GR, ascorbic acid (AsA) and APX activities increased irrespective to the nature of wheat cultivar [68]. After 10 days of salinity stress at 100 mM of NaCl showed significant higher POD and SOD contents and non-significant increase in the CAT and APX contents with a decrease in GR and DHAR contents in wheat seedlings [67].

3.2.4 Ion imbalance in wheat

Salinity stress also causes an imbalance in ion uptake and ion toxicity in the plant. Na^+ absorption varies from nature of wheat cultivars against salinity stress [68]. Salinity increase the intake of Na^+ and Cl^- and reduced the K^+ and Ca^{2+} uptake along with the lower accumulation of NO_3^- and PO_4^{3-} in wheat seedlings under 125 mM of NaCl level for one-week exposure, and decreased the K^+/Na^+ ratio in wheat shoots at 120 mM of NaCl [11, 65, 66]. Similarly at high EC 15–16 dS m^{-1}, K^+ accumulation significantly decreased, and under medium salinity stress, Na^+ and Cl^- accumulation increase and decreased the uptake of K^+, Ca^{2+}, and Zn^{2+} [64, 65, 72].

3.3 Salinity and maize (*Zea mays* L.)

Maize is an important cereal crop which is being cultivated over a large area under a wide spectrum of edaphic and climatic conditions. It is categorized as a C4 plant of the *Poaceae* family and is moderately sensitive to salinity [73]; nevertheless, a considerable intraspecific genetic potential against salinity also exists in the maize. The threshold level of salinity for maize is 0.25 mM NaCl or 1.8 dS m^{-1}, and a further increase in salinity may stunt growth and cause severe damages [74].

3.3.1 Salinity and maize growth

Salinity significantly induces the detrimental changes in growth and development of maize, but the response of maize varies with the crop growth stage and degree of stress. The short term exposure to salinity may influence the growth of maize plants duet to osmotic stress without causing the ionic toxicity. The germination and early seedling stages of maize are more sensitive to salinity than later developmental stages. Generally, salinity during germination period delays the initiation, reduces the rate, and increases the dispersion of germination phases [75]. Salinity induces the detrimental impact on seed germination; (a) by sufficiently reducing the osmotic potential of the soil, leading to retard the water absorption by seed, and (b) by inducing Na^+ or Cl^- or both ions toxicity to the seed embryo. Therefore, hyper-osmotic effects and toxic stress of Na^+ and Cl^- ions on germinating seeds under saline conditions may delay or reduce germination [75]. Maize as a salt-sensitive crop, the shoot growth in maize is sharply reduced during the osmotic stress phase [76]. However, Schubert et al. [77] proved that it was cell wall extensibility, which limited the cell extension growth during osmotic stress phase than turgor in the cells. In crux, salinity-induced growth reduction in maize is primarily due to the suppressed leaf initiation and expansion, as well as internode growth and also by increased leaf abscission. Additionally, Salinity reduced the grain number and weight, leading to low grain yield of maize. This reduction was due to the limitation of the sink and reduced activity of acid inverses in developing maize grains lead to poor kernel setting as well as reduced grain numbers.

3.4 Salinity and cotton (*Gossypium hirsutum*)

Cotton is grown as the most important fiber oilseed crop, providing 35% of the total fiber used globally [78]. About 29.5 million hectares of cotton were grown during 2016–2017 with a total production reaching to 106.49 million bales during 2017 [79] worldwide. *Gossypium hirsutum* is giving over 90% of the world cotton crop annually, after spreading from its origin in Mesoamerica to more than 50 countries in Northern and Southern hemispheres.

3.4.1 Effects of salinity on cotton plant

Cotton is mostly grown in arid and semi-arid regions of the world, where water shortage is a dominant factor [80]. In general, salinity severely hinders cotton growth and development, including the reduced plant height, fresh and dry weights of shoot and roots, leaf area index, node number, canopy development, photosynthesis, transpiration rate, stomatal conductance, yield, fiber quality, and root development [81]. However, cotton is considered a moderately salt tolerant crop which can withstand EC up to 7.7 dS m^{-1} [34]. Generally, salinity effects on cotton at all ontogenetical levels, from molecular to organismal, which lead towards the reduced plant growth, economic yield, and fiber quality. But these effects depend on the timing and intensity of salt stress, the plant growth stage, and the species. Therefore, seed germination and early seedling stage of cotton are considered as the most sensitive stages to salinity [1]. It has been advocated that plants having a higher tolerance to salinity generally maintain lower Na^+/K^+ ratio in their tissues [82]. Furthermore, Wang et al. [83] found that soil ECe and sodium absorption ratio (SAR) values of root zone were significantly and linearly correlated with the final germination percentage of the cotton. The FG% was adversely affected by increasing EC and SAR. These results also show that the vulnerability of cotton plants towards salinity increases with increase in plant age. Therefore, cotton plant is more sensitive to the salinity during peak flowing period, leading to less number of bolls, boll weight, and lint yield [84]. Many studies [34, 85] also reported up to 50% yield reduction when the salinity level was increased from 7.7 to 17.0 dS m^{-1}. Soil salinity also induces a wide range of morpho-physiological and biochemical changes that adversely affect the cotton growth and productivity. Additionally, plant biomass accumulation and the final output are pre-determined by the rate of photosynthesis, salinity induced a direct impact on both stomatal and mesophyll conductance [86].

3.4.2 Salinity and fiber quality

The production of higher fiber quality is a key objective of cotton breeding and genetics programs globally [87]. However, salinity induced lower lint percentage and fiber quality parameters, including fiber length, strength, and micronaire [84].

However, salinity during the flowering season imposed no detrimental impacts on fiber quality, but salinity after flowering resulted in reduced fiber quality.

3.5 Effects of salinity on sorghum [(*Sorghum bicolor* L.) Moench]

Sorghum is monocot species, and C4 plant with high photosynthetic capability and productivity has a spot with *Poaceae* family. The most of the sorghum species found in Australia and the rest of the world (Asia, Africa, Mesoamerica, India, and Pacific Oceans). Sorghum is the extremely beneficial yield, which can be used for essentialness source, human sustenance (grain), domesticated animals feed (grain and biomass), and mechanical reason (fiber or paper and treatment of natural side-effect). The sorghum biomass is used as fuel (ethanol generation) and sugar substrate through aging (methane creation) [88].

3.5.1 Effects of salinity on sorghum

The sorghum plant has an extraordinary adjustment potential to abiotic stresses, particularly high salinity, which is significant for genotypes developing in an extreme environment [89, 90]. By and large, sorghum is considered as a respectable salinity tolerant species with genotypic varies from cultivar to cultivar. The threshold level of salinity for grain sorghum is (6.8 dS m^{-1}), and the reduction reaches 25% and 50% at 7 and 10 dSm^{-1} respectively [34]. Salinity also influences the sorghum plant's physiological procedures, for example, seed germination rate, K^+ take-up, net photosynthesis rate (Pn), biomass amassing, and biochemical qualities (chlorophyll substance or electrolyte leakage). In sorghum plants, a notable salinity induced phenotype of plant growth was observed after 4 days of exposure of 200 mM NaCl salinity stress [91]. Similarly, in sweet sorghum, salinity increase the duration of germination and reduced germination percentage [92].

3.5.2 Effects of salinity on ROS production in sorghum

Under salinity stress, the production of reactive oxygen species (ROS) and an increase in the antioxidant enzymatic activity is a vital component of salt tolerance capacity of the plant. Salinity stress is linked with associated with enhanced antioxidant activity. Salinity decreased superoxide dismutase (SOD), ascorbate peroxidase (APX), catalase (CAT), peroxidase (POX), and glutathione reductase (GR), and total antioxidant and phenol contents of tissues in sorghum cultivars [93]. The stem yield and soluble carbohydrate contents decreased as salinity level increased in sweet sorghums cultivars such as Keller and Sofra, and in one-grain sorghum cultivar Kimia, whereas it is also reported that at the higher salinity stress the sorghum cultivar 'Keller' showed high sucrose contents and stem yield [90].

3.5.3 Salinity and ion toxicity in sorghum

The large aggregation of toxic ions such as Na^+ and Cl^- causes unsettling influence in ion uptake and K^+ status of plant tissues. In this manner, it is the high K^+/Na^+ perception and the conservation of low Na^+/K^+ ratio in plant tissues, which describe as salt-tolerant genotypes [94]. The Na^+ content in sorghum plant's tissue enhanced with excessive Na^+ contents, and as a result of significant contrasts in Na^+ contents of root and shoot among genotypes. Lesser accumulation of Na^+ in the shoot might be due to lower Na^+ uptake by the root or from the variation in the Na^+ transfer rate to the shoot. For example, salt-tolerant sorghum variety (Jambo) amassed less Na^+ concentration in the root and shoot tissues than the salt-sensitive genotypes and kept up lower Na^+/K^+ ratios both in the root and shoot [95]. Particular testimony of Na^+ ions in the shoot depends on leaf base [96], and enhancing levels of Ca^{2+} in the control condition increased plant growth and brought down Na^+ take-up of sorghum plants [97]. The high Ca^{2+} accumulation in leaf and root tissues were observed in the salt-tolerant genotype Jambo than the salt sensitive varieties, Payam and Kimia [98].

3.6 Effects of salinity on sugarcane (*Saccharum* sp.)

Sugarcane is a key commercial and irrigated crop of the tropical and subtropical areas of the world [99]. Sugarcane is propagated further by setts from the stem cuttings of mature plants (one-year-old crop). Sugarcane is an important source of sugar in Asia and Europe. It also supplied the basic raw material for the production of jaggery (Gur), white sugar, and khandsari. Further, sugarcane juice is widely being used for drinking and beverage purposes.

3.6.1 Salinity and sugarcane production

The salinity is a major environmental concern, responsible for a significant decline in sugarcane yield [100]. The sugarcane production is low under less fertile soil caused by salinity stress. This plant is categorized as a moderately salt sensitive species which can withstand the ECe up to 1.7 dS m^{-1}. But, a further increase in EC could induce the adverse effects on its production. The detrimental impacts of salinity at germination or bud emergence stage mainly varied across the different species. Akhta et al. [101] reported a significant reduction in sprout emergence at different days after sowing under moderate and severe salinity stress depends on the nature of cultivars.

Under severe salinity stress conditions, growth could be significantly influenced by the accumulation of active oxygen species [102]. Vasantha et al. [103] observed the reduced leaf area index (LAI) of sugarcane by 36% during Formative Growth

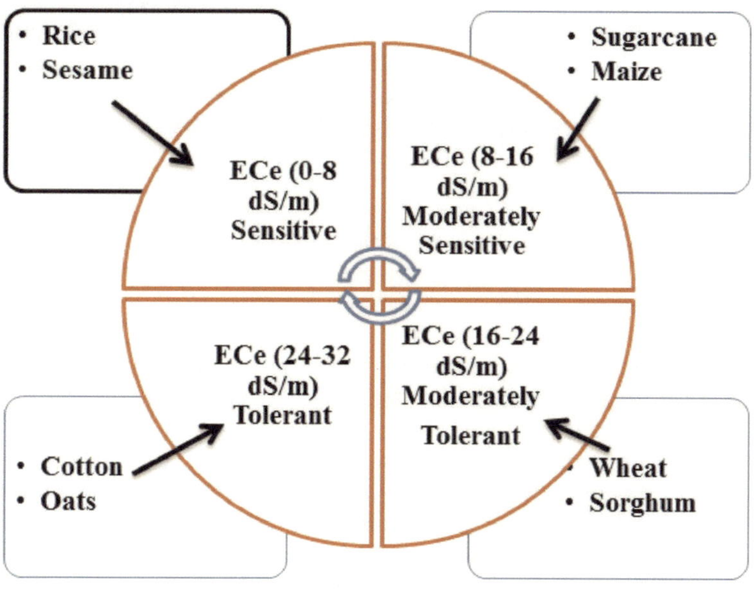

Figure 2.
Classification of field crops subjected to salinity stress. Extracted from Maas and Grattan [105].

Phase (FGP) and by 21% during Grand Growth Period (GGP). Additionally, they observed it decreased in biomass accumulation by 44% during FGP and 32% during GGP. The significant reduction in shoot and root biomass accumulation in sugarcane sprouts with increasing salinity level from normal to 120 mM NaCl [101]. Similarly, the increasing NaCl level resulted in a reduction of the shoot, root length, root volume, and leaf area of sugarcane seedlings by 36–41, 29–42, and 52–66%, and chlorophyll contents by 20.0–45.0% respectively [104]. The other factors which directly reflect the depletion of growth of sugarcane are linked with alterations in gas exchange parameters, and reduced transpiration and photosynthetic rates due to stomatal closure. As concern sugarcane yield and related traits, the sucrose juice (6%) of sugarcane was significantly reduced induced during Grand Growth Period (GGP) and so also the brix [103]. Similar to the millable canes (MC) and cane yield were reduced drastically under salinity. The MC decreased by 8.0–100% by exposing under salinity. Additionally, salinity caused negative impacts on cane yield, cane length, and single cane weight. Hence, the different field crops showed a different level of response to salinity stress depends on their genetic nature and as EC increased from 32 dS m^{-1}, the yield is unacceptable from the most of the field crops (**Figure 2**) [105].

4. Management strategies

There are two groups of management strategies against salinity, first one natural adaptation responses towards salinity, and second are human-made management strategies to handle the salinity stress in field crops or plants. Tolerance or resistance of rice plant to salt stress involves many adaptive responses at molecular, cellular, and physiological levels. Among the natural management strategies by the plants to salinity stress based on three strategies: (i) exclusion of Na^+ from the cytoplasm due to low uptake, or pumping out of the ion from the cell by active mechanisms, (ii) requisitioning of Na^+ into the vacuole and (iii) preferential accumulation in the leaf tissues. However, the genotypes with high leaf Na contents proved to be generally salt sensitive, and only those can tolerate high tissue concentrations, which can sequester Na^+ into the vacuoles of leaf cells. The essential processes leading to plant adaptation to high salinity include ionic, metabolic, and osmotic adjustments. The salt-resistant genotypes can successfully cope with osmotic and ionic stresses caused by the excess of NaCl; they can effectively reduce the oxidative damage and can detoxify the harmful metabolites [106].

4.1 Natural adaptation responses towards salinity by plant

4.1.1 Osmotic adjustment

Osmotic adjustment is the best and favorable plant physiological strategy to endure concentration of toxic ion (Na^+ and Cl^-) in cytoplasm and compartmentalization in vacuoles, and define the salinity tolerance limits for plant [107]. Under osmotic stress, accumulation of free sugar, glycine betaine, organic solutes, and the proline in the plant's cytoplasm is also an important strategy to cope with the salinity stress [108]. This phenomenon is important to handle the antagonistic abiotic stresses, including salinity and maintain the homeostasis in osmotic or ionic signaling [17]. Similarly, leaf area or leaf architecture is also an important trait of the plant, which can reduce the excessive amount of Na^+ in leaves through dilution effects and the transpiration force [109].

4.1.2 Closure of stomata

The ultimate response of plant subjected to salt stress is the closure of stomata [110]. The carbon dioxide assimilation decrease, as EC level increase (0–20 dS m^{-1}) which results in plant growth reduction as well as the closure of stomata. This closure of stomata decreased the intracellular (Ci) CO_2 partial pressure leading to hampering the Pn [5]. High salinity stress in rhizosphere decrease the transpiration rate (Tr), reduce the root water potential. Salinity stress enhances the biosynthesis of abscisic acid (ABA) and closes the stomata after reaching the guard cells. ABA passage from root to shoot causes closure of stomata and save the leave tissue from dehydration [54, 111]. Mostly, salinity hinders Pn in various crop plants. However, the sound reasons for lower Pn are stomatal closure, lower sink activity, reduced efficiency of rubisco, dislocation of vital cations from the membrane structure of leaf which lead to changes in permeability, and swelling and inefficiency of the grana [112], or might be due to the direct effects of salinity on conductance of stomata through a decrease in guard cell turgidity and CO_2 partial pressure within plant cell [113]. Closure of stomata plays a vital role to survive with salinity stress. Chen and Gallie [114] studied that the ascorbate or ascorbic acid (AsA) redox state controls the transpiration rate and conductance of stomata. Stomatal guard cells control through Na^+ which control transpiration rate according to the concentration of salt presented in soil environment [115].

4.2 Agronomic practices to cope salinity

Salinity occurs because of excessive accumulation of soluble salts via soil chemical properties and irrigated water. As a result of salinity stress and ion (Na^+ and Cl^-) toxicity, the disturbance of ion imbalance occurs. By adopting some measures, these problems can manage plant growth by adopting some agronomic strategies such as water and nutrient management to improve soil health, plant growth, and input use efficiency (IUE) under salinity [116].

4.2.1 Water and nutrient management strategies

Irrigation water with high electrical conductivity (EC), sodium adsorption ratio (SAR), residual sodium carbonate (RSC), and pH value also causes of salinity stress and plant growth reduction [117]. For the better survival of plant against salinity stress, a wise water management strategy is indispensable. Availability of good quality irrigated water is very vital for the survival of crop plant and yield [118]. The usage of good quality water is a good option to drain or leeched down the soluble salts from the root zone for better soil management and plant growth [118]. The canal water is a good replacement of brackish underground water for irrigation of field crops. If canal water is unavailable, the use of gypsum with brackish underground water is the best option, and it increased 25–294% rice yield and 182% wheat production under salinity [119]. Similarly application of canal water with 100% gypsum help to lower the ECe, pH value, and SAR of soil at 0–30 cm depths than saline water with 100% gypsum in field crops [120]. In case of less availability of good quality water, then the 25% gypsum amendment with unfit irrigating water is the best option. The wise use of less good quality is than never mix up with the unfit underground water or tub well water, and follows the irrigation scheduling [119].

The management of salinity stress by nutrient management is the wise use of calcium (Ca^{2+}) source in the form of gypsum ($CaSO_4 \cdot 2H_2O$) and to improve soil water infiltration and better plant growth. Application of gypsum (100%), a combination of gypsum + farmyard manure (FYM) + H_2SO_4, $CaSO_4{\cdot}2H_2O$ + FYM + chiseling, pyrite, and humic acid (HA), in rice, wheat, sorghum crop improved soil properties, plant biomass, and yield [118, 121]. Application of N, P, K, S, Zn, B, and Mn separately or with different combination increased rice total above ground biomass and grain production under salt-affected soils [122, 123]. Humic acid improves nutrients availability by chelating with unavailable nutrients (P, K, Ca, Fe, Zn, and Cu) and buffered pH value, and enhanced soil microbial, enzymatic and physiological activities, and plant growth under salinity [121, 124]. The combined effect of humic acid (HA) with gypsum (24 and 48 kg/ha) in rice was higher than the alone effect of HA on ECe and SAR due to its chelating effect with other nutrients subjected to salinity [125]. The use by-product of sugarcane (press-mud), green manure, poultry manure, and Sesbania as a cover crop for amendment of soil to reduce the effect of salinity which is a source of macro and micronutrients especially Zn and S in crops are also good options [126]. There are some other useful agronomic practices to reduce the effect of salt stress on the plant are periodic use of fresh water, subsoiling, deep tillage, sanding, and application of organic and inorganic fertilizers, and adopting crop rotation [118].

4.2.2 Application of hormones regulators

The hormonal imbalance is one of the salinity effects on plants. There are many plant growth regulators being used as hormones regulator or plant growth regulators such as aminoethoxyvinylglycine (AVG), ethephon, and 1-methylcyclopropene (1-MCP) for ethylene inhibitor under salt stress and enhance the boles and spikelets development in rice and cotton respectively [4]. Similarly, exogenous applications of abscisic acid (ABA), brassinosteroids (BRs) or their analogs (D-31, D-100, etc.) are good option to improve plant performance under salinity [127, 128].

4.2.3 Traditional breeding for salt tolerance

To meet the demand for food and livelihood of the increasing population on the globe, the increase in the agriculture production is indispensable. Therefore, many efforts have been made to improve salinity tolerance capacity of the crops through conventional plant breeding and biotechnology [129]. Salinity tolerance is a complex trait both at the genetic and physiological level and controlled by polygenes. It has been speculated that salinity tolerance seems to be regulated by independent genes at different growth stages [130]. Traditional breeding has been considered as a more promising and efficient approach to improve the salt tolerance. Conventional breeding involves identification of QTLs using closely linked markers along with their phenotypic evaluation. One of the best-studied QTL for salt tolerance; saltol was identified by the conventional breeding approach in rice [131]. This QTL was found to control shoot Na^+/K^+ ratio at the seedling stage. So, the identification of new QTLs and later pyramiding of these QTLs would lead to the development of the more promising salt tolerant line. Marker-assisted backcrossing (MAB), which is one of the best traditional breeding approaches that involve the transfer of the specific allele at target locus from donor to recipient parent, can be used for this purpose. Traditional breeding mainly relied on the use of diverse germplasm resources to identify the landraces showing salt tolerance and then map the locus responsible for salt tolerance. This can be seen as an advantage as well as a disadvantage. Salt tolerance is an outcome of involvement of diverse cellular processes like ion transport and homeostasis, osmoregulation, and oxidative stress protection. Identification and characterization of key genes for salt tolerance would need the collective application of advanced molecular mapping, genomics, transcriptomics, and proteomics approaches.

4.2.4 Molecular breeding to improve salt tolerance

Many salt tolerance genes have been discovered by using traditional breeding techniques, such as subtractive hybridization, differential hybridization, and through genetic information from the model organism. Furthermore, protein crystallography, a proteomic study has enabled researchers to the exploration of the protein's structure and function for salt tolerant genes. After salt tolerance gene identification, many latest techniques for foreign gene transformation to the desired plant can help to improve field crop production. Such as CRISPR CAS9, PEG-mediated gene transfer, electroporation, partial or the micro projectile bombardment, microinjection, and Agrobacterium-mediated gene transfer. These techniques are available for many crops.

5. Conclusions and future perspectives

Salinity stress is the one of the key growth hampering agents for field crops. Salinity not only effects the plant growth but also affect yield by creating osmotic, ionic, and oxidative stresses. From this chapter, it is concluded that, the rice (sensitive), sugarcane and maize (moderately sensitive, wheat and sorghum (moderately tolerant), and cotton (tolerant) subjected to salinity. There are many management strategies, including traditional soil, water, and nutrient management strategies as well as genetic modification and by using molecular breeding, tools are suitable for producing salt tolerance cultivars. The bunch of information in this chapter wills able the scientific community to understand the role of salinity stress in field crops and their management options [115].

Acknowledgements

Sajid Hussain is thankful to China National Rice Research Institute for the necessary facilities and to Chinese Academy of Agricultural Sciences, China for the award of Postdoctoral Fellowships. SH is giving special thanks to Prf. Dr. Qianyu Jin and Junhua Zhang for motivated me to write this chapter and provided my friendly environment to complete this task. We are also highly thankful to Dr. M. Ashraf and M. Shaukat, for contributing in chapter write up and providing us supporting material.

Author details

Sajid Hussain[1], Muhammad Shaukat[2], Muhammad Ashraf[3], Chunquan Zhu[1], Qianyu Jin[1] and Junhua Zhang[1*]

1 State Key Laboratory of Rice Biology, China National Rice Research Institute, Hangzhou Zhejiang, China

2 Department of Agronomy, University of Agriculture, Faisalabad, Pakistan

3 Department of Soil Science, Bahauddin Zakariya University, Multan, Pakistan

*Address all correspondence to: zhangjunhua@caas.cn

References

[1] Munns R, Tester M. Mechanisms of salinity tolerance. Annual Review of Plant Biology. 2008;**59**:651-681

[2] FAO: High Level Expert Forum—How to Feed the World in 2050. Rome, Italy: Economic and Social Development, Food and Agricultural Organization of the United Nations; 2009

[3] Ghassemi F, Jakeman AJ, Nix HA. Salinisation of Land and Water Resources: Human Causes, Extent, Management and Case Studies. Sydney, Australia, and CAB International, Wallingford, UK: UNSW Press; 1995

[4] Hussain S, Bai Z, Huang J, Cao X, Zhu L, Zhu C, et al. 1-Methylcyclopropene modulates physiological, biochemical, and antioxidant responses of Rice to different salt stress levels. Frontiers in Plant Science. 2019;**10**:124. DOI: 10.3389/fpls.2019.00124

[5] Amirjani MR. Effect of NaCl on some physiological parameters of rice. European Journal of Biological Sciences. 2010;**3**(1):6-16

[6] Flowers TJ. Improving crop salt tolerance. Journal of Experimental Botany. 2004;**55**(396):307-319

[7] Ye WW, Liu JD, Fan BX, Hu QM. The effect of salt on the fiber characteristics in Upland cotton. China Cottons. 1997;**24**:17-18

[8] Ye WW, Liu JD, Fan BX, Hu QM. The effect of salt on the fiber characteristics in Upland cotton. China Cottons. 1997;**24**:17-18

[9] Parihar P, Singh S, Singh R, et al. Effect of salinity stress on plants and its tolerance strategies: A review. Environmental Science and Pollution Research. 2015;**22**:4056

[10] Rengasamy P. World salinization with emphasis on Australia. Journal of Experimental Botany. 2006;**57**(5): 1017-1023

[11] Jamal Y, Shafi M, Bakht J, Arif M. Seed priming improves salinity tolerance of wheat varieties. Pakistan Journal of Botany. 2011;**43**:2683-2686

[12] Ghafoor A, Qadir M, Murtaza G. Salt-Affected Soils: Principles of Management. Urdu Bazar, Lahore, Pakistan: Allied Book Centre; 2014. ISBN: 969-547

[13] Bohn HL, BL MN, O'Connor GA. Soil Chemistry. 2nd ed. New York: John Wiley and Sons; 1985

[14] Shereen A, Mumtaz S, Raza S, Khan MA, Solangi S. Salinity effects on seedlings growth and yield components of different inbred Rice lines. Pakistan Journal of Botany. 2005;**37**(1):131-139

[15] Amirjani MR. Effect of salinity stress on growth, sugar content, pigments and enzyme activity of rice. International Journal of Botany. 2011;7: 73-81

[16] Munns R. Comparative physiology of salt and water stress. Plant, Cell & Environment. 2002;**25**:239-250

[17] Kumar K, Manu K, Seong-Ryong K, Hojin R, Yong-Gu C. Insights into genomics of salt stress response in rice. The Rice a Springer Open Journal. 2013; **6**:27

[18] Hussain S, Zhang J, Zhong C, Zhu L, Cao X, Yu S, et al. Effects of salt stress on rice growth, development characteristics, and the regulating ways: A review. Journal of Integrative Agriculture. 2017;**16**(11):2357-2374

[19] Nishimura T, Cha-um S, Takagaki M, Ohyama K. Survival percentage, photosynthetic abilities and growth characters of two indica rice (*Oryza sativa* L. spp. indica) cultivars in response to isosmotic stress. Spanish Journal of Agricultural Research. 2011;**9**: 262-270

[20] Munns R. Genes and salt tolerance: Bringing them together. The New Phytologist. 2005;**167**:645-663

[21] James RA, Blake C, Byrt CS, Munns R. Major genes for Na^+ exclusion, Nax1 and Nax2 (wheat HKT1;4 and HKT1;5), decrease Na+ accumulation in bread wheat leaves under saline and waterlogged conditions. Journal of Experimental Botany. 2011;**62**(8): 2939-2947

[22] Siringam K, Juntawong N, Cha-um S, Kirdmanee C. Salt stress induced ion accumulation, ion homeostasis, membrane injury and sugar contents in salt-sensitive rice (*Oryza sajtiva* L. spp. indica) roots under isoosmotic conditions. African Journal of Biotechnology. 2011;**10**:1340-1346

[23] Jung JY, Shin R, Schachtman DP. Ethylene mediates response and tolerance to potassium deprivation in Arabidopsis. The Plant Cell. 2009;**21**: 607-621

[24] Abdelgadir EM, OKA M, Fujiyama H. Nitrogen nutrition of rice plants under salinity. Biologia Plantarum. 2005;**49**(1):99-104

[25] Garcia MJ, Lucena C, Romera FJ, Alcántara E, Perez-Vicente R. Ethylene and nitric oxide involvement in the up-regulation of key genes related to iron acquisition and homeostasis in Arabidopsis. Journal of Experimental Botany. 2010;**61**:3885-3899

[26] Demiral T, Turkan I. Comparative lipid peroxidation, antioxidant defense systems and proline content in roots of two rice cultivars differing in salt tolerance. Environmental and Experimental Botany. 2005;**53**:247-257

[27] Pedranzani H, Racagni G, Alemano S, Miersch O, Ramirez I, Pena-Cortes H, et al. Salt tolerance tomato plants show increased levels of Jasmonic acid. Plant Growth Regulation. 2003;**41**: 149-158

[28] Godfray HCJ, Beddington JR, Crute IR, Haddad L, Lawrence D, Muir JF, et al. Food security: The challenge of feeding 9 billion people. Science. 2010; **327**(5967):812-818. DOI: 10.1126/ science.1185383

[29] Maggio A, Hasegawa PM, Bressan RA, Consiglio MF, Joly RJ. Review: Unravelling the functional relationship between root anatomy and stress tolerance. Functional Plant Biology. 2001;**28**(10):999-1004

[30] Munns R, Termaat A. Whole-plant responses to salinity. Functional Plant Biology. 1986;**13**(1):143-160

[31] Asana RD, Kale VR. A study of salt tolerance of four varieties of wheat. Indian Journal of Plant Physiology. 1965; **8**:5-22

[32] Venkateswarlu J, Ramesam M, Murali Mohan Rao GV. Salt tolerance in rice varieties. Indian Society of Soil Science. 1972;**20**:169-173

[33] Maas EV, Donovan TT, Francois LE. Salt tolerance of irrigated guayule. Irrigation Science. 1988;**9**:199-211

[34] Maas EV, Hoffman GJ. Crop salt tolerance—Current assessment. Journal of the Irrigation and Drainage Division. 1977;**103**(IR2):115-134

[35] Bernstein L, Ford R. Salt tolerance of field crops. In: United States Salinity Laboratory Report to Collaborators; Riverside, CA. 1959. pp. 34-35

[36] Dev G, Bajwa MS. Studies on salt tolerance of sugarcane. Indian Sugar (Calcutta). 1972;**22**:723-726

[37] Joseph B, Jini D, Sujatha S. Biological and physiological perspectives of specificity in abiotic salt stress response from various rice plants. Asian Journal of Agricultural Sciences. 2010; **2**(3):99-105

[38] Lou WP, Wu L, Chen HY, Ji ZW. Assessment of rice yield loss due to torrential rain: A case study of Yuhang Country, Zhejiang Province, China. Natural Hazards. 2012;**60**:311-320

[39] Calpe C. Rice International Commodity Profile. Food and Agricultural Organization of the United Nations: Rome; 2006

[40] IRRI (International Rice Research Institute). Rice: Hunger or Hope? MCPO box 3217, Makati 1271, Philippines: IRRI; 1998

[41] Zhenhua Z, Qiang L, Hai-Xing S, Xiang-min R, Ismail AM. Responses of different rice (*Oryza sativa* L.) genotypes to salt stress and relation to carbohydrate metabolism and chlorophyll content. African Journal of Agricultural Research. 2012;**7**(1):19-27

[42] Grattan SR, Zeng L, Shannon MC, Roberts SR. Rice is more sensitive to salinity than previously thought. California Agriculture. 2002;**56**:189-195

[43] Kazemi K, Eskandari H. Effects of salt stress on germination and early seedling growth of rice (*Oryza sativa*) cultivars in Iran. African Journal of Biotechnology. 2011;**10**(77):17789-17792

[44] Mostofa GM, Saegusa D, Fujita M, Phan TL. Hydrogen sulfide regulates salt tolerance in rice by maintaining Na+/K+ balance, mineral homeostasis, and oxidative metabolism under excessive salt stress. Frontiers in Plant Science. 2015;**6**:1055

[45] Smet ID, White PJ, Bengough AG, Dupuy L, Parizot B, Casimiro I, et al. Analyzing lateral root development: How to move forward. The Plant Cell. 2012;**24**:15-20

[46] Kronzucker JK, Britto TD. Sodium transport in plants: Acritical review. New Phytologist. 2011;**189**:54-81

[47] Gong HJ, Randall DP, Flowers TJ. Silicon deposition in the root reduces sodium uptake in rice (*Oryza sativa* L.) seedlings by reducing bypass flow. Plant Cell Environment. 2006;**29**: 1970-1979

[48] Jiang XJ, Zhang S, Miao L, Tong T, Liu Z, Sui Y. Effect of salt stress on rice seedling characteristics, effect of salt stress on root system at seedling stage of rice. North Rice. 2010;**40**:21-24 (in Chinese)

[49] Hasamuzzaman M, Fujita M, Islamm MN, Ahamed KU, Kamrin N. Performance of four irrigated rice varieties under different levels of salinity stress. International Journal of Integrative Biology. 2009;**6**:85-89

[50] Ramezani MR, Mohammad JS, Gholamreza M, Mohammad HS. Effect of salinity and foliar application of iron and zinc on yield and water use efficiency of Ajowan (*Carum copticum*). International Journal of Agriculture and Crop Sciences. 2012;**4**(7):421-426

[51] Munns R, James RA, Lauchli A. Approaches to increasing the salt tolerance of wheat and other cereals. Journal of Experimental Botany. 2006; **57**:1025-1043

[52] Cha-umi S, Kanyaratt S, Chalermpol K. Comparative effects of salt stress and extreme pH stress combined on glycinebetaine accumulation, photosynthetic abilities and growth characters of two Rice genotypes. Rice Science. 2009;**16**(4): 274-282

[53] Tester M, Davenport R. Na^+ tolerance and Na^+ transport in higher plants. Annals of Botany. 2003;**9**: 503-527

[54] Zeinolabedin J. The effects of salt stress on plant growth. Technical Journal of Engineering and Applied Sciences. 2012;**2**(1):7-10

[55] Razzaque AM, Nur MT, Tofazzal IM, Kumar DR. Salinity effect on mineral nutrient distribution along roots and shoots of rice (*Oryza sativa* L.) genotypes differing in salt tolerance. Archives of Agronomy and Soil Science. 2011;**57**(1):33-45

[56] Wimmer MA, Muehling KH, Lauchli A, Brown PH, Goldbach HE. Interaction of salinity and boron toxicity in wheat (*Triticum aestivum* L.). In: Horst WJ, Schenk MK, Burkert A, Claassen N, Flessa H, Frommer WB, Goldbach H, Olfs HW, Romheld V, editors. Plant Nutrition: Food Security and Sustainability of Agro Ecosystems Through Basic and Applied Research. Netherlands: Springer; 2001. pp. 426-427. ISBN: 978-0-7923-7105(2)

[57] Amanullah I. Dry matter partitioning and harvest index differ in rice genotypes with variable rates of phosphorus and zinc nutrition. Rice Science. 2016;**23**:78-87

[58] Hasanuzzaman M, Nahar K, Rahman A, Anee TI, Alam MU, Bhuiyan TF, et al. Approaches to Enhance Salt Stress Tolerance in Wheat. 2017:151-187. http://dx.doi.org/10.5772/67247

[59] Ghiyasi M, Seyahjani AA, Tajbakhsh M, Amirnia R, Salehzadeh H. Effect of osmopriming with polyethylene glycol (8000) on germination and seedling growth of wheat (*Triticum aestivum* L.) seeds under salt stress. Research Journal of Biological Sciences. 2008;**3**:1249-1251

[60] Akbarimoghaddam H, Galavi M, Ghanbari A, Panjehkeh N. Salinity effects on seed germination and seedling growth of bread wheat cultivars. Trakia Journal of Sciences. 2011;**9**:43-50

[61] Fuller MP, Hamza JH, Rihan HZ, Al-Issawi M. Germination of primed seed under NaCl stress in wheat. ISRN Botany. 2012;**167801**:5. DOI: 10.5402/2012/167804

[62] Hasanuzzaman M, Nahar K, Fujita M, Ahmad P, Chandna R, Prasad MNV, et al. Enhancing plant productivity under salt stress—Relevance of poly-omics. In: Ahmad P, Azooz MM, Prasad MNV, editors. Salt Stress in Plants: Omics, Signaling and Responses. Berlin: Springer; 2013. pp. 113-156

[63] Afzal I, Rauf S, Basra SMA, Murtaza G. Halopriming improves vigor, metabolism of reserves and ionic contents in wheat seedlings under salt stress. Plant Soil and Environment. 2008;**54**:382-388

[64] Asgari HR, Cornelis W, Damme PV. Salt stress effect on wheat (*Triticum aestivum* L.) growth and leaf ion concentrations. International Journal of Plant Production. 2012;**6**:195-208

[65] Afzal I, Basra SMA, Cheema MA, Farooq M, Jafar MZ, Shahid M, et al. Seed priming: A shotgun approach for alleviation of salt stress in wheat. International Journal of Agriculture and Biology. 2013;**15**:1199-1203

[66] Wahid A, Perveen M, Gelani S, Basra SMA. Pretreatment of seed with H_2O_2 improves salt tolerance of wheat seedlings by alleviation of oxidative damage and expression of stress proteins. Journal of Plant Physiology. 2007;**164**:283-294

[67] Zou P, Li K, Liu S, He X, Zhang X, Xing R, et al. Effect of sulfated chitooligosaccharides on wheat seedlings (*Triticum aestivum* L.) under salt stress. Journal of Agricultural and Food Chemistry. 2016;**64**:2815-2821

[68] Mandhania S, Madan S, Sawhney V. Antioxidant defense mechanism under salt stress in wheat seedlings. Biologia Plantarum. 2006;**50**:227-231

[69] Lv DW, Zhu GR, Zhu D, Bian YW, Liang XN, Cheng ZW, et al. Proteomic and phosphoproteomic analysis reveals the response and defense mechanism in leaves of diploid wheat *T. monococcum* under salt stress and recovery. Journal of Proteomics. 2016; **143**:93-105

[70] Tanou G, Molassiotis A, Diamantidis G. Induction of reactive oxygen species and necrotic death-like destruction in strawberry leaves by salinity. Environmental and Experimental Botany. 2009;**65**:270-281

[71] Sairam RK, Roa KV, Srivastava GC. Differential response of wheat genotypes to long term salinity stress in relation to oxidative stress, antioxidant activity and osmolyte concentration. Plant Science. 2002;**163**: 1037-1046

[72] Guo R, Yang Z, Li F, Yan C, Zhong X, Liu Q, et al. Comparative metabolic responses and adaptive strategies of wheat (*Triticum aestivum*) to salt and alkali stress. BMC Plant Biology. 2015; **15**:170. DOI: 10.1186/s12870-015-0546-x

[73] Chinnusamy V, Jagendorf A, Zhu JK. Understanding and improving salt tolerance in plants. Crop Science. 2005; **45**:437-448

[74] Menezes-Benavente L, Kernodle SP, Margis-Pinheiro M, Scandalios JG. Salt-induced antioxidant metabolism defenses in maize (*Zea mays* L.) seedlings. Redox Report. 2004;**9**:29-36

[75] Farsiani A, Ghobadi ME. Effects of PEG and NaCl stress on two cultivars of corn (*Zea mays* L.) at germination and early seedling stages. World Academy of Science, Engineering and Technology. 2009;**57**:382-385

[76] Wakeel A, Sümer A, Hanstein S, Yan F, Schubert S. In vitro effect of Na+/K+ ratios on the hydrolytic and pumping activity of the plasma membrane H+-ATPase from maize (*Zea mays* L.) and sugar beet (*Beta vulgaris* L.) shoot. Plant Physiology and Biochemistry. 2011;**49**:341-345

[77] Schubert S, Neubert A, Schierholt A, Sumer A, Zorb C. Development of salt resistant maize hybrids: The combination of physiological strategies using conventional breeding methods. Plant Science. 2009;**177**:196-202

[78] USDA-ERS: Cotton and Wool Outlook. August 2017 [Accessed: 29 August 2018]

[79] USDA-ERS: Cotton and Wool Outlook: 2017 [Accessed: 29 August 2018]

[80] Penna JCV, Verhalen LM, Kirkham MB, Mcnew RW. Screening cotton genotypes for seedling drought tolerance. Genetics and Molecular Biology. 1998;**21**:545-549

[81] Loka DM, Derrick M, Oosterhuis DM, Ritchie GL. Water-deficit stress in cotton. In stress physiology in cotton. In: Oosterhuis DM, editor. Number Seven the Cotton Foundation Book Series. National Cotton Council of America; 2011. pp. 37-72

[82] Gorham J, Wyn Jones RG, McDonnell E. Some mechanisms of salt tolerance in crop plants. Plant and Soil. 1985;**89**:15-40

[83] Wang R, Wan S, Sun J, Xiao H. Soil salinity, sodicity and cotton yield parameters under different drip irrigation regimes during saline wasteland reclamation. Agricultural Water Management. 2018;**209**:20-31

[84] Ashraf M, Ahmad S. Influence of sodium chloride on ion accumulation, yield components and fiber

characteristics in salt-tolerant and salt-sensitive lines of cotton (*Gossypium hirsutum* L.). Field Crops Research. 2000;**66**:115-127

[85] Ashraf M. Salt tolerance of cotton: Some new advances. Critical Reviews in Plant Sciences. 2002;**21**:1-30

[86] Chaves MM, Flexas J, Pinheiro C. Photosynthesis under drought and salt stress: Regulation mechanisms from whole plant to cell. Annals of Botany. 2009;**102**:551-560

[87] May OL, Jividen GM. Genetic modification of cotton fiber properties as measured by single- and high-volume instruments. Crop Science. 1999;**39**: 328-333

[88] Tari I, Laskay G, Takacs Z, Poor P. Responses of sorghum to abiotic stresses: A review. Journal of Agronomy and Crop Science. 2013;**1999**(4):264-274. https://doi.org/10.1111/jac.12017

[89] Almodares A, Hadi MR, Dosti B. The effects of salt stress on growth parameters and carbohydrate contents in sweet sorghum. Research Journal of Environmental Sciences. 2008;**2**: 298-304

[90] Almodares A, Hadi MR, Ahmadpour H. Sorghum stem yield and soluble carbohydrates under different salinity levels. African Journal of Biotechnology. 2008;7:4051-4055

[91] Swami AK, Alam SI, Sengupta N, Sarin R. Differential proteomic analysis of salt stress response in *Sorghum bicolor* leaves. Environmental and Experimental Botany. 2011;**71**:321-328

[92] Almodares A, Hadi MR, Dosti B. Effects of salt stress on germination percentage and seedling growth in sweet sorghum cultivars. Journal of Biological Sciences. 2007b;7:1492-1495

[93] Kafi M, Nabati J, Masoumi A, Mehrgerdi MZ. Effect of salinity and silicon application on oxidative damage of sorghum [*Sorghum bicolor* (L.) Moench.]. Pakistan Journal of Botany. 2011;**43**:2457-2462

[94] Hasegawa PM, Bressan RA, Zhu JK, Bohnert HJ. Plant cellular and molecular responses to high salinity. Annual Review of Plant Physiology and Plant Molecular Biology. 2000;**51**:463-499

[95] Bavei VB, Shiran B, Khodambashi M, Ranjbar A. Protein electrophoretic profiles and physiochemical indicators of salinity tolerance in sorghum (*Sorghum bicolor* L.). African Journal of Biotechnology. 2011;**10**:2683-2697

[96] Lacerda CF, Cambraria J, Oliva MA, Ruiz HA, Prisco JT. Solute accumulation and distribution during shoot and leaf development in two sorghum genotypes under salt stress. Revista Brasileira de Fisiologia Vegetal. 2003;**49**:107-120

[97] Asghar D, Khan AA, Collins JC. Effects of salinity on growth, ionic relations and solute content of *Sorghum bicolor* L. Moench. Journal of Plant Nutrition. 2009;**32**:1219-1236

[98] Bavei V, Shiran B, Ahmad A. Evaluation of salinity tolerance in sorghum (*Sorghum bicolor* L.) using ion accumulation, proline and peroxidase criteria. Plant Growth Regulation. 2011; **64**:275-285

[99] Vasantha S, Venkataramana S, Rao PNG, Gomathi R. Long term salinity effect on growth, photosynthesis and osmotic characteristics in sugar cane. Sugar Tech. 2010;**12**:5-8

[100] Santana MJ, Carvalho JA, Souza AMG, Sousa KJ, Vasconcelos CL, Andrade LAB. Efeitos da salinidade da água de irrigação na brotação e desenvolvimento inicial da cana-de-açúcar (Saccharum spp) eemsolos com

diferentes niveis texturais. Cienciae Agrotecnologia. 2007;**31**:1470-1476

[101] Akhtar S, Wahid A, Rasul E. Emergence, growth and nutrient composition of sugarcane sprouts under NaCl salinity. Biologia Plantarum. 2003; **46**:113-116

[102] Willadino L, Oliveira Filho RA, Silva EA, Gouveia Neto AS, Camara TR. Estresse salino em duas variedades de canade-açúcar: Enzimas do sistema antioxidativo e fluorescência da clorofila. Revista Ciência Agronômica. 2011;**42**:417-422

[103] Vasantha S, Gomathi R, Brindha C. Growth and nutrient composition of sugarcane genotypes subjected to salinity and drought stresses. Journal Communications in Soil Science and Plant Analysis. 2017;**48**: 989-998

[104] Anitha R, Mary PCN, Savery MAJR, Sritharan N, Purushothaman RS. Differential responses of sugarcane (*Saccharum officinarum* l.) varieties exposed to salinity under a hydroponic system. Plant Archives. 2015;**15**:817-822

[105] Maas EV, Grattan SR. Crop yields as affected by salinity. In: Skaggs RW, van Schilfgaard J, editors. Agricultural Drainage, Agronomy Monograph 38. Madison, WI: Am. Soc. Agron.; 1999

[106] Zhu JK. Plant salt tolerance. Trends in Plant Science. 2001;**6**:66-71

[107] Asch F, Wopereis MCS. Responses of field grown irrigated rice cultivars to varying levels of floodwater salinity in a semi-arid environment. Field Crops Research. 2001;**70**:127-137

[108] Jampeetong A, Brix H. Effects of NaCl salinity on growth, morphology, photosynthesis and proline accumulation of Salvinia natans. Aquatic Botany. 2009;**91**:181-186

[109] Akita S, Cabuslay GS. Physiological basis of differential response to salinity in rice cultivars. Plant and Soil. 1990; **123**:277-294

[110] Foad M, Abdelbagi MI. Responses of photosynthesis, chlorophyll fluorescence and ROS-scavenging systems to salt stress during seedling and reproductive stages in rice. Annals of Botany. 2007;**99**:1161-1173

[111] Zhang J, Jia W, Yang J, Ismail AM. Role of ABA in integrating plant responses to drought and salt stresses. Field Crops Research. 2006;**97**:111-119

[112] Flowers TJ, Yeo AR. Variability in the resistance of sodium chloride salinity within rice (*Oryza sativa* L.) varieties. New Phytologist. 1981;**88**: 363-373

[113] Dionisio-Sese ML, Tobita S. Antioxidant responses of rice seedlings to salinity stress. Plant Science. 2000; **135**:1-9

[114] Chen Z, Gallie DR. The ascorbic acid redox state controls guard cell signaling and stomata movement. The Plant Cell. 2004;**16**:1143-1162

[115] Ashour N, Serag MS, Abd El-Haleem AK, Mandour S, Mekki BB, Arafat SM. Use of the killar grass (*Leptochloa fusca* L.) Kunth. In saline agriculture in arid lands of Egypt. Egyptian Journal of Agronomy. 2002; **24**:63-78

[116] USSLS (US Salinity Laboratory Staff). Diagnosis and improvement of saline and alkali soils. USDA Handbook No. 60. Washington, DC, USA; 1954

[117] Murtaza G, Ghafoor A, Qadir M. Irrigation and soil management strategies for using saline-sodic water in a cotton-wheat rotation. Agriculture Water Management. 2006;**81**:98-114

[118] Ezeaku PI, Ene J, Joshua AS. Application of different reclamation methods on salt affected soils for crop production. American Journal of Experimental Agriculture. 2015;**9**:1-11

[119] Zaka MA, Schmeisky H, Hussain N, Rafa HU. Utilization of brackish and canal water for reclamation and crop production. In: The International Conference on Water Conservation in Arid Regions; ICWCAR09.2009; Qatar. pp. 1-16

[120] Mehdi SM, Sarfraz M, Qureshi MA, Rafa HU, Ilyas M, Javed Q, et al. Management of high RSC water in salt affected conditions under rice and wheat cropping system. International Journal of Scientific and Engineering Research. 2013;**4**:684-698

[121] Manzoor A, Khattak RA, Dost M. Humic acid and micronutrient effects on wheat yield and nutrients uptake in salt affected soils. International Journal of Agriculture and Biology. 2014;**16**: 991-995

[122] Pandeyl SK, Pathak LP, Pathak RK. Effect of some nutrients in rice plant under sodic soils. International Journal of Technical Research and Applications. 2013;**1**:1-6

[123] Dash AK, Singh HK, Mahakud T, Pradhan KC, Jena D. Interaction effect of nitrogen, phosphorus, potassium with sulphur, boron and zinc on yield and nutrient uptake by rice under rice-rice cropping system in inceptisol of coastal Odisha. International Research Journal of Agricultural Science and Soil Science. 2015;**5**:14-21

[124] Khattak RA, Haroon K, Muhammad D. Mechanism(s) of humic acid induced beneficial effects in salt affected soils. Scientific Research and Essays, Academic Journals. 2013;**8**: 932-939

[125] Shaaban M, Abid M, RAI A-S. Amelioration of salt affected soils in rice paddy system by application of organic and inorganic amendments. Plant Soil Environment. 2013;**59**:227-233

[126] Ismail AM. Development of Technologies to Harness the Productivity Potential of Salt-Affected Areas of the Indo-Gangetic, Mekong, and Nile River basins, CPWF Project Report. CGIAR Challenge Program on Water and Food; Colombo, Sri Lanka. 2009. pp. 1-6

[127] Singh NK, Larosa PC, Handa AK, Hasegawa PM, Bressan RA. Hormonal regulation of protein synthesis associated with salt tolerance in plant cells. Proceedings of the National Academy of Sciences of the United States of America. 1987;**84**:739-743

[128] Ashraf M, Akram NA, Arteca RN, Foolad MR. The physiological, biochemical and molecular roles of brassinosteroids and salicylic acid in plant processes and salt tolerance. Critical Reviews in Plant Sciences. 2010; **29**:162-190

[129] Silva EN, Ribeiro RV, Ferreira-Silva SL, Viegas RA, Silveira JAG. Comparative effects of salinity and water stress on photosynthesis, water relations and growth of Jatropha curcas plants. Journal of Arid Environments. 2009;**74**:1130-1137

[130] Le HL, Linh TH, Xuan TD, Ham LH, Ismail AM, Khanh TD. Molecular breeding to improve salt tolerance of rice (*Oryza sativa* L.) in the red river delta of Vietnam. International Journal of Plant Genomics. 2012;**2012**:1-9. DOI: 10.1155/2012/949038

[131] Thomson MJ, de Ocampo M, Egdane J, et al. Characterizing the saltol quantitative trait locus for salinity tolerance in rice. Rice. 2010;**3**:148-160

Chapter 13

Capacity Development for Scaling Up Climate-Smart Agriculture Innovations

Elliot Mahlengule Zwane

Abstract

Climate change and climate variability are creating negative impacts to agriculture. It affects both food security and crop and livestock production. In the process, it affects the livelihood of communities. Climate-smart agriculture is seen as an alternative to mitigate the challenges of climate change. Literature studies were obtained from journal articles on capacity development. The problem investigated is that climate-smart agriculture (CSA) is a recent concept which needs to be understood with climate change, and the extension advisors do not have requisite skills. Ethical tea partnership singles out tea farmers and advisors in the tea sector. The findings included the definition of climate-smart agriculture (CSA) and the linkage with climate change. It further identified key issues involved in CSA, adaptation and mitigation, and identified the technologies that need to be scaled up in order to mitigate against climate change. The study recommended the area of competency required to serve farmers by advisory services by showing the needed factors that will serve as strategies in order to scale up the needed technologies useful in managing climate-smart agriculture. The chapter recommends capacity development for extension advisors and concludes with a series of mitigation steps.

Keywords: climate change, mitigation, impact, climate-smart agriculture

1. Introduction

It is acknowledged that climate change and climate variability are creating negative impacts to agriculture. In the process it affects the livelihood of communities. Experts have attempted to explain the challenges that the continents will face due to climate change and also predicted that unless something is done, it has potential to destroy the environment which promotes food production [1, 2]. It creates a lot of effects to both food security and animal and crop production. Both developed and developing countries are going to be affected. Studies have shown that researchers can contribute by describing the potential role innovative agricultural practices and technologies can play in climate change [3, 4].

Effort have been directed to the understanding of climate change, its definition, causes and mitigation. However, not much has been done in technologies that can be used in the circumstances of climate change. According to [5] innovative agricultural practices and technologies can play an important role in climate change.

If it can be mitigated and adaptation be encouraged, they can make a difference. This chapter focuses on climate-smart technologies with these specific objectives:

- To define climate smart and how it is linked to climate change.
- To discuss the impact of climate change in agriculture.
- To indicate the impact of climate change to crops and livestock as well as the mitigation strategies that can be used.
- To identify climate-smart technologies and strategies to scale up by extension advisors.
- To indicate steps that extension advisors can adopt in scaling it.

2. Background and motivation

Evidence suggests that climate change has become a big matter in the world with organizations pulling their efforts in an attempt to find solution, for example, according to [6]. Climate change will exert increasing pressure on our ability to meet other major challenges, with feeding the world's growing population paramount 9.6 billion by 2050; see [7]. Over the next 40 years, the need to increase global cereal production by a minimum of 60–70% [8], the question to be asked is what is our understanding of climate change, what are the causes, and can these be mitigated? Researchers and scientists have been grappling with these questions, and some have given their understanding. Climate change plays a negative role to agriculture and causes excessive gases in the atmosphere, and the existence of "high levels of CO_2 would have a catastrophic effect on the planet's ecosystems" [7].

Rising air temperatures trigger several important secondary effects. Increased global day- and nighttime temperatures are causing changes to seasonality. Warmer air temperatures are melting the polar ice caps, northern latitude ice shields, and high-altitude glaciers worldwide, leading to changes in the timing and volume of freshwater discharge and rising sea levels [7].

In order to slow down this process, human beings should be helped to understand some of the steps that need to be taken. One would interrogate as to what is the link between climate change and CSA. It can be argued that while climate change increases, the vulnerability to agriculture which results in variability of temperature and reduced rainfall or in some instances brings out flood. On the other hand, Juvadi [9] identified three critical contributions that come as a result of CSA, namely: (a) CSA reduces agriculture contribution to climate change; (b) strengthens resilience to climate change and variability; and (c) sustainably increases productivity and income [9].

International organizations, such as International Crops Research Institute for Semi-Arid Tropics (ICRISAT), saw the importance of CSA to an extent that it has established projects in Africa reaching out to countries such as Ethiopia, the Horn of Africa, and Sudan with one aim to capacitate the extension functionaries in order to best serve farmers with climate-smart agriculture. The project will develop CSA technologies and CSA farms in vulnerable regions of Ethiopia and Sudan. These technologies include improved crop varieties and land management, improved soil fertility management, integrated pest and weed management, agroforestry, and improved livestock systems. These CSA farms will also serve as research and training sites for students from the universities and be used as demonstration sites. Gender equality and the promotion of female farmers is a core activity of the project.

3. Rationale for capacity development in CSA

It is the opinion of the author that the need for climate change knowledge on both adaptation and mitigation is a must in order to assist in the solving the challenges posed by the negative effects of climate change in farmers especially the smallholder one because in most cases, they are not well positioned in terms of knowledge as well as financial muscle to handle the threat posed by climate change.

For example, areas that receive flood change the ecosystems leading to new host of pathogenic organisms, or in some cases the environment is altered, and new diseases emerge which threatens the survival of these smallholder farmers. Authors have developed manuals to assist agricultural advisors to cope with such situations, see [6, 9–12].

4. The effect of climate change in agriculture

Productivity in agriculture and the quality and quantity of produce will be low. The sector of agriculture will not be able to meet the demand of the increase in high population due to climate change. Maize production will be severely affected because there will be limited rainfall for irrigation.

There will be no products of maize that are exported to other countries, and the GDP of the country will be low [13]. It will result in poor economic growth that will lead to low level of employment.

Agriculture practices will be affected; certain cultivation method will not be considered effective because of climate change [14]. Deep cultivation method cannot be adopted because of low rainfall received [15]. Minimal cultivation will be introduced to try to meet the requirement of the crops [16]. Farmers will sell their produce at higher price trying to meet daily operational cost of the farm. There will low level of employment rate in agriculture, and other permanent workers will lose their jobs [17].

There will be an introduction of other methods of cultivation that are neglected currently [18]. The mulching method will be used because of the following advantages; it promotes the water retention in the soil; the level of water that is being lost will be minimized. It also reduces the growth of weeds. There will be newly full adaptation of irrigation systems; the drip irrigation will be fully used [19].

It is because of the following advantages the total amount of water is applied directly to the root zone of maize plants, which will help in the full development of the plant [20]. There will be less evaporation rate due to the direct application of the water in the root zone. Other methods of irrigation that waste the usage of water will not be used. The disadvantage of the furrow irrigation is the high infiltration rate and evaporation of water that will result in a great loss [21].

Agriculture inputs will be ineffective; climate change will cause other species of insects to mutate. It will affect those insecticides that are registered now; they will not be able to control the pest population. It will pose a negative impact to the yield of farmers [22] leading to farmers not being able to meet daily operational cost of the farm.

Environmental effect of climate change in the soil will be negative [21]. There will be high leaching of nutrients due to unpredictable receivable rainfall. Soil that has good base status of the soil will result in having low base status. Other soil nutrients will not be available to the root zone of the crops. It will cause the plant to suffer from nutrient deficiency. Farmers will need to supplement those leached-out nutrients [23].

There will be a high soil erosion and a washing away of the top soil by rainfall [24]. There are different types of soil erosion that will occur. The gullies and the dongas which will affect the top soil of farmers. The topsoil is rich in plant nutrients that are essential to different kinds of crops; see [25]. There will be a loss in different types of crop due to the effect of climate change. Other diversification method cannot be used due to the lack of rainfall received [25].

The potential land in agriculture will be lost because 20% of the land will be turned into desertification [26]. In the field of agriculture, there will be new species that will become more resistant to a number of chemicals, and those insects will affect the quality of production in agriculture. Shortage of scientific knowledge in climate change causes farmers not to understand fully what is actually occurring around the globe [27]. Agronomists have concluded that there will be shortage of food by 2030. Climate change affects the entire farming systems and food security. Climate change will affect stable food of other countries. There will be shortages of substitutes in the markets of stable products [28].

5. Identification of key issues

There are different approaches that could be used to gather data; however, for this exercise a literature review was adopted. Different search engines were used to search climate change and climate smart as well as capacity building in climate-smart agriculture (CSA). A number of documents were consulted, and related papers in journals and books were found and were consulted. All these documents were found useful in terms of expanding the frontier of knowledge on climate smart in agriculture. The findings are presented next.

6. Discussion on key issues of climate change

6.1 Definition of climate smart

Literature is full of definition of the concept of climate smart. Climate-smart agriculture (CSA) is defined as agriculture that sustainably increases productivity and resilience and reduces or removes greenhouse gases while enhancing the achievement of national food security and development goals; see [8]. This concept is supported by international multilateral agencies such as the World Bank, International Fund for Agricultural Development (IFAD), and Consultative Group for International Agricultural Research (CGIAR), and practices have been documented; see [22].

It should be realized that prior to the implementation of CSA, one has to first understand what is climate change prior to talking of climate-smart agriculture. The United Nations Framework Convention on Climate Change (UNFCCC) defines climate change as "a change of climate that is attributed directly or indirectly to human activity that alters the composition of the global atmosphere and that is in addition to natural climate variability observed over comparable time periods," see [7]. It is viewed as the greatest environmental challenge the world is facing in this century; see [29].

6.2 Capacity development training for the officers and training for farmers

It has been realized that it is imperative that technical advisors should be capacitated in the area of climate change. Literature shows that developing countries are

Adaptation needed	Mitigation needed to be done
• Improve resilience of social and physical infrastructure • Change of clones and agricultural practices • Improve water and soil management • Prepare for future pests and diseases • Manage existing environmental threats	• Low carbon energy source • Energy efficiency • Change of agricultural practices, e.g., judicious fertilizer use • Change in consumer behavior
Source: Cracknell [10].	

Table 1.
Climate change adaptation and mitigation.

putting efforts in this regard [6], while there are also pockets of areas where similar efforts are being done in the African continent. It is noted that extension functionaries have to be capacitated in this area with competencies that include communication, farming, science, social, technical, and methodological. In the past extension, advisors were not trained in the relevant areas that could be used to tackle climate change-related aspects.

Although some of these aspects may seem to be obvious, the effects of climate change may confuse the farmer and threaten his survival. It is this reason that a need for continuous training is found to be important. One example is that of tea farming in Kenya which the small-scale farmers have found themselves facing challenges such as having erratic rainfall, poor quality tea due to high temperatures, and other related challenges [10]. A training manual has been developed in order to equip extension technicians in skills that would be needed to mitigate the impact of climate change. Some of the examples that an extension advisor is expected to deliver in the case of challenges of tea farmers are shown in **Table 1**.

Table 1 shows examples of the knowledge needed by the agricultural advisor to assist small-scale tea farmers. He will be expected to understand the steps needed to help farmers to adapt as well as to mitigate on the other hand. It should be noted that the knowledge will play a huge role in helping farmers to make profit in their farming enterprise. A training like this was provided in the Horn of Africa by different partners as included in this work; see [9].

According to [12], such training was provided to technical advisors in order to enable them to have adequate technical knowledge and tools to better advice and train farmers and thus enhance their capacity to adapt to the effects of variability and climate change. The project will develop CSA technologies and CSA farms in vulnerable regions of Ethiopia and Sudan. The CSA farms will focus on integrating promising CSA technologies and creating synergies between the different technologies [30]. Training manuals have been developed [10, 31].

There are many organizations that contribute towards training technical advisors, and they have formed partnerships, for example, ICRISAT working in the Horn of Africa has formed partnerships with organizations like International Relief Development (IRD), Malian Agency for the Environment and Sustainable Development (AEDD), and Building Resilience and Adaptation to Climate Extreme and Disasters Program (BRACED) [30, 32].

These technologies include improved crop varieties and land management, improved soil fertility management, integrated pest and weed management, agroforestry, and improved livestock systems. These CSA farms will also serve as research and training sites for students from the universities and be used as demonstration sites.

6.3 Types of innovations

Different types of technologies exist such as those that are related to water which may be called water-smart technologies. There are five types of technologies of which two are discussed, while the other three are presented in **Table 1**; see [33]. According to [33] these are the kind of interventions that reduce water requirements to produce the same or a higher level of yield.

Rainwater may need to be managed such as harvesting it either from rain or from the runoff by using different tools. Water management is another issue that needs more attention. The impact of 2015/2016 drought in Western Cape Province of South Africa has led to the major dams to become dry. The dam levels were estimated to be 30–40% [34]. In situ rainwater storage in rice paddies with 20–25 cm bunds can also be regarded as technology. It is believed that this technique is for rice only [33].

Another technology is called laser land leveling. In this technology the land is leveled with a laser this is done by a laser leveler. This kind of technology has been found suitable in rice fields which is a system of rice intensification; in this technology 7- to 10-day-old seedlings are transplanted at 20 cm spacing with 1–2 seedlings per hill. The other technologies are presented in **Table 2**.

These technologies include improved crop varieties and land management, improved soil fertility management, integrated pest and weed management, agroforestry, and improved livestock systems [12]. It should be taken seriously that these technologies are not exhaustive, but they can form a base to create awareness in the manner how farmers can benefit from them, if the advisor can use them to guide the farmers. The challenge is that advisors may not see their usefulness until a farmer is threatened by the effects of negative climate change. In this case an advisor needs to be trained to quickly identify the scenarios which he can make his positive contribution to solve farmer's problems; see [10].

6.4 Strategies of scaling up of approaches

In order to increase our understanding and actions about the technologies, one has to link a number of facts. Few of the challenges are discussed next.

6.4.1 Policy match

Earlier studies have shown that the capital cost of the technology has a great bearing on technology adoption [35]. Thus, if the cost of adoption is totally private, the technology will be implemented if the private returns from investment are more than the private costs. If this is not the case, the adoption of technologies may be deferred until the benefits exceed the cost.

6.4.2 Financial support

Most of the small and marginal farmers may not have access to the formal credit system. Studies have established that increased access to credit helps farmers overcome short-term liquidity constraints and increase technology adoption.

6.4.3 Strengthening capacity of institution

Access to information is a key element in the adoption of new technologies; a farmer will adopt a technology that will maximize his food security. It is therefore

Type of technology	Definition
1. Energy-smart technologies	Technologies that help reduce energy consumption during land preparation without affecting yield levels. These also help reduce water requirements for crops
• Direct-seeded rice	Dry seeds are sown either by broadcasting or drilling in line
• Zero tillage/minimum tillage	The crop is seeded through a seeder in an untilled field, and the crop residue is incorporated into the soil. At present, this technique is limited to wheat only
2. Energy-smart technologies	Technologies that help reduce energy consumption during land preparation without affecting yield levels. These also help reduce water requirements for crops
• Direct-seeded rice	Dry seeds are sown either by broadcasting or drilling in line
• Zero tillage/minimum tillage	The crop is seeded through a seeder in an untilled field, and the crop residue is incorporated into the soil. At present, this technique is limited to wheat only
3. Nutrient-smart technologies	Technologies that save/supplement/avoid chemical fertilizer use for crops and enrich carbon in the soil
• Green manure	Cultivation of legumes in a cropping system. This practice improves nitrogen economy and soil health/quality
• Integrated nutrient management	Integrated use of organic and chemical fertilizers to partially (25–50%) reduce NPK (nitrogen, phosphorus, and potassium) requirements without affecting productivity and improve soil health
• Leaf color chart	Standardized color charts are used to identify nutrient deficiency to estimate fertilizer doses in different field locations
4. Weather-smart instruments	Interventions that provide services related to financial security and weather advisories to farmers
• Crop insurance	Crop-specific insurance to compensate income loss due vagaries of weather
• Weather advisories	Information and communication technology-based forecasting about the weather
5. Introduction of stress-tolerant crops and diversification	Tolerant crops withstand biotic and abiotic stresses, and crop diversification reduces water demands and helps in harnessing nutrients from different soil layers
• Drought-tolerant variety	Seed variety that is tolerant to drought or relatively dry weather conditions
• Crop diversification (maize-wheat cropping)	Rice is replaced by maize on part of the land to economize on water use

Source: Taneja et al. [33].

Table 2.
Selected technology options for choice experiment.

recommended to create institutions to build capacity among technology developers, disseminators, and farmers [10]. Farmers and other role players should be made aware that technologies should be adopted [32].

6.5 Mitigation

According to [36] mitigation refers to the efforts undertaken to "reduce anthropogenic [greenhouse gas] emissions or to enhance natural sinks of greenhouse gases" [36]. In agriculture, mitigation generally refers to the sequestration of atmospheric carbon dioxide (CO_2) in plant tissue through photosynthesis and its storage in soil organic matter and the reduction in direct emissions from fossil fuel usage and energy intensive inputs.

Efforts to reduce the impact of climate change, on the other hand, put it simple, mitigation is defined as activities aimed at "avoiding or minimizing sources of pollution that can have a deleterious effect on levels of greenhouse.

Gases (GHGs), global warming and climate change. Contributions towards reducing the levels of anthropogenic greenhouse gas production need to be actively encouraged. This includes fossil fuel-related activities, methane and nitrous oxide emissions [37].

A number of steps were identified as facts to be considered when mitigating against climate change; see [10]. In terms of livestock, the herd sex, age, and breed should be optimized in order to allow the national herd to be reduced while maintaining the same level of production. Supplementing the feed with high protein forage would reduce the methane production from enteric fermentation and increase productivity.

Extending feedlot manure management to include anaerobic digestion and the collection and use of the methane gas produced would improve the emission of negative gases in the atmosphere. The following would make a positive impact, for example, promoting the use of game in place of beef production, avoiding the burning of agricultural residues, including those from sugar cane plantations, even where such methods are accepted management practice, reduction of the frequency of fires by enhanced fire management practices, promoting savannah thickening over substantial areas, effectively managing soil organic matter, adopting minimum tillage methods, and exploring synergies between adaptation and mitigation measures in the areas of agricultural product diversification and the application of more socially beneficial agro-technologies such as permaculture to provide sustainable livelihoods [37].

The use of insurance can be piloted. Risk mitigation may involve a variety of private and government policy and institutions. The government-supported crop insurance in developed countries has often been highly subsidized [5], whereas in developing countries the situation is different. Insurance remains an individual responsibility. It has been indicated that in India, farmers were prepared to buy weather index insurance products even when these products are not subsidized [5].

6.6 Impacts of climate change to farmers

Different categories of farmers suffer the impacts of climate change. For example, the negative effects of climate change affect food security of the targeted households and smallholder farmers. Furthermore, the most vulnerable to the expected impacts of climate change are developing countries and their citizens who have a lower resilience to climate change impacts due to limited financial and technical resources to support adaptation. Commercial farmers in different fields like crops grapes and livestock also suffer the impacts [38].

Climate change and climate variability have been found to be presenting a negative influence to crops [39], for example, where irrigation is insufficient, crops wither and die thus reducing the yield. In a study conducted in Kenya checking the suitability of places to be used in the future, a number of problems were found. The problems include the reduced yield could further mean reduced profit and increased poverty. However certain steps need to be taken in order to mitigate against this situation. It is believed that if the advisors are conscientised about the impact, they will realize the need to be capacitated in climate smart in agriculture. Advisors have to be in the forefront of knowledge in climate-smart agriculture if they have to be meaningful to farmers.

7. Conclusions and recommendations

7.1 Conclusions

Climate smart in agriculture has its basis in the climate change and its variability. Key issues playing a role were identified as well as the rationale for training in climate change agriculture. It is important to have in mind the meaning of both concepts, namely, climate change and climate smart in agriculture (CSA). While the causes of climate change are known, i.e., both nature- and man-induced activities, it can be concluded that efforts of mitigation should become a norm. The study has identified the impacts of climate change in agriculture, and furthermore, the chapter has identified some of the technologies that are needed in CSA.

These technologies were discussed as well as how they can be scaled out based on three strategies, namely, policy matching, capacitating the extension functionaries in CSA, and financial support. CSA should be seen as an engine of green growth and a provider of environmental services.

7.2 Recommendations

Based on the findings, it can be stated that the following should serve as recommendations:

In order for the technology to be effectively promoted, there is a need of the following: policy financial support and the willingness of farmers to adopt such technologies on conditions the benefits outweigh the costs of implementing it.

On the other hand, there is a need for a policy within the various institutions that accommodate agricultural advisors, to craft policies that among other will encourage in service training among the advisors in keeping abreast of climate change issues, biased to women disabled and youth in farming.

There is a need to train the advisors in climate-smart principles to enable them to be ahead of their farmers with knowledge in climate change and climate smart in agriculture. A climate change module should be part of the training in tertiary institutions, as well as in high schools in order to create awareness within the changing environment of farmers.

Acknowledgements

I would like to thank the University of Limpopo and the research directorate for providing funding to this work. I extend my thanks to my student Kethikuthula M. Maseko who provided some of the background information in this chapter.

Conflict of interest

I declare that there is no conflict of interest in this work.

Acronyms

UNFCCC	United Nations Framework Convention on Climate Change
CO_2	carbon dioxide
GHGE	greenhouse gas emission

CSA	climate-smart agriculture
BRACED	Building Resilience and Adaptation to Climate Extreme and Disasters Program
AEDD	Malian Agency for the Environment and Sustainable Development
IRD	International Relief Development
IFAD	International Fund for Agricultural Development
CGIAR	Consultative Group for International Agricultural Research
ICRISAT	International Crops Research Institute for the Semi-Arid Tropics

Author details

Elliot Mahlengule Zwane
Centre for Rural Community Empowerment (CRCE), Department of Agricultural Economics and Animal Production, University of Limpopo, Polokwane, South Africa

*Address all correspondence to: elliot.zwane@ul.ac.za

References

[1] IPCC. Climate Change 2007: Synthesis Report. Contributions of Working Groups I, II, and III to the Fourth Assessment Report of the Intergovernmental Panel on Climate Change. Geneva: IPCC; 2007

[2] Swanson BE. Global Review of Good Agricultural Extension and Advisory Service Practices. Rome: Food and Agriculture Organization; 2008

[3] Nelson G. Climate Change: Impact on Agriculture and Costs of Adaptation. Washington, DC: IFPRI; 2009

[4] SEO S. Is an integrated farm more resilient against climate change? A micro econometric analysis of portfolio diversification in African agriculture. Food Policy. 2010;**35**(1):32-40

[5] Lybbert TJ, Sumner DA. Agricultural technologies for climate change in developing countries: Policy options for innovation and technology diffusion. Food Policy. 2012;37(1). DOI: 10.1016/j.foodpol.2011.11.001 [Accessed: 31 July 2017]

[6] Simpson BM, Burpee CG. Adaptation under the "New Normal" of Climate change: The future of Agricultural Extension and Advisory Services. In: MEAS Discussion Paper Series on Good Practices and Best Fit Approaches in Extension and Advisory Service Provision. MEAS Discussion Paper No. 3. USAID; 2014. Available from: www.meas-extension.org [Accessed: 15 May 2017]

[7] United Nations Department of Economic and SOCIAL Affairs (UNDESA). Population Division. World Population Prospects: The 2012 Revision. New York, NY: United Nations; 2013

[8] FAO. Climate-Smart Agriculture—Policies, Practices and Financing for Food Security. Adaptation and Mitigation. Rome: FAO; 2010

[9] Juvadi DP. Capacity Building in Extension: Key to Climate Smart Agriculture. Hyderabad: Centre for Good Governance

[10] Cracknell R. What is climate change? Extension officers training manual. Adapting to climate change in the tea sector. Climate change adaptation. Ethical Tea Partnership. 2011

[11] Maponya P, Mpadeli S. The role of extension services in climate change adaptation in Limpopo Province, South Africa. Journal of Agricultural Extension and Rural Development. 2013;**5**(7):137-142

[12] Mwala M, Muyinda K. Capacity building of extension workers on climate smart. Research and capacity building in climate smart agriculture in the Horn of Africa. Research and capacity building in climate smart agriculture in the Horn of Africa. Available from: www.icrisat.org/capacity-building-of-extension-workers [Accessed: April 2017]

[13] Mortier S, Faict D, Lameire NH, De Vriese A. Benefits of switching from a conventional to a low-GDP bicarbonate/lactate-buffered dialysis solution in a rat model. Kidney International. 2005;**67**(4):1559-1565

[14] Parry ML, Rosenzweig C, Iglesias A, Livermore M, Fischer G. Effects of climate change on global food production under SRES emissions and socio economic scenarios. Global Environmental Change. 2004;**14**(1):53-67

[15] Berglund K. Agricultural improvement of cultivated organic soils II. Effects of liming and deep

cultivation on soil properties and root development. Soil Use and Management. 1996;**12**(4):176-180

[16] Adelberg EA, Myers JW. Modification of the penicillin technique for the selection of auxotrophic bacteria. Journal of Bacteriology. 1953;**65**(3):348

[17] Mills P, Newell G, Johnson D. Testicular cancer associated with employment in agriculture and oil and natural gas extraction. The Lancet. 1984;**323**(8370):207-210

[18] Xia C, Zhang J, Zhang W, Hu B. A new cultivation method for microbial oil production: Cell pelletization and lipid accumulation by Mucor circinelloides. Biotechnology for Biofuels. 2011;**4**(1):15

[19] Steinmetz Z, Wollmann C, Schaefer M, Buchmann C, David J, Tröger J, et al. Plastic mulching in agriculture. Trading short-term agronomic benefits for long-term soil degradation. Science of the Total Environment. 2016;**550**:690-705

[20] Kang S, Zhang J. Controlled alternate partial root-zone irrigation: Its physiological consequences and impact on water use efficiency. Journal of Experimental Botany. 2004;**55**(407):2437-2446

[21] Yohannes F, Tadesse T. Effect of drip and furrow irrigation and plant. Agricultural Water Management. 1998;**35**:201-207. DOI: 10.1016/S0378-3774(97)00039-5

[22] Scherr SJ, Shames S, Friedman R. From climate smart agriculture to climate smart landscapes. Agriculture and Food Security. 2012;**1**:12

[23] Hansen EM, Djurhuus J. Nitrate leaching as influenced by soil tillage and catch crop. Soil and Tillage Research. 1997;**41**(3-4):203-219

[24] Renard KG, Foster GR, Weesies GA, McCool DK, Yoder DC. Predicting Soil Erosion by Water: A Guide to Conservation Planning with the Revised Universal Soil Loss Equation (RUSLE). Vol. 703. Washington, DC: United States Department of Agriculture; 1997

[25] Romig DE, Garlynd MJ, Harris RF, McSweeney K. How farmers assess soil health and quality. Journal of Soil and Water Conservation. 1995;**50**(3):229-236

[26] Rosenthal JL. The fruits of revolution: Property rights, litigation, and French agriculture, 1700-1860. The Journal of Economic History. 1990;**50**(2):438-440

[27] Mengistu DK. Farmers' perception and knowledge on climate change and their coping strategies to the related hazards: Case study from Adiha, Central Tigray, Ethiopia. Agricultural Sciences. 2011;**2**(02):138

[28] Keeton JT. Low-fat meat products—Technological problems with processing. Meat Science. 1994;**36**(1-2):261-276

[29] Department of Environmental Affairs and Tourism (DEAT). Greenhouse Gas Inventory South Africa 1990 to 2000: National Inventory Report [online]. 2009. Available from: http://www.pmg.org.za/files/docs/090812greenhouseinventory.pdf [Accessed: 13 July 2018]

[30] Tabo R. Training in Climate change. Research Program Director—West and Central Africa and Country Representative, ICRISAT—Mali. Available from: www.icrisat.org/capacity-building-of-extension-workers [Accessed: 20 June 2017]

[31] Brooks DR, Hoberg EP. How will global climate change affect

parasite—Host assemblages. Trends in Parasitology. 2007;**23**(12):571-574

[32] Wright ML, Prokopy JG, Arbuckle JR, Ingles M, Thelen R, Bellm D, et al. Climate Change and Agricultural Extension; Building Capacity for Land Grant Extension Services to Address the Agricultural Impacts of Climate Change and the Adaptive Management Needs of Agricultural Stakeholders. Technical Report Series: Findings and Recommendations of the Climate and Corn-based Cropping Systems Coordinated Agricultural Project; Vol. 3. CSCAP Publication no. CSCAP-0192-2016. 2016

[33] Taneja G, Depal B, Pramod K, Joshi PK, Agarwal PK, Tyang INK. Farmers' Preferences for Climate-Smart Agriculture an Assessment in the Indo-Gangetic Plain. IFPRI Discussion Paper 01337 New Delhi Office; 2014

[34] Du Preez J. Western Cape drought declaration: 'crisis for farmers'—Farmers Weekly. 2017. Available from: www.farmersweekly.co.za>south-africa [Accessed: 24 May 2017]

[35] Garrido A. Using good economic principles to make irrigators true partners of water and environment policies. In: Paper presented at OECD Workshop on Agriculture and Water Sustainability, Markets and Policies; 14-18 November; Adelaide, Australia. 2005

[36] Intergovernmental Panel on Climate Change (IPCC). Summary for policymakers. In: Field CB, Barros V, Stocker TF, Qin D, Dokken DJ, Ebi KL, et al., editors. Managing the Risks of Extreme Events and Disasters to Advance Climate Change Adaptation, A Special Report of Working Groups I References. Page | iv and II of the Intergovernmental Panel on Climate Change. Cambridge and New York, NY: Cambridge University Press; 2012. pp. 1-19

[37] Department of Agriculture, Forestry and Fisheries (DAFF). Draft Climate Change Sector Plan. Government Gazette notice 7 of 2013 [online]. 2013. Available from: www.gov.za/documents/download.php?f=181645 [Accessed: 22 November 2013]

[38] Swart C. Dialogue. Drought 201 and 2016. Challenges Facing Commercial Producers. AGRI Western Cape; 2016. [Accessed: 12 July 2017]

[39] Afful DB. Public extension agents' need for new competencies: Evidence from a climate variability study in Limpopo Province, South Africa. South African Journal of Agricultural Extension. 2016;**44**(2):59-70. DOI: 10.17159/2413-3221/2016/v44n2a387